Explainable AI with Python

Antonio Di Cecco • Leonida Gianfagna

Explainable AI with Python

Second Edition

Antonio Di Cecco
University of Chieti-Pescara
Pescara, Italy

Leonida Gianfagna
Cyber Guru
Rome, Italy

ISBN 978-3-031-92228-2   ISBN 978-3-031-92229-9   (eBook)
https://doi.org/10.1007/978-3-031-92229-9

© The Editor(s) (if applicable) and The Author(s), under exclusive license to Springer Nature Switzerland AG 2021, 2025

This work is subject to copyright. All rights are solely and exclusively licensed by the Publisher, whether the whole or part of the material is concerned, specifically the rights of translation, reprinting, reuse of illustrations, recitation, broadcasting, reproduction on microfilms or in any other physical way, and transmission or information storage and retrieval, electronic adaptation, computer software, or by similar or dissimilar methodology now known or hereafter developed.

The use of general descriptive names, registered names, trademarks, service marks, etc. in this publication does not imply, even in the absence of a specific statement, that such names are exempt from the relevant protective laws and regulations and therefore free for general use.

The publisher, the authors and the editors are safe to assume that the advice and information in this book are believed to be true and accurate at the date of publication. Neither the publisher nor the authors or the editors give a warranty, expressed or implied, with respect to the material contained herein or for any errors or omissions that may have been made. The publisher remains neutral with regard to jurisdictional claims in published maps and institutional affiliations.

This Springer imprint is published by the registered company Springer Nature Switzerland AG
The registered company address is: Gewerbestrasse 11, 6330 Cham, Switzerland

If disposing of this product, please recycle the paper.

Welcome to a journey that transforms opaque algorithms into transparent, reliable tools empowering you to implement, evaluate, and trust AI models in real-world applications.

Foreword

Explainable AI with Python is not merely a technical manual, but a journey into one of the most fascinating and transformative areas of modern artificial intelligence. At a time when algorithms have moved from being deterministic sets of rules to complex, data-driven systems, the need for clarity and transparency has never been more critical. This book emerges in response to that need, guiding the reader through the landscape of machine learning with a particular focus on demystifying the "black box" nature of contemporary models. It provides a comprehensive description of the methods and tools that enable us to understand, interpret, and ultimately have confidence in the decisions made by sophisticated AI systems.

The evolution of artificial intelligence over recent decades has led to systems that perform with remarkable accuracy yet remain opaque in their inner workings. Traditional software engineering relied on explicit instructions, ensuring that every output could be directly traced to a specific part of the code. In contrast, modern machine learning models, especially deep neural networks, learn by example and are inherently complex, making it challenging to explain how they arrive at a particular conclusion. Explainable AI (XAI) addresses this challenge by developing methods that bridge the gap between high-performing models and human comprehension. It aims to provide explanations that are not only technically accurate but also intuitively clear, thereby enhancing system reliability and easing understanding.

This book stands as a robust guide for both practitioners and theorists who seek to navigate this complex field. It begins by laying a solid foundation in the principles of machine learning and the intrinsic need for model transparency. From the fundamentals of supervised, unsupervised, and reinforcement learning to nuanced discussions on the limitations of black-box models, every chapter is designed to build a coherent narrative. Through detailed theoretical explanations and hands-on Python examples, readers are invited to explore diverse methodologies employed in making AI systems explainable—be it through model-agnostic approaches that treat the AI as a black box or intrinsic methods that weave interpretability into the model's architecture.

What truly distinguishes this work is its holistic approach to the subject. Rather than merely cataloguing techniques, the authors critically explore why explainability matters. In many high-stakes fields such as healthcare, finance, and legal systems, AI decisions can have profound implications for individuals and society. An AI system that offers clear and accessible explanations is more likely to gain acceptance from regulators, practitioners, and the public alike. By ensuring that every prediction or classification can be audited and understood, the book fosters a new era of responsible and ethical AI deployment, advocating for transparency as a core principle rather than an optional feature.

The second edition builds upon the strong foundation of its predecessor by addressing rapid advancements in the field. Notably, it now investigates the world of large language models and multimodal systems. These additions reflect current trends in artificial intelligence where models process not only text but also integrate visual, auditory, and sensor data to form richer, more nuanced representations of the world. The expansion into these areas brings with it a discussion of how such complex systems can be made clear and interpretable without compromising performance. Furthermore, the enhanced treatment of additive models provides a more extensive exploration of their role in improving model interpretability. These updates ensure that the book remains an up-to-date and indispensable resource for anyone interested in the ethical, technical, and practical dimensions of explainable AI.

Throughout its pages, the book maintains a balance between rigorous technical detail and accessible narrative. The authors invite readers not only to learn the techniques but also to understand the underlying principles that make these methods effective. The blend of theory and practical application creates a journey that is both intellectually stimulating and immediately useful, equipping readers with the skills needed to implement XAI methods in real-world scenarios.

In an era where debates around AI ethics and accountability are intensifying, this book challenges the conventional wisdom that there must always be a trade-off between performance and interpretability. Instead, it demonstrates that through careful design and innovative techniques, high performance can go hand in hand with explanations that are comprehensible to humans. By reading this book, you gain insights into state-of-the-art XAI methods while also contributing to a broader movement towards more responsible artificial intelligence.

Explainable AI with Python is an invitation to embrace a future where technology serves humanity in an open and ethical manner. Whether you are an experienced practitioner looking to refine your skills or a newcomer eager to grasp the complexities of modern AI, this book offers a comprehensive exploration of the field. It is a vital resource that empowers you to build AI systems whose decision-making processes can be clearly understood, ensuring that technological progress is aligned with principles of fairness and accountability.

Università di Pisa/CNR   Carlo Metta

Preface

This book emerged during the COVID-19 pandemic, which, despite its challenges, provided us time to contemplate the vital role of transparency in artificial intelligence.

The team at Springer Nature deserves special recognition for their commitment to this project during global uncertainty, providing essential guidance and expertise throughout the process.

Our diverse professional backgrounds have shaped our approach to explainable AI. Both of us share a foundation in theoretical physics, which has given us a unique perspective that spans from scientific discovery to industrial applications. Antonio contributes perspectives from his experience at Sony, his academic journey from La Sapienza to Università di Pescara, and his work founding an AI for Good association and as an AI communicator. Leonida brings insights from his background at IBM and his current work in cybersecurity, focusing on the security implications of AI systems.

This book represents our shared commitment to developing AI technologies that are powerful yet accountable and trustworthy. Created through virtual collaboration across lockdowns, it addresses the practical need for explaining complex models to stakeholders—a necessity we've both encountered repeatedly as data scientists.

We hope to provide you with tools and frameworks to develop AI systems that can be understood, trusted, and deployed effectively. In our algorithm-driven world, demystifying AI isn't just a technical skill—it's fundamental to responsible innovation.

In this revised edition, we further explore XAI methods beyond SHAP and introduce a dedicated section on additive models in Chapter 6, accompanied by a supporting GitHub repository. Additionally, given the transformative impact of LLMs since the first edition, we have extensively analyzed the relationship between LLMs, XAI, and the potential emergence of AGI.

We extend our deepest gratitude to our wives, whose unwavering support sustained us through countless evenings of writing and coding.

Pescara, Italy Antonio Di Cecco
Rome, Italy Leonida Gianfagna
March 2025

Competing Interests The authors have no competing interests to declare that are relevant to the content of this manuscript.

Introduction

In an era where machine learning models achieve remarkable accuracy yet often operate as inscrutable "black boxes," our project, *Explainable AI with Python*, emerges as both a practical guide and a philosophical manifesto. This book is dedicated to demystifying the inner workings of modern AI systems, making them transparent and trustworthy. Our approach is built on two key principles:

1. **Visual Reinforcement**
   We believe that clear understanding is best achieved through visual cues. That's why we repeat key images throughout the book. These visuals are not just included once "just in case" they might be useful—they are presented "just in time" to capture and maintain the reader's attention exactly when needed.
2. **Transparency and Trust**
   As AI models grow more sophisticated, it is essential to not only achieve high performance but also to understand how these systems make decisions. Our goal is to provide practical techniques that reveal the inner workings of AI models, ensuring they are both accountable and secure.

March 2025
University of Chieti-Pescara Antonio Di Cecco,
Pescara, Italy Leonida Gianfagna,
Cyber Guru
Rome, Italy

Contents

1	**The Landscape**		1
	1.1 Examples of What Explainable AI Is		2
		1.1.1 Learning Phase	2
		1.1.2 Knowledge Discovery	4
		1.1.3 Reliability and Robustness	4
		1.1.4 What Have We Learnt from the Three Examples	5
	1.2 Machine Learning and XAI		6
		1.2.1 Machine Learning Tassonomy	7
		1.2.2 Common Myths	10
	1.3 The Need for Explainable AI		11
	1.4 Explainability and Interpretability: Different Words to Say the Same Thing or Not?		13
		1.4.1 From World to Humans	13
		1.4.2 Correlation Is Not Causation	14
		1.4.3 So What Is the Difference Between Interpretability and Explainability?	17
	1.5 Making Machine Learning Systems Explainable		19
		1.5.1 The XAI Flow	19
		1.5.2 The Big Picture	21
	1.6 Do We Really Need to Make Machine Learning Models Explainable?		23
	1.7 Summary		25
	References		25
2	**Explainable AI: Needs, Opportunities and Challenges**		27
	2.1 Human in the Loop		28
		2.1.1 Centaur XAI Systems	28
		2.1.2 XAI Evaluation from "Human in the Loop Perspective"	30
	2.2 How to Make Machine Learning Models Explainable		33

		2.2.1	Intrinsic Explanations	36
		2.2.2	Post Hoc Explanations	40
		2.2.3	Global or Local Explainability	41
	2.3	Properties of Explanations		44
	2.4	Summary		46
	References			46
3	**Intrinsic Explainable Models**			47
	3.1	Loss Function		48
	3.2	Linear Regression		51
	3.3	Logistic Regression		60
	3.4	Decision Trees		70
	3.5	K-Nearest Neighbors (KNN)		77
	3.6	Summary		80
	References			81
4	**Model-Agnostic Methods for XAI**			83
	4.1	Global Explanations: Permutation Importance and Partial Dependence Plot		84
		4.1.1	Ranking Features by Permutation Importance	85
		4.1.2	Permutation Importance on the Train Set	88
		4.1.3	Partial Dependence Plot	89
		4.1.4	Accumulated Local Effects (ALE)	93
		4.1.5	Properties of Explanations	96
	4.2	Local Explanations: XAI with Shapley Additive Explanations		98
		4.2.1	Shapley Values: A Game-Theoretical Approach	99
		4.2.2	The First Use of SHAP	100
		4.2.3	Properties of Explanations	102
	4.3	The Road to KernelSHAP		103
		4.3.1	The Shapley Formula	103
		4.3.2	How to Calculate Shapley Values	104
		4.3.3	Local Linear Surrogate Models (LIME)	105
		4.3.4	KernelSHAP Is a Unique Form of LIME	106
	4.4	KernelSHAP and Interactions		107
		4.4.1	The NewYork Cab Scenario	107
		4.4.2	Train the Model with Preliminary Analysis	107
		4.4.3	Making the Model Explainable with KernelShap	110
		4.4.4	Interactions of Features	111
	4.5	A Faster SHAP for Boosted Trees		112
		4.5.1	Using TreeShap	113
		4.5.2	Providing Explanations	113
	4.6	A Naïve Criticism to SHAP		115
	4.7	Summary		116
	References			118

5 Explaining Deep Learning Models ... 119
5.1 Agnostic Approach ... 120
5.1.1 Adversarial Features ... 120
5.1.2 Augmentations ... 122
5.1.3 Occlusions as Augmentations ... 122
5.1.4 Occlusions as an Agnostic XAI Method ... 124
5.2 Neural Networks ... 126
5.2.1 The Neural Network Structure ... 126
5.2.2 Why the Neural Network Is Deep? (vs Shallow) ... 129
5.2.3 Rectified Activations (and Batch Normalization) ... 130
5.2.4 Saliency Maps ... 131
5.3 Opening Deep Networks ... 131
5.3.1 Different Layer Explanation ... 131
5.3.2 CAM (Class Activation Maps) and Grad-CAM ... 132
5.3.3 DeepShap/DeepLift ... 134
5.3.4 Integrated Gradients ... 137
5.3.5 TracIn: Tracing Training Data Influence ... 140
5.4 A Critic of Saliency Methods ... 142
5.4.1 What the Network Sees ... 142
5.4.2 Explainability Batch Normalizing Layer by Layer ... 143
5.5 Unsupervised Methods ... 144
5.5.1 Unsupervised Dimensional Reduction ... 144
5.5.2 Dimensional Reduction of Convolutional Filters ... 145
5.5.3 Activation Atlases: How to Tell a Wok from a Pan ... 146
5.6 Summary ... 149
References ... 150

6 Additive Models for Interpretability ... 153
6.1 Additive Models: When Interpretability Meets Predictive Power ... 153
6.2 Generalized Additive Models (GAM): Clarity in Complex Decisions ... 154
6.3 GAM with Interactions: Caruana's Innovation (GA^2M) ... 156
6.4 GAM for Fairness: Explicit Bias Mitigation ... 157
6.5 Explainable Boosting Machines (EBM): The Bridge Between Precision and Comprehensibility ... 159
6.6 Tree Ensemble Additive Model (TEAM): Didactic Interpretability ... 161
6.7 Neural Additive Models (NAM): When Neural Networks Become Transparent ... 164
6.8 Kolmogorov-Arnold Networks (KAN): The Frontier of Advanced Interpretability ... 166
6.9 Summary ... 169
References ... 170

7 Adversarial Machine Learning and Explainability ... 173
- 7.1 Adversarial Examples (AE) Crash Course ... 174
  - 7.1.1 Hands-on Adversarial Examples ... 183
- 7.2 Doing XAI with Adversarial Examples ... 186
- 7.3 Defending Against Adversarial Attacks with XAI ... 189
- 7.4 Summary ... 192
- References ... 192

8 Explainability of Language Models (XAI and LLM) ... 195
- 8.1 Introduction ... 195
- 8.2 Evolution of Sequential Models: From RNNs to GRUs ... 196
  - 8.2.1 The Challenge of Sequentiality ... 196
  - 8.2.2 Recurrent Neural Networks: An Artificial Memory ... 196
  - 8.2.3 The Vanishing Gradient Problem ... 197
  - 8.2.4 LSTM: Selective Memory ... 197
  - 8.2.5 GRU: Simplifying Without Losing Power ... 198
  - 8.2.6 Explainability and Interpretation ... 199
  - 8.2.7 Limitations and Transition to Transformers ... 200
- 8.3 The Attention Mechanism: A Conceptual Revolution ... 201
  - 8.3.1 From Sequential Processing to Attention ... 201
  - 8.3.2 The Intuition Behind Attention ... 201
  - 8.3.3 The Mathematics of Attention ... 202
  - 8.3.4 Multi-Head Attention: Parallel Attention ... 204
  - 8.3.5 Self-Attention: Introspective Attention ... 205
  - 8.3.6 Explainability of Attention ... 206
  - 8.3.7 Limitations and Considerations ... 208
  - 8.3.8 The Future of Attention ... 208
- 8.4 Transformer Architecture: Beyond Sequentiality ... 209
  - 8.4.1 The Transformer Revolution ... 209
  - 8.4.2 General Architecture ... 209
  - 8.4.3 The Encoder ... 210
  - 8.4.4 The Decoder ... 211
  - 8.4.5 Key Components for Explainability ... 212
  - 8.4.6 Layer Normalization ... 213
  - 8.4.7 Explainability of the Transformer ... 214
  - 8.4.8 Variants and Innovations ... 215
  - 8.4.9 Explainability Considerations ... 216
- 8.5 BERT and Encoder-Based Models: Understanding Context ... 217
  - 8.5.1 The BERT Innovation ... 217
  - 8.5.2 Architecture and Pre-training ... 217
  - 8.5.3 BERT Explainability ... 219
  - 8.5.4 Evolution of Encoder-Based Models ... 220
  - 8.5.5 Applications to Explainability ... 221
  - 8.5.6 Comparative Model Analysis ... 221
  - 8.5.7 Explainability for End-Users ... 222

	8.6	Challenges and Future Directions	222
	8.7	Vision Transformers: From NLP to Computer Vision	223
		8.7.1 The Vision Transformer Revolution	223
		8.7.2 From Images to Patches	223
		8.7.3 Key Differences from CNNs	224
		8.7.4 Explainability of ViT	224
		8.7.5 Variants and Innovations	225
		8.7.6 Applications to Explainable Computer Vision	226
		8.7.7 Challenges and Future Directions	226
	8.8	Multimodal Models: Integrating Text, Images, and Audio	227
		8.8.1 The Emergence of Multimodal Architectures	227
		8.8.2 Explainability Techniques for Multimodal Models	230
		8.8.3 Challenges in Multimodal Explainability	232
		8.8.4 Future Directions in Multimodal Explainability	233
		8.8.5 Conclusion	234
	8.9	Explainability Techniques for Language Models	234
		8.9.1 The Challenge of LLM Explainability	234
		8.9.2 Techniques for Internal Analysis	235
		8.9.3 Techniques for External Analysis	237
		8.9.4 Metrics and Evaluation	238
		8.9.5 The Future of LLM Explainability	239
		8.9.6 Practical Applications in Different Domains	241
		8.9.7 Ethical Considerations in LLM Explainability	242
		8.9.8 Conclusion and Future Outlook	243
	8.10	Case Studies and Practical Applications	243
		8.10.1 Introduction to Case Studies	243
		8.10.2 Case Study 1: Clinical Decision Support Systems	244
		8.10.3 Case Study 2: Financial Risk Assessment	245
		8.10.4 Case Study 3: Educational Content Generation	247
		8.10.5 Comparative Analysis and Lessons Learned	249
		8.10.6 Future Directions	251
		8.10.7 Conclusion	252
	References		252
9	**Making Science with Machine Learning and XAI**		**259**
	9.1	Scientific Method in the Age of Data	260
	9.2	Ladder of Causation	263
	9.3	Discovering Physics Concepts with ML and XAI	269
		9.3.1 The Magic of Autoencoders	269
		9.3.2 Discover the Physics of Damped Pendulum with ML and XAI	273
		9.3.3 Climbing the Ladder of Causation	277
	9.4	Expanding the Role of Machine Learning in Scientific Discovery: The Impact of Large Language Models	278

	9.5	Rethinking the Damped Pendulum: From Prediction to Discovery with LLMs	281
	9.6	Science in the Age of ML and XAI	283
	9.7	Summary	284
	References		285
10	**AGI, LLM, XAI**		287
	10.1	Defining Intelligence and AGI	288
		10.1.1 Evolving Definitions of AGI: From Theoretical Foundations to Practical Constraints	289
	10.2	LLM: Evolution and Architectures	291
	10.3	Emergent Properties and Fragility of LLMs	293
		10.3.1 The Case for Emergence in LLMs: Sparks of AGI	293
		10.3.2 The Case Against Emergence: Is It a Mirage?	295
	10.4	LLMs Are Fragile	297
	10.5	The Role of XAI in AGI Powered by LLMs	300
		10.5.1 LLM Toy Model Example: Using SHAP for Explainability	302
	10.6	Conclusions and Future Perspectives	305
	10.7	Summary	306
	References		306
11	**A Proposal for a Sustainable Model of Explainable AI**		309
	11.1	The XAI "Fil Rouge"	310
	11.2	XAI and GDPR	311
		11.2.1 FAST XAI	313
	11.3	Conclusions	317
	11.4	Summary	321
	References		321
Appendix A			323

Chapter 1
The Landscape

> *Everyone knows that debugging is twice as hard as writing a program in the first place. So if you're as clever as you can be when you write it, how will you ever debug it?*
>
> Brian Kernighan

This Chapter Covers:

- What is Explainable AI in the context of Machine Learning
- Why do we need Explainable AI
- The big picture of how Explainable AI works

For our purposes we place the birth of AI with the seminal work of Alan Turing (Turing, 1950) in which the author posed the question "*Can machines think?*" and the later famous mental experiment proposed by Searle called the *Chinese Room*.

The point is simple: suppose to have a "black box"-based AI system that pretends to speak Chinese in the sense that it can receive questions in Chinese and provide answers in Chinese. Assume also that this agent may pass a Turing Test, which means it is indistinguishable from a real person that speaks Chinese. Would we be fine on saying that this AI system is capable of speaking Chinese as well? Or do we want more? Do we want the "black box" to explain itself clarifying some Chinese language grammar?

So, the root of Explainable AI was at the very beginning of Artificial Intelligence, albeit not in the current form as a specific discipline. The key to trust the system as a real Chinese speaker would be to make the system less "opaque" and "explainable" as a further requirement besides getting proper answers.

Jumping to our days, it is worth mentioning the statement of GO champion Fan Hui commenting on the famous 37th move of AlphaGo, the software developed by Google to play GO, that defeated in March 2016 the Korean champion Lee Sedol with a historical result: "It's not a human move, I've never seen a man playing such a move" (Metz, 2016). GO is known as a "computationally complex" game, more complex than Chess and before this result, the common understanding was that it was not a game suitable for a machine to play successfully. But for our purposes and start this journey, we need to focus on Fan Hui statement quoted. The GO champion could not make sense of the move even after having looked at all the match, he recognized it as brilliant, but he had no way to provide an explanation. So, we have an AI system (AlphaGo) that performed very well (defeating the GO champion) but no explanation of how it worked to win the game that is where "Explainable AI" inside the wider Machine Learning and Artificial Intelligence starts to play a critical role.

Before presenting the full landscape, we will give some examples less sensationalistic but more practical in terms of understanding what we mean by the fact that most of the current machine learning models are "opaque" and not "explainable." And the "fil rouge" of the book will be to learn in practice, leveraging different methods, how to make ML models explainable—that is, to answer the questions "What," "How," and "Why" about their results—especially in light of the paradigm shift introduced by large language models (LLM) marked by the public release of ChatGPT in late 2022, as we will explore in the new dedicated sections of this revised edition.

1.1 Examples of What Explainable AI Is

Explainable AI (aka XAI) is more than just a buzz word, but it is not easy to provide a definition that includes the different angles to look at the term. Basically speaking, XAI is a set of methods and tools that can be adopted to make ML models understandable to human beings in terms of providing explanations on the results provided by the ML models elaboration.

We'll start from some examples to get into the context. In particular, we will go through three easy cases that will show different but fundamental aspects of explainable AI to keep in mind for the rest of the book:

- the first one is about the **learning phase**;
- the second example is more on **knowledge discovery**
- the third introduces the argument of **reliability and robustness** against external attacks to the ML model.

1.1.1 Learning Phase

One of the most brilliant successes of modern Deep Learning techniques against the traditional approach comes from Computer Vision. We can train a Convolutional Neural Network (CNN) to understand the difference between different classes of labelled pictures. The applications are probably infinite: we can train a model to

1.1 Examples of What Explainable AI Is

discriminate between different kinds of pneumonia RX pictures or teach it to translate sign language into speech. But are the results truly reliable?

Let's follow a famous toy-task in which we have to classify pictures of wolves and dogs (Fig. 1.1).

After the training, the algorithm learned to distinguish the classes with remarkable accuracy: only a misclassification over 100 images! But if we use an Explainable Ai method asking the model "Why have you predicted wolf?" The answer will be with a little of surprise "because there is snow!" as shown in Fig. 1.2 (Ribeiro et al., 2016).

So maybe giving a model the ability to explain itself to humans can be an excellent idea. An expert in machine/deep learning can immediately see the way a model goes wrong and make a better one. In this case, we can train the network with

Fig. 1.1 ML Classification of wolves and dogs (Singh, 2017)

Fig. 1.2 (a, b) Classification mistake (Ribeiro et al., 2016)

(a) Husky classified as wolf (b) Explanation

occlusions (part of the images covered) as possible augmentations (variations of the same pictures) for an easy solution.

1.1.2 Knowledge Discovery

The second case deals with Natural Language Processing (NLP) models like Word Embedding can learn some sort of representation of the semantics of words. Words are embedded in a linear space as vectors and making logic statement becomes simple as adding two vectors.

Man is to King as Woman is to Queen becomes a formula like:

$$Man - King = Woman - Queen$$

Very nice indeed! But on the same dataset, we find misconceptions such as **Man is to Programmer as Woman is to Housekeeper.** As we say "garbage in garbage out" the data were biased, and the model has learned the bias. A good model must be Fair, and Fairness is also one of the goals of Explainable AI.

1.1.3 Reliability and Robustness

Now let's look at Fig. 1.3 below, you may see a guitar a penguin and two weird patterns (labelled again with Guitar and Penguin) with some numbers below. It is a result of an experiment in which state of the art Deep Neural Network has been trained to recognize guitars and penguins.

Fig. 1.3 ML classification of guitars and penguins with strange patterns (Nguyen et al., 2015)

1.1 Examples of What Explainable AI Is

The numbers below each image are the confidence levels that Machine Learning system assigns to each recognition (e.g. say that the ML system is pretty sure that the first image is a guitar with 98.90% confidence level). But you may see that also the second image is recognized as a Guitar with 99.99% CL and the fourth as a penguin with 99.99% CL. What is happening here?

This is an experiment conducted to fool the Deep Neural Network (DNN): the engineers maintained in the second and fourth image only the elements that the systems use to recognize a guitar and a penguin and changed all the rest so that the system still "see" them like a guitar and a penguin. But these characteristics are not the ones that we, as humans, would use to do the same task, said in other way these elements are not useful as an explanation to make sense of why and how the DNN recognizes some images (Nguyen et al., 2015).

1.1.4 What Have We Learnt from the Three Examples

As promised, let's critically think about the three examples above to see how they introduce different angles to look at explainable AI capturing various aspects:

- **Example 1 about wolves classification**: Accuracy is not enough as an estimator of a good machine learning model, without explanations we would not be able to discover the bad learning that caused the strong relation between snow and wolfs.
- **Example 2 about Natural Language Processing**: There is the need for checking the associations against bias to make the process of knowledge discovery fair and more reliable. And as we will see in the following, knowledge discovery is among the main applications of XAI.
- **Example 3 about penguins**: this is trickier, in this case, the engineers did a kind of reverse engineering of the ML model to hack it, and we need to keep this in mind to see the relation between explainable AI and making the ML models more robust against malicious attacks.

These experiments are trivial in terms of impacts on our life, but they can be easily generalized to cases in which a DNN is used to recognize tumor patterns or take financial decisions.

In these critical cases, we won't rely only on the outcome, but we will also need the rationale behind any decision or recommendation coming from the DNN, to check that the criteria are reliable, and we can TRUST the system.

For an easy reminder, we'll require every Explainable AI model to be **F.A.S.T.** as in **Fair** and not negatively biased, **Accountable** on its decisions, **Secure** to outside malevolent hacking and **Transparent** in its internals. Rethinking at the examples, the second one needs more fairness and the last one more security.

This is precisely what Explainable AI (XAI) as an emerging discipline inside Machine Learning tries to do: make the ML systems more transparent and interpretable to build trust and confidence in their adoption. To understand how XAI works, we need to do a step back and place it inside Machine Learning. Let's clarify from this very beginning that we are using terms interchangeably like interpretable and explainable. We will provide better definitions in the following sections starting from Sect. 1.4 of this chapter.

1.2 Machine Learning and XAI

Without going through a historical digression on how machine learning was born inside the broader context of AI, it is useful to recall some concepts for proper positioning of Explainable AI in this field and understand from a technical point of view how the need of explainability stands out.

Let's start with Fig. 1.4 below as a visual representation to place Machine Learning in the right landscape.

Among the vast number of definitions for ML, we will base on this simple but effective one that captures very well the core:

> Machine Learning is the field of study that gives computers the ability to learn without being explicitly programmed.
> (Samuel, 1959).

For our purposes, we need to focus on "without being explicitly programmed." In the old world of software, the solution to whatever problem was demanded to an algorithm. The existence of an algorithm guarantees by itself full explainability and full transparency of the system. Knowledge of the algorithm directly provides the explainability in terms of "Why," "What," and "How" questions. Algorithms are a process or set of rules to be followed to produce an output given an input; they are not opaque to human understanding, all the knowledge around a problem is translated in the set of steps needed to produce the output.

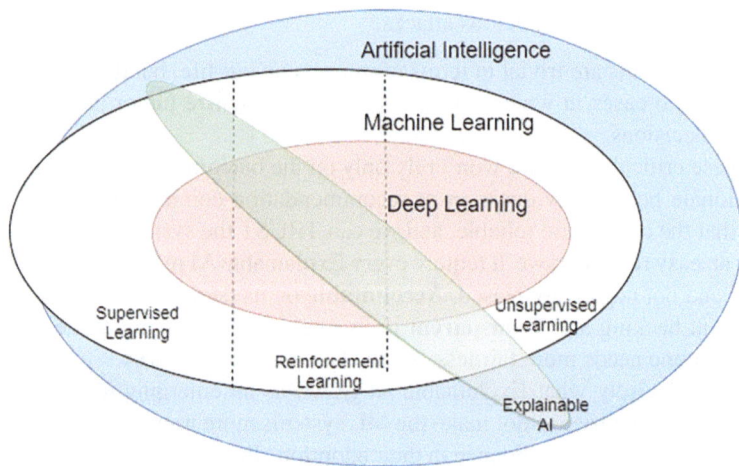

Fig. 1.4 Different areas of Artificial Intelligence in their mutual relations as Venn diagram

1.2 Machine Learning and XAI

The age of algorithms is being replaced by the **age of data** (Fig. 1.5 below), the current Machine Learning systems learn the rules from data during the learning phase and then can produce the expected outputs based on the given inputs.

However, you don't have direct access to the algorithmic steps that were followed, which means you might not be able to explain why a particular output was produced.

1.2.1 Machine Learning Tassonomy

Are all machine learning systems opaque by definition? No, it is not like that; let's have a quick categorization of Machine Learning systems before getting into details. There are three different main categories of Machine Learning systems based on the type of training that is needed (Fig. 1.4):

Supervised Learning: The system learns a function to map inputs to outputs (A to B) based on a set of data that include the solutions (labels) on which the system is trained. At present Supervised Learning is the most used form of Machine Learning for the wide range of its possible applications. You can model as A to B correspondence spam filtering, ad-clicks prediction, stock prediction, sales prediction, language translation, and so on with a multitude of techniques e.g. linear regressors, random trees, boosted trees, and neural networks.

Unsupervised Learning: The training data are not labelled, they do not contain the solutions, and the system learns to find patterns autonomously (e.g. KMeans, PCA, TSNE, autoencoders). This is the part of ML that is more affine to GAI (General Artificial Intelligence) because we have a model that autonomously labels the data without any human intervention. A typical example of unsupervised learning are reccomender systems like the ones used to suggest movies based on user's preferences.

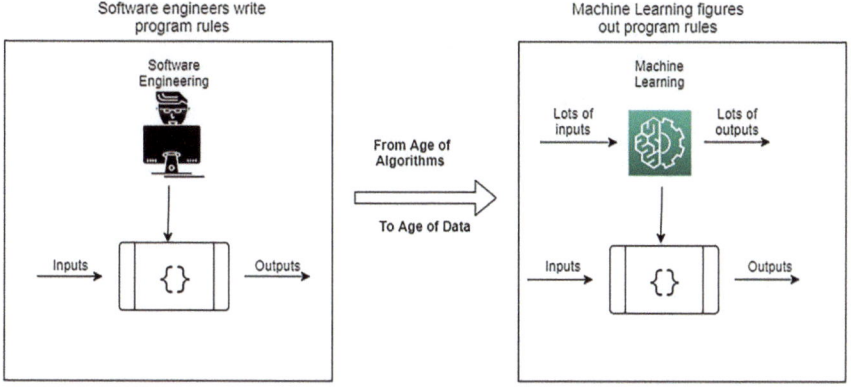

Fig. 1.5 Transition from standard software engineering driven by algorithms to sw engineering driven by data

Reinforcement Learning: is different from the two previous categories; there is no training on existing data. The approach here is to have an agent that performs actions in an environment and get rewards that are specific to each action. The goal is to find a policy that is a strategy to maximize the rewards. (e.g. deep learning networks and Montecarlo tree search to play games like AlphaGo) We can think of an RL model like at the intersection of Supervised and Unsupervised systems for the model generates its own examples, so it learns in an unsupervised manner how to generate examples exploring the example's space and learning from them in a supervised way.

Looking at Fig. 1.4, **Deep Learning** is a subset of machine learning that does not fall in a unique category in terms of type of learning, the term deep refers specifically to the envisioned architecture of the neural networks that are implemented with multiple hidden layers making the neural network hard to interpret in terms of how it is producing results. Deep Neural Networks (DNN) are the machine learning systems that are producing the most successful results and performance.

Given the categories above (the three different types of learning and the deep learning), there is not a unique mapping or set of rules to say that a specific category may need explainability more than the other in relation to the interpretability of the algorithms that belong to that category.

Explainable AI (XAI) is an emerging and transversal need (as pictured above) across the different AI domains. Let's make an example in Supervised Learning. Say we want to take a loan and the bank says "NO." We want to know the reason and the system cannot respond. What if people eligible for a loan could be classified as in Fig. 1.6 below?

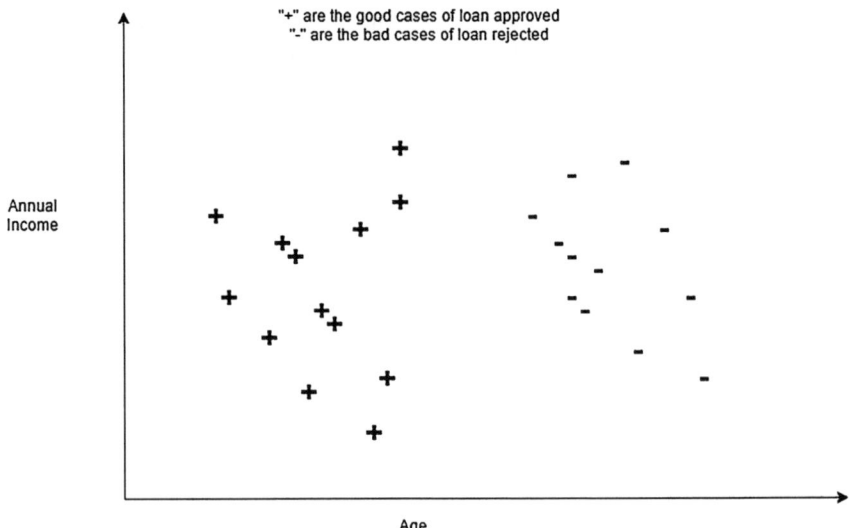

Fig. 1.6 Loan Approval, good and bad cases

1.2 Machine Learning and XAI

Here the axes are relevant features of the model like age and annual income of the borrower. As can you see, we can solve the problem with a linear classifier, and the model shows the suitable range of values for features to get the approval for the loan (Fig. 1.7).

But in a more complex model we must face a trade-off problem, look at Fig. 1.8 we may see the striking difference between the outcome of a simple linear classifier (Fig. 1.7) and a more sophisticated (most used) nonlinear one (Fig. 1.8).

In this last case, it is not easy to explain the outcomes of the ML model in terms of approved and non-approved loans.

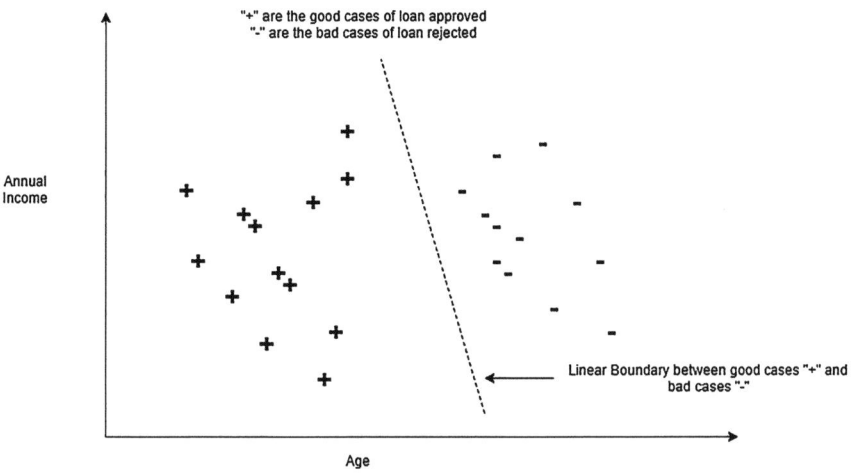

Fig. 1.7 Loan approval, good and bad cases with linear boundary

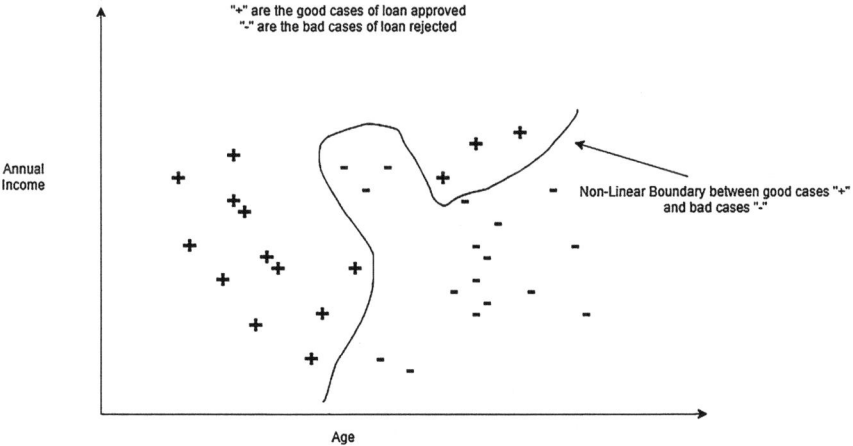

Fig. 1.8 Loan approval with nonlinear boundary

In linear models, you can easily say the effect of an increase or a decrease of a specific feature that is not generally possible for nonlinear cases. If you have a linear boundary to separate the two sets, you can explain how (in this specific case) age and annual income determine the approval or rejection of the loan. But with a nonlinear boundary, the two same features don't allow a straightforward interpretation of what's going on. We will go into details of how nonlinearity makes XAI harder starting from Chap. 2, for now, it is enough to note that the use of more complex models is unavoidable to achieve greater performance and inside Machine Learning, as we said, DNNs are the models that are most successful at all. This takes us to start demystifying two widespread beliefs.

1.2.2 Common Myths

The first common myth is that there is a strict need of explainability only for Deep Learning Systems (DNNs) that are opaque as constituted by many complex layers. This is only partially true in the sense that we will show that the need for explainability may also arise in the very basic Machine Learning models like Regression and it is not necessarily coupled with the architecture of deep learning systems. The question that may occur at this point is if there is any relation between the fact the DNN is seen both as the most successful ML systems in terms of performance (as noted previously) and the ones to need more explainability?

We will detail the answer in the following chapters, but it is useful to have a general idea right now of the standard answer in the field (somehow unjustified).

Indeed, **the second myth is that there is an unavoidable trade-off between machine learning system performance** (in terms of accuracy that is for example how well images are recognized) and interpretability, you cannot get both at the same time, the better the ML systems perform, the more it is opaque becoming less explainable. We can make a graph to visualize it.

The two sets of points represent the qualitative trends for today and the expectation for the future (Fig. 1.9). We will come back to this picture, but as for now it is already important to keep in mind that XAI is not for depressing performance to the advantage of explanations but to evolve the curve both in terms of performance and explainability at the same time, which means that given a model like DNN is performing very well today but not so explainable, the expectation is to move it toward increasing explainability but keeping (even improving) the performance. The trade-off between explainability and performance is on a single curve, but the overall trend is to push the curve for an overall improvement.

We will learn better across this book how these common beliefs have some validity, but they cannot be taken as general truths. Keeping this in mind, in the next two sections we will go into more details about the need of XAI beside the introductory examples and provide definitions of the terms we have been using to give them a reliable and operational meaning. As you may have noted we have been using so far

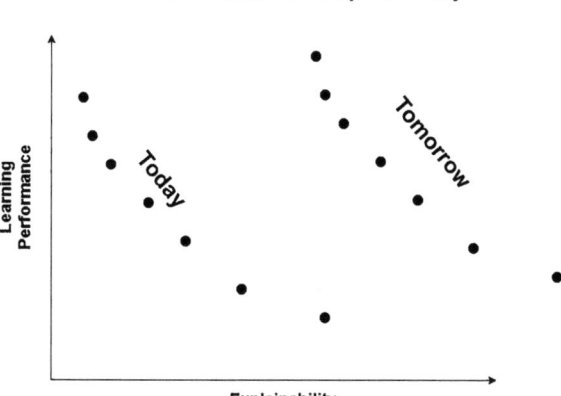

Fig. 1.9 Qualitative trend and relation of learning performance vs. explainability

terms like explainable, interpretable, and "less opaque" interchangeably while they refer to slightly different aspects of the same concept to be clarified.

1.3 The Need for Explainable AI

The picture that is emerging from the what explained so far should allow to easily understand at this point why do we need Explainable AI. From one side it may be obvious to answer given the above examples, but it could be fair enough to use Machine Learning model to do a task given a high level of accuracy and performance without getting into details of making the model more transparent.

Let's try to understand the need of explainability in more general terms. As argued by Karim et al. (2018), a single metric like a classification accuracy may not wholly describe and provide an answer to our real-world problem, getting the answer of "the what" might be useless without the addition of "the why" that is the explanation of how the model get the answer, in such cases the prediction itself is only a partial solution.

In fact, classification accuracy is not always a reliable metric. Generally, human expertise in selecting appropriate evaluation measures remains indispensable.

Science fiction often features Artificial General Intelligence agents that, regardless of their assigned task, will do whatever it takes to achieve maximum accuracy. In 2001: A Space Odyssey, HAL—the spaceship's computer—decided to eliminate the human crew, reasoning that "this mission is too important for me to allow you to jeopardize it".

The three main applications that are often coupled with the prediction of a machine learning model and that introduce the need of explainability are (Du et al., 2019):

Model Validation: It is connected to the properties of fairness, lack of bias and privacy in relation to the ML model. Explanations are needed to check whether the machine learning model has been trained on a "Biased" data set that may produce discriminations on a specific set of people. If a person is excluded from a loan, you need the possibility of looking into the black box to examine and produce the criteria that have been adopted for the decision. At the same time, the privacy of sensitive information is a must in specific cases (legal or medical among the others).

Model Debugging: To guarantee reliability and robustness, the ML model should ensure some level of debugging, which means the possibility to look behind the scenes into the machinery that produces the outputs. A small change in the inputs should not produce a huge change in the outputs to reduce the exposure to malicious attacks aimed at fooling the ML system and provide some level of robustness. Transparency and interpretability are needed to allow debugging in case of misbehavior and weird predictions.

Knowledge Discovery: This is the most complex application to comment, being related to situations in which ML models are not just used to make predictions but to increase the understanding and knowledge of a specific process, event or system. The extreme case that we will discuss further in the book is the adoption of ML models to gain scientific knowledge in which prediction is not enough without also providing explanations and causal relations. An infamous example that is helpful to understand the relation between explainability and trust is a rule-based model used to predict the mortality risks for patients with pneumonia. The unexpected result was that having asthma could decrease the risk of dying because of pneumonia. But the truth was that the patients with asthma were provided stronger medical treatments with better overall results; long story short, the need of explainability is deeply coupled with the level of trust we can have on a ML model.

It is easy to guess how the arguments connected to fairness, privacy and trust represent fundamental factors that might strong limit the adoption of ML systems in case explainability would not be guaranteed. We have not explicitly mentioned yet, in addition to the items above, the legal need of explainability because of regulations like GDPR in Europe, which make explainability a must for AI, but this will be fully discussed in Chap. 8.

Are there any cases in which explainability might not be needed? The general answer is that the ML model may be treated just like black boxes only in the cases in which the model is not expected to produce any significant impact. And this is pretty evident if we look at the AI adoption speed in the consumer market in which there is a large diffusion of recommender systems and personal assistants to be compared to the still low adoption of AI in regulated industries.

1.4 Explainability and Interpretability: Different Words to Say the Same Thing or Not?

"If a lion could speak, we could not understand him" and not because of different languages but because of two different worlds or better two different "language games." We start with a quote from L. Wittgenstein to set the context and expectation of this section that is a bit more philosophical than the rest of the book. Wittgenstein worked in the domain of philosophy of language to investigate the conditions that make whatever statement understandable. For our purpose, we want to build ML models explainable, but we need to clarify before starting the journey into the real world of techniques what we mean or at least agree on what we mean for explainability and interpretability As in case of lion, the language we use is strongly coupled with the world of our experience, the language is inherited from the world but also builds the world itself. So, we need to pay attention to the specific language we are searching for providing explanations about the opaque ML models. And, we need to be sure about the domains, the "language games" in which explainability and interpretability terms are used to avoid misunderstandings.

1.4.1 From World to Humans

Figure 1.10 makes more evident the concept: the layer "interpretability methods" lives between the black box and humans and two layers above the real world and the data with an increasing level of abstraction. We start from the bottom with grapes and wine is not a case, which will be elements of the real case scenario we will handle in Chap. 3.

Interpretability Methods should bridge the gap between predictions or decisions generated by the opaque ML models and humans to make them trust the predictions through explanation and interpretation (Fig. 1.10).

It is challenging to define the appropriate type and level of explanation, as it depends both on the source—which varies across different machine learning models—and the audience, which typically consists of individuals with diverse backgrounds, knowledge, and experiences. There is no single, universally accepted or quantitative definition of interpretability. Moreover, the terms 'explainability' and 'interpretability' are often used interchangeably.

Following Doshi-Velez and Kim (2017), we define interpretability in the context of machine learning as *"the ability to explain or to present in understandable terms to a human."* The definition is fuzzy and nonoperational, but it is used as a practical starting point for our analysis of the different meaning of interpretation and explanation. We want to state again that although these arguments may appear somehow abstract and philosophical, they are needed to set the stage before going into hands-on through the different techniques.

Fig. 1.10 From world to humans through Machine Learning

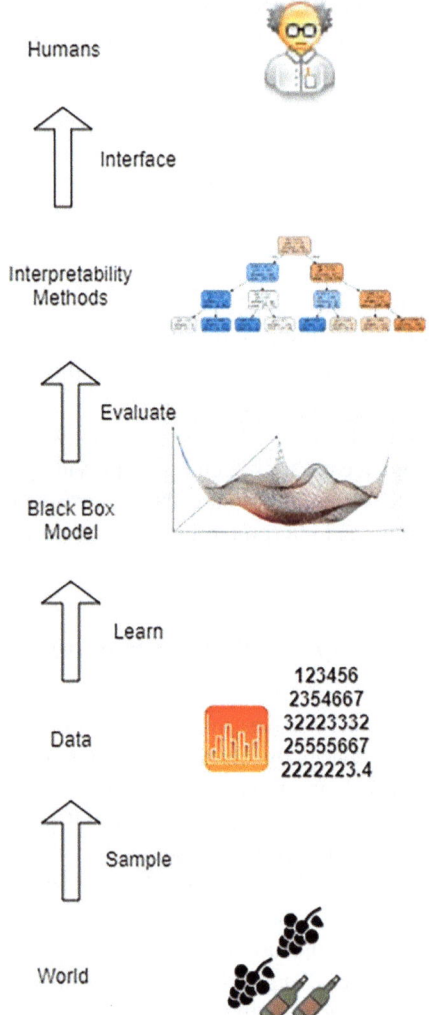

1.4.2 Correlation Is Not Causation

To get the proper view on these ambiguities, we need to start from one of the applications Explainable AI mentioned above, which is knowledge discovery. In the age of data, the solid boundary between correlation and causation is fading away. There is no need to go into mathematical details to highlight the basic difference between the two terms; correlation is a statistical tool to check the connection between two or more items, so to evaluate their coupling (the fact that change together).

But correlation doesn't imply causation, which is something stronger and means that one variable causes the change in the other, one thing makes the other to happen

1.4 Explainability and Interpretability: Different Words to Say the Same Thing or Not?

(cause and effect). If you see a correlation between doughnuts sales and the number of homicides in a certain area you are probably just seeing a coincidence without any underlying relation of causation between the two events. But when you have a lot of data and the possibility to learn from them to make predictions that is what machine learning does, you are almost relying on correlation to find patterns. But you are not going into the direction of building new knowledge that requires explanations, causal relations and not just coincidences.

This is an exciting stream of research inside the field of machine learning aimed at understanding if the classical scientific method based on models may be replaced by a brute analysis of the data to find patterns and generate predictions. For our purposes, it is important to emphasize again the concept that making predictions with machine learning is basically doing correlation, and these predictions need to be **interpreted and explained** to generate knowledge (Fig. 1.11).

We explicitly said "interpreted and explained" because these are two different actions. Most of the techniques and tools that we will study and discuss through this book are tools aimed at providing interpretations of the opaque ML models, but this could not be enough to get explanations. Interpretations are an element of an explanation but do not exhaust it.

To get an analogy from science, the same theory, for example, quantum mechanics, you can use it to build predictions that work well without understanding absolutely anything of it, then you may come to interpretations that are not just one but many (Copenhagen interpretations with collapsing wave functions, many-worlds interpretation) that are tested by predictions in a continuous loop to generate a full

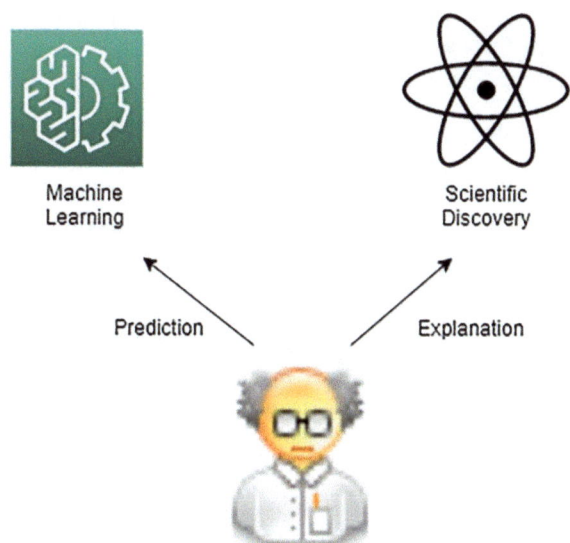

Fig. 1.11 Machine Learning, prediction vs. explanation

theory **an explanation that is composed of interpretations expressed through formalisms and predictions** (Fig. 1.12).

The same argument can be more effectively illustrated using a practical and concrete example from machine learning. Figure 1.13 demonstrates how various high-performing models can be developed to assess the risk of granting a loan to a customer, based on income and interest rate variables.

If we model the error as a mountain landscape each choice of parameters makes a different model and we prefer those that minimize the error (loss) function. So, every local minimum on the landscape of the error function makes, to some performance, a good choice of the model and each different model would generate a

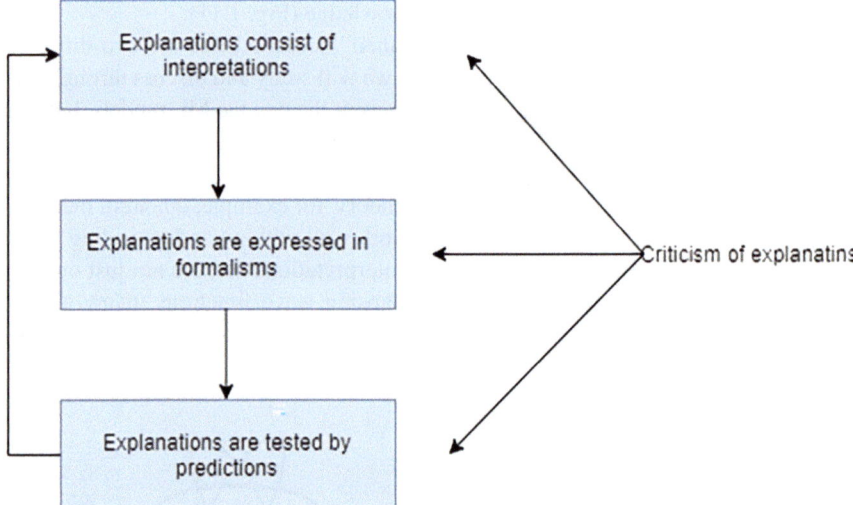

Fig. 1.12 Explanations decomposed (Deutsch, 1998)

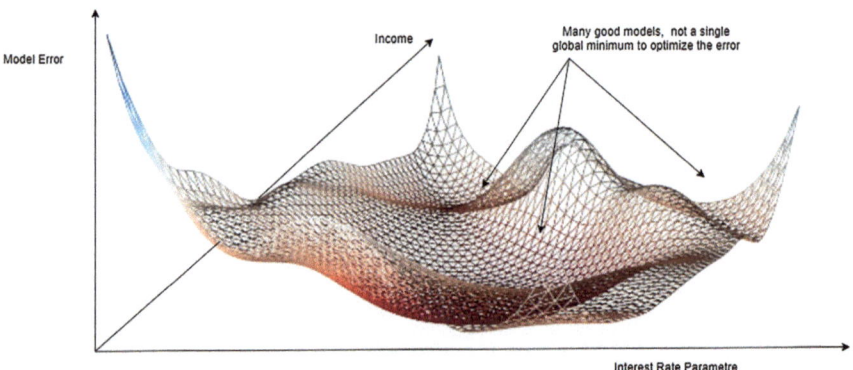

Fig. 1.13 An Illustration of the error surface of machine learning model

1.4 Explainability and Interpretability: Different Words to Say the Same Thing or Not?

different set of interpretations. But you don't have a global model that comprises the overall phenomenon, the kind of explanation we expect from a scientific theory.

1.4.3 So What Is the Difference Between Interpretability and Explainability?

To provide a further visual example of this distinction between interpretability and explainability, let's think about the boiling water, the temperature increases with time steadily until the boiling point after it will stay stable. If you just rely on data before the boiling point, the obvious prediction with the related interpretation would be that temperature rises continuously. Another interpretation may make sense of data taken after the boiling point with a steady temperature.

But if you search for a full explanation, a full theory of water "changing state," this is something deeper that exceeds the single good interpretations and predictions, in the two different regimes. ML would be good at predicting the linear trend and the flat temperature after the boiling point, but the physics of the phase transition would not be explainable to predict otherwise unobservable data (Fig. 1.14).

Interpretability would be to understand how the ML systems predict temperature with passing time in the normal regime; explainability would be to have a ML model that takes into account also the changing state that is a global understanding of the

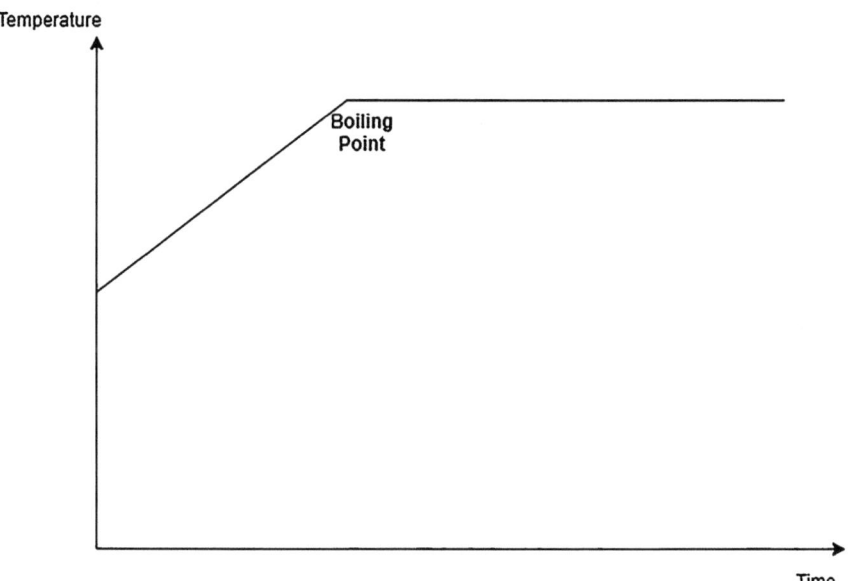

Fig. 1.14 Water phase transition diagram with two different trends for temperature before and after the boiling point

phenomenon more related to the application of knowledge discovery already mentioned.

To summarize, with the risk of an oversimplification of the discussion above but getting the core, **we will consider interpretability as the possibility of understanding the mechanics of a machine learning model but not necessarily knowing why**.

To provide an operational approach to how to distinguish interpretability from explainability, we summarize what stated above with a table (Table 1.1) that differentiates the two terms relying on the different questions that may be answered in each of the two scopes.

Explainability is for us something more in terms of being able to answers questions about what happens in case of new data, "What if I do x, does it affect the probability of y" and counterfactual cases to know what would have changed is some features (or values) would not have occurred, explainability is a theory that deals also with unobserved facts toward a global theory while interpretability is limited to make sense of what is already present and evident.

As stated by Gilpin et al. (2018): "We take the stance that interpretability alone is insufficient. For humans to trust black-box methods, we need explainability—models that can summarize the reasons for neural network behavior, gain the trust of users, or produce insights about the causes of their decisions. **Explainable models are interpretable by default, but the reverse is not always true**."

Keeping in mind these distinctions, we will have our path through the different techniques to make Machine Learning models more explainable, in most of the cases the methods and systems we will present are aimed at providing interpretability and not explainability as we have been discussing in this section and we will make this evident.

And it won't be important for most of the time to come back again on the difference between interpretability and explainability until the end of the book in which we will touch the argument of knowledge discovery (Chap. 6 focused how

Table 1.1 Difference between Interpretability and Explainability in terms of the questions to answer for the two different scopes

Question	Interpretability	Explainability
Which are the most important features that are adopted to generate the prediction or classification?	✓	✓
How much does the output depend on the input? How sensitive is the output to small changes in the input?	✓	✓
Is the model relying on a good range of data to select the most important features?	✓	✓
What are the criteria adopted to come across the decision?	✓	✓
How would the output change if we put different values in a feature not present in the data?	✗	✓
What would happen to the output if some feature or data had not occurred?	✗	✓

1.5 Making Machine Learning Systems Explainable

we may do science with ML and XAI) again and try to present a framework for a standard approach to AI (Chap. 8) in which this distinction will rise again.

1.5 Making Machine Learning Systems Explainable

The landscape of Explainable AI should be clear at this point; we provided a high level description of what Explainable AI is and why it is needed in the broader context of machine learning. Also, we tried to present better and clarify the terms and buzzwords that are used in this field. This section aims to depict a global map of the Explainable AI (XAI) system and process that may be used to get in on shot the big picture and as orientation for the rest of this work.

1.5.1 The XAI Flow

The best starting point is to locate the XAI inside the classical machine learning pipeline, which briefly consists of three phases (we are not getting into details of data preparation and optimization for our purposes, Fig. 1.15):

1. Training data
2. Machine Learning Process
3. Learned function that generates predictions, decision, or recommendation.

The main point of XAI is to make sense of the output producing explanations and interpretations that can be understood by the humans, that is to make the model explainable and provide an explanation interface open to the users.

The two blocks inside the red ellipse that are the Explainable Model and Explainable Interface represent the core content of this book. They are further

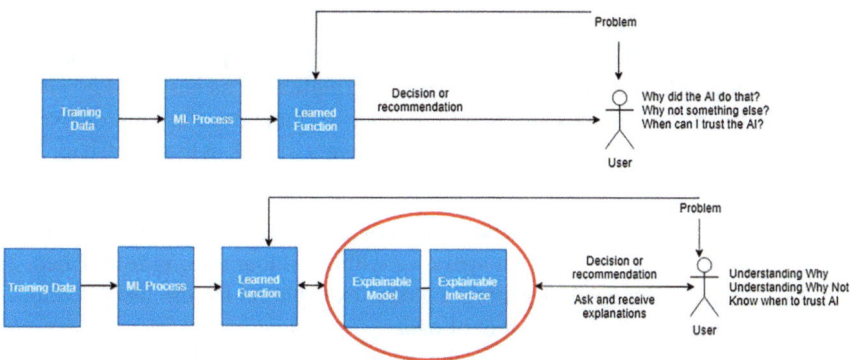

Fig. 1.15 Machine Learning Pipeline with focus on XAI, blocks inside the ellipse

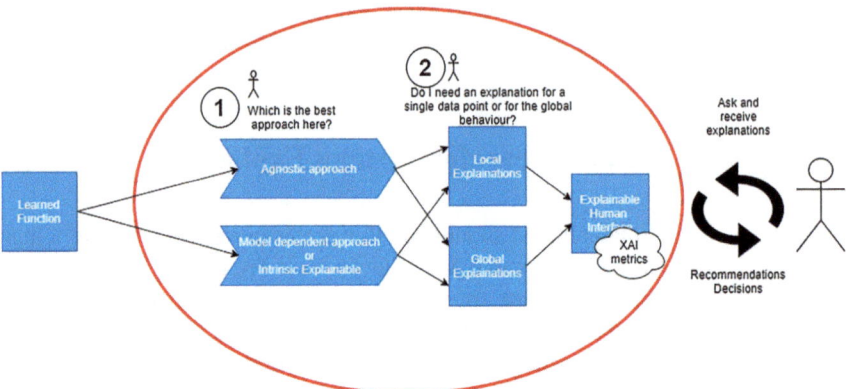

Fig. 1.16 Different approaches to make an ML model explainable

expanded in Fig. 1.16 below, note that if exploded the two blocks become methods and systems not a system only. This means that we are not making a ML system explainable just changing his inner components, but most of the times the ML is left untouched and we make sense of it from the external through the proper techniques.

As per the picture below the XAI mental model is a flow that, given a ML model, provides the proper options and techniques to make it explainable (Fig. 1.16).

The map now should be easily readable but let's just emphasize the main points. We have a given ML process (existing or built from scratch) that is providing outputs through the learned function and we need to make it explainable. The "Explainable Model" of Fig. 1.15 is exploded in Fig. 1.16 into the different techniques and approaches that, given the original Learned Function, drive to the Explainable Human Interface and the human-readable XAI metrics. Note that there are two main decision points in the flow: in the first one, the "human" may choose the main XAI approach to adopt that is an agnostic approach or a model-dependent one. These techniques will be deep dived starting from Chap. 2; for now it is enough to get the main points below:

- **Agnostic approach** means that the XAI works with the ML model as a "black box" without assuming any knowledge of the internals to produce explanations.
- **Model dependent approach or Intrinsic Explainable** means that knowledge of the ML model internals are used to produce explanations. As sub-case we put here also the "intrinisic explainable" models (we will look at them in Chap. 3) in which the model parameters provide explanations directly.

The decision point number "2" is about the choice of a global vs. local explanation: this depends from case to case (but again we will detail that starting from Chap. 2) and it is related to the need of getting a global explanation for the full ML model behavior or only for a specific subset of data.

Then we can do cycles until the produced explanations (XAI metrics) are satisfactory (Fig. 1.17).

1.5 Making Machine Learning Systems Explainable

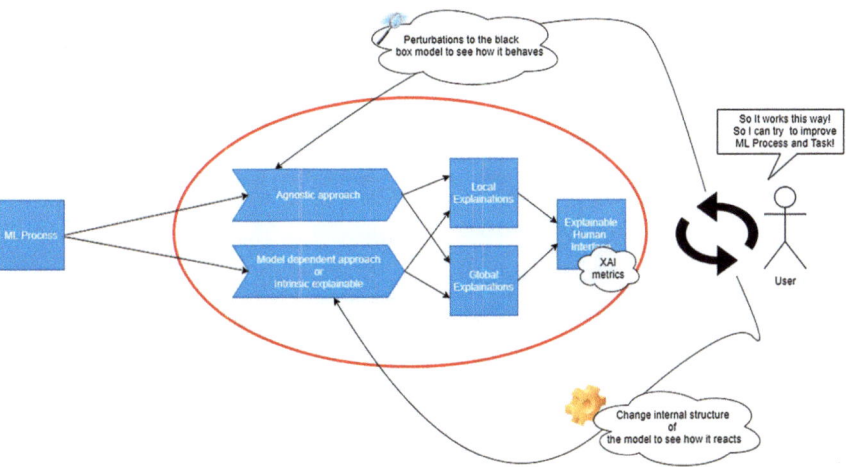

Fig. 1.17 Different approaches to make an ML model explainable with focus on the feedback from the external to further improve XAI

Figure 1.17 shows the two main type of loops to provide feedback to the system and improve explanations:

"Black box perturbations" is to feed the model with artificial data, even strange data, to test its response while in case of model-dependent approach the internal structure of the model might be changed to see how it reacts and improve the level of the produced explanations (local for some data or global ones).

"Change internals" and observe system's reaction is more suitable for the model-dependent approach in which we have full access to the specific model and we may play with the internal parameters to see how it works and generate explanations (local for some data or global ones).

1.5.2 The Big Picture

The spirit of the book is not only to provide techniques to deal with XAI if needed and produce model interpretations but mostly to complement the practical examples with critical thinking to understand the real reasons behind the adopted techniques, the urge to adopt them, and avoid false expectations. The technology rapidly changes so in terms of trading off we prefer to try to share a mindset, a way of thinking with practical methods instead of just a list of references to XAI methods and tools. Each XAI technique has its own domain of applicability and limitations. Every technique (global or local, agnostic or model dependent as in Fig. 1.17) needs to be carefully chosen and tailored toward our objective. We will use our mental model across all the book to keep always the focus on the global XAI picture and quickly position the specific technique we will deep dive into the proper context. So, Fig. 1.18 below

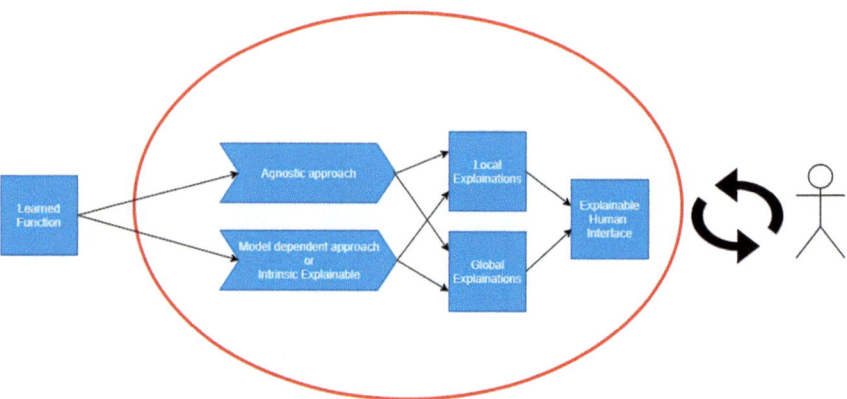

Fig. 1.18 XAI main flow

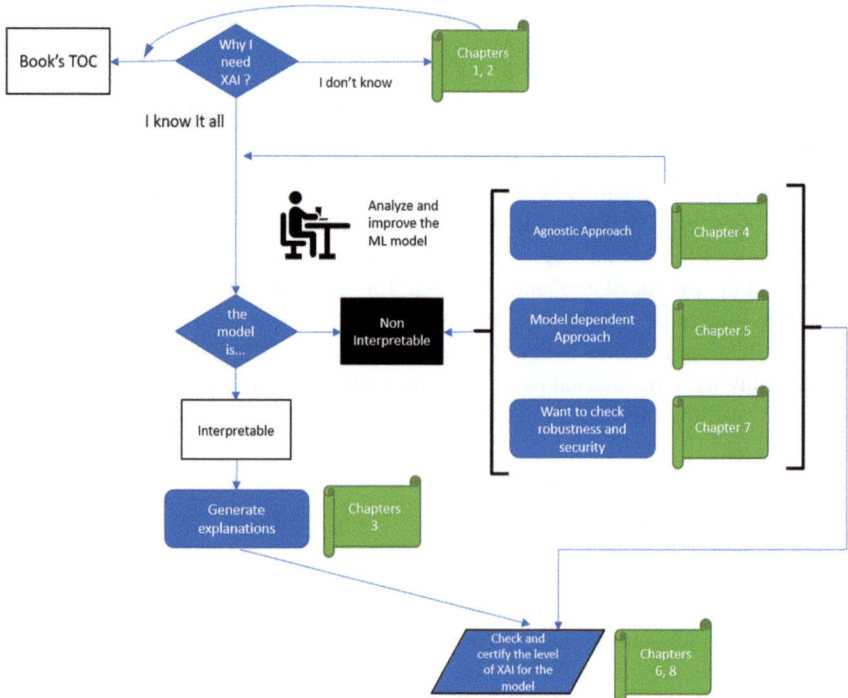

Fig. 1.19 Visual map of the main concepts and topics across the book

(same flow but without annotations to keep it simple) will appear again and again to see where we are during our XAI journey.

While Fig. 1.18 represents the high-level internals to make an ML model explainable, it might be useful to expand the flow to have a map of the book with the pointers to the different chapters (Fig. 1.19).

As discussed, there are different levels of possible explanations that may be required. Reflecting critically on these levels can help readers progress toward a standardized certification path for an XAI system, if necessary (as outlined at the bottom of the flow).

To have another look at the flow, we will have another chapter (Chap. 2) still focused on theory and landscape before putting hands on code to use XAI methods on real-life scenarios. Real-life scenarios here means to put in the shoes of an XAI scientist that is asked to provide explanations on the predictions provided by a ML model. We will learn how to use these methods to answer "What," "Why,"and "How" questions on the predictions from the ML models.

We will also deep dive into the robustness and security of ML models from an XAI perspective to prevent (or to be aware) of attacks aimed at fooling the ML models to change their predictions. The ultimate goal is to look at XAI from different perspectives to successfully deal with problems coming from real file to trust and certify ML models as explainable.

1.6 Do We Really Need to Make Machine Learning Models Explainable?

In this section we want to play the role of devil's advocate and challenge the real need of explainability beside the arguments already discussed and before jumping into the real world of XAI methods of the next chapters.

The mantra around XAI is that to trust Machine Learning we must be able to explain how the decisions or predictions have been generated. But let's think again about the core of machine learning, basically we are talking about systems that solve problems relying on learning by examples (experience) instead of algorithms (explanation of how things work).

And the best cases in which machine learning shows its full strength are the cases in which providing all the rules is not feasible (e.g. describe how to recognize a cat with an algorithm instead of training the system with millions of cat images).

Thus, could we say that you might get explainability at the price of losing the power of machine learning? This would mean that you will get explainable models only in the cases in which machine learning is not producing any huge benefit having already a complete algorithm to solve the problem.

To make these arguments more practical, let's suppose that we are determined to get the secrets of how our favorite football player kicks the ball (Fig. 1.20).

Would you try to get the physics of the kick or try to learn by experience looking at him and doing the same again and again? I would guess that the physics model would be useless and you would learn by experience. In this analogy relying on the physics model to learn how to kick is the same that using XAI to understand how a deep machine learning model works.

If you remember one of the original streams of AI, the idea was to make neural networks to mimic how human's brain learns and work. And up to some extent,

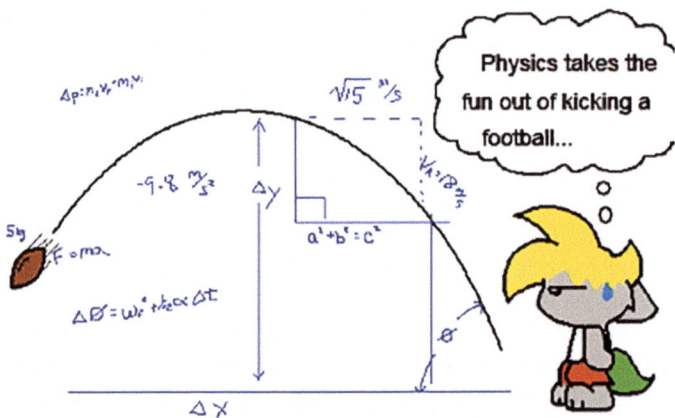

Fig. 1.20 Is knowing Physics really necessary to play football? (ChaosKomori, 2003)

deep neural networks are doing exactly that: learning from experience to do complex tasks that does not necessarily mean having an "acceptable" explanation of how to perform these tasks.

And this is totally acceptable for humans in the sense that you would never rely on a deep-dive in neuroscience to get the neurons activity and provide such an explanation of how a person is performing a mathematical task. In case of a student, in order to check if his learning of math, you would test him with questions or problems that the student should be able to solve given the required level of knowledge. But you would not care about "how" the student derived his answers. And this is exactly what happens when you transfer a Deep NN to work on test data after training.

So, is this to say that XAI is not really needed and close the book? Not at all, the message is again to rely on critical thinking to understand the context. Much of misunderstanding comes from not knowing or from misinterpreting the background of your problem and the related needs.

Let's get back to the main applications that need XAI: model validation, model debugging and knowledge discovery. For the last one, knowledge discovery, there is no sense in having predictions with machine learning without explanations. In this case we are trying to understand a phenomenon by relying on a brute machine learning approach on data and without any abstract modelling. Also, assuming we have the right predictions, it won't be any progress in the knowledge without explanations (because of the difference between correlation and causation).

In this specific case we strongly need details on how the system works also at the risk of decreasing the interpretability, because we want to answer questions about system behavior that are in the scope of new data and counterfactual examples (unobserved data). But in the other two cases, we are targeting different properties that are fairness, lack of bias, reliability, and robustness.

What we may search for, if not full explainability in terms of causal relations, is some degree of interpretability that can be provided by a variety of methods (discussed in later chapters). **The important concept to get here is that interpretability, as a lighter version of explainability, can be fulfilled also without an algorithmic equivalent of the ML system (complete set of rules) but with artifacts like local approximation of the systems, weights of different features in generating the outcomes, or generation of rules through decomposition.**

It is normal that these terms are not clear at this point of the book, but it is enough to get now the basic idea: using a black box ML model to solve a problem that doesn't fit an algorithmic resolution. Even though this black box might not be suitable for full explainability we can do things (like approximate the black box with an interpretable model somehow) to reach the needed level of interpretability and start to trust the model itself. Also we will see how asking for a model to be interpretable and/or explainable will help on building a better model as well with best practices that on one side may guarantee the explanations and on the other avoid problems with overfitting and bias among the others.

1.7 Summary

- Understand what is meant by "Machine Learning systems are getting "opaque" to human understanding."
- Use practical examples to show how the trust in ML may be reduced without the possibility of answering "Why" questions about the output and adopted criteria.
- Explainable AI is an emerging discipline inside machine learning to make ML models more interpretable. You need XAI to generate trust on ML.
- Explainable AI might be critical for the application of ML in regulated industries like finance and health. Without XAI, ML application might be strongly reduced in scope.
- Distinguish and understand the primary applications of XAI in terms of model validation, model debugging and knowledge discovery.
- Explainability and Interpretability are often used interchangeably, but they mean different things.

In the next chapter, we detail the different approaches to XAI depending on the specific context (which ML model needs to be interpreted) and the needs (which level of explanation I'm looking for).

References

ChaosKomori. (2003). *The physics of football*. DeviantArt. Retrieved from https://www.deviantart.com/chaoskomori/art/The-Physics-of-Football-1870988

Deutsch, D. (1998). *The fabric of reality*. Penguin UK.

Doshi-Velez, F., & Kim, B. (2017). *Towards a rigorous science of interpretable machine learning*. arXiv preprint arXiv:1702.08608.

Du, M., Liu, N., & Hu, X. (2019). Techniques for interpretable machine learning. *Communications of the ACM, 63*(1), 68–77.

Gilpin, L. H., Bau, D., Yuan, B. Z., Bajwa, A., Specter, M., & Kagal, L. (2018). Explaining explanations: An overview of interpretability of machine learning. In *2018 IEEE 5th International Conference on data science and advanced analytics (DSAA)* (pp. 80–89). IEEE.

Karim, A., Mishra, A., Newton, M. A., & Sattar, A. (2018). Machine learning interpretability: A science rather than a tool. *arXiv preprint arXiv:1807.06722*.

Metz, C. (2016). *How Google's AI viewed the move no human could understand*. Retrieved from https://www.wired.com/2016/03/googles-ai-viewed-move-no-human-understand/

Nguyen, A., Yosinski, J., & Clune, J. (2015). Deep neural networks are easily fooled: High confidence predictions for unrecognizable images. In *Proceedings of the IEEE Conference on computer vision and pattern recognition* (pp. 427–436). IEEE.

Ribeiro, M. T., Singh, S., & Guestrin, C. (2016). Why should I trust you? Explaining the predictions of any classifier. In *Proceedings of the 22nd ACM SIGKDD International Conference on knowledge discovery and data mining* (pp. 1135–1144). ACM.

Samuel, A. L. (1959). Some studies in machine learning using the game of checkers. *IBM Journal of Research and Development, 3*(3), 210–229.

Singh, S. (2017). *Explaining black-box machine learning predictions*. Presented at #H2OWorld 2017 in Mountain View, CA. Retrieved from https://youtu.be/TBJqgvXYhfo

Turing, I. B. A. (1950). Computing machinery and intelligence-AM Turing. *Mind, 59*(236), 433.

Chapter 2
Explainable AI: Needs, Opportunities and Challenges

> *One could put the whole sense of the book perhaps in these words: What can be said at all, can be said clearly; and whereof one cannot speak, thereof one must be silent.*
>
> Ludwig Wittgenstein

This Chapter Covers:

- What is an explanation and how to evaluate it
- Subtleties on the need of making a ML model explainable
- High level overview of the different XAI methods and properties.

This chapter is a bridge between the high-level overview of XAI presented in Chap. 1 and the hands-on work with XAI methods that we will start in Chap. 3. The chapter will introduce a series of key concepts and a more complete terminology as you will find in literature and papers.

The examples in Chap. 1 made evident the cases in which XAI needs to be coupled with ML model predictions to make them useful and trusted. We also saw what is really meant by words like explainability and interpretability.

The core of this chapter is to provide a general presentation of XAI methods with the proper taxonomy. In order to do this, we will rely on a practical example in which the goal is to forecast sales of a product depending on the age of the customers.

Before jumping to a general presentation of XAI methods, we will start this chapter with some discussion about how to evaluate an explanation from a human standpoint and the role that humans may play into ML and XAI pipeline we presented in the previous chapter.

2.1 Human in the Loop

In Chap. 1, we briefly touched on the ambiguities that emerge when using different words like explainability and interpretability. The position adopted in this book is that explainability is a stronger request than interpretability. Interpretability is a first step to gain explainability in which methods are adopted to get some hints on how the ML model is producing a certain output. But explainability requires a full understanding of the ML model, the possibility to answer "Why" questions at the point of being able to anticipate the outcomes.

The tricky point in this path is how to check that an explanation is good enough and who is the audience for the "good enough" explanations. You may easily understand that a good explanation for a ML practitioner in terms of technical details might not work at all for people coming from outside of ML world. People not in the field might not increase their trust in the model with such kind of technical assessment only.

2.1.1 Centaur XAI Systems

If you think about a XAI system as a "Centaur" combining a Machine Learning model with a human trying of making sense of the explanations coming from the model, we can use it as a model for the "Human in the Loop"-XAI paradigm.

In Fig. 2.1 we have portrayed a typical task such as an AI Classifier which outputs a "dog vs. wolf" classification with some probability.

If the Classifier is not confident enough, it can ask for help from the human. The human can add new annotations to the pictures or makes fine-tunings to the model.

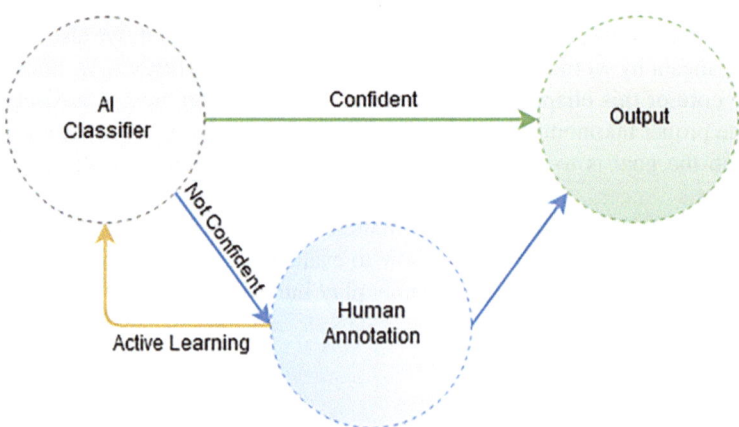

Fig. 2.1 The Human in the loop improves the performance of an AI Classifier taking part in the training process

2.1 Human in the Loop

This form of collaboration between AI systems and humans is aimed at further improvement of performance. Of course, humans can use active learning even in earlier steps by choosing a best-suited model for the Classifier or building smarter features the model can use in the training process.

The typical example to understand this is to think about what has been happening in chess. The common belief that followed the victory of IBM's Deep Blue against Gary Kasparov in 1997 was that there was not further room for humans to play chess against machines. But it was the same Kasparov to raise the doubt, 1 year later in 1998, that maybe a human might not defeat an AI system anymore but a "human in the loop" could successfully collaborate with a machine to play chess against a human or a machine with better performance.

Long story short, freestyle chess tournaments were organized and the results of these Centaurs systems (humans + AI) have been impressive against the common belief: a strong computer + a strong human consistently beat the strongest computers on their own. But it is even more than that; it has been shown how three weak computers + two young chess amateurs players won against Hydra, one of the most powerful chess AI at the time. Where the three weak computers had different recommendations, the humans may interact with the system to do further analysis and take a decision. In a way, the interaction between humans and machine models is more than the mere sum of the parts. The use of machine learning models exposes some human planning ability that artificial machine models still lack.

This example that comes from chess can be extended in a more general way. **Putting humans in the loop is not only needed for XAI but also to achieve better results.** This can be effectively summarized **in a variation of Pareto's 80:20 rule**.

> The usual call for Pareto principle (V. Pareto was an Italian economist that set the 80/20 rule in 1896 at University of Lausanne) is that for a variety of events the 80% of effects usually come from 20% of the causes and it is usually adopted as a mantra in the corporations to say that 80% of sales come from 20% of clients, or that 80% of the software quality is achieved fixing 20% of the most common defects. The mathematical root of this outcome is to be searched in the underlying power-law probability distribution that produces the phenomena.

In this case, an ideal Machine Learning system is one in which 80% of the results are AI-driven, but in order to further boost accuracy, 20% of the efforts should come from humans (Fig. 2.2).

Humans have an active role not only as in the XAI domain to receive and produce explanations but also as active players to enhance the overall performance as in Centaur collaboration models. **Machines are good at providing answers, but humans are often better at finding the right questions** to take critical decisions or to interpret results for rare cases outside of the learning dataset.

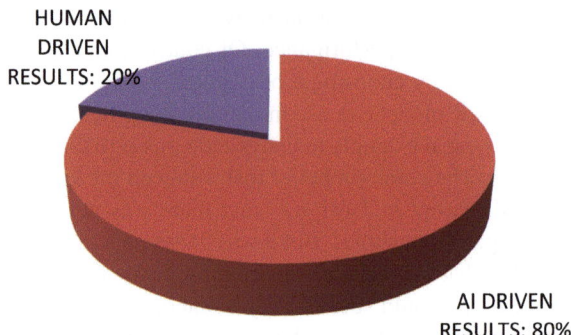

Fig. 2.2 The Pareto principle suggests that an efficient ML system must have 20% of creative (human) effort

Also, models are shockingly weak when it comes to answer questions such as "What is more beautiful?" or "What is the right thing to do?" or to answer those cases arising from explainability. For example, it is an open question how to create a model that will be Fair and not affected by bias like the "Man is to Woman as Programmer is to … Home maker" problem we talked about in Chap. 1.

But pragmatically a model can ask human collaborators, "What answers are fair and which ones are not fair?" and then attempt to learn from the answers. For such reasons in XAI humans can contribute to all four F.A.S.T. aspects of explaining a model (remember that he F.A.S.T. acronym is a reminder for the attributes fairness, accountability, security, and transparency).

Do you remember the common belief about the need to trade-off between explainability and performance we already talked about? This is another angle to demystify it. Performance comes together with "humans in the loop" boosting performance and explainability at the same time.

For example using some field expertise, we can build smarter features making simpler and more explainable models while boosting performance.

Back to the specific XAI scope but having in mind this pattern of collaboration, XAI systems put a "human in the loop" to make sense of the output by adding explanations and interpretations that can be understood by the humans with an interface open to the users.

2.1.2 XAI Evaluation from "Human in the Loop Perspective"

As outlined by Gilpin et al. (2018), an explanation can be assessed by two main features that are its interpretability and its completeness.

The main objective of **interpretability** is to provide a set of descriptions that allows a person to understand what the ML model is doing, up to the extent of gaining meaningful knowledge and trust in the system, according to a person's specific needs.

2.1 Human in the Loop

Completeness is the accuracy of the description of the systems up to the possibility of anticipating the results of the model. So a description of the model is complete if it can distill all the knowledge of the model in human-understandable language.

In this sense, it could be difficult to trade-off between interpretability and completeness because the most interpretable explanations are usually simple and a description of the model that is complete can be as complex as the model itself. As argued by Gilpin: "Explanation methods should not be evaluated on a single point on this tradeoff, but according to how they behave on the curve from maximum interpretability to maximum completeness" (Gilpin et al., 2018).

This curve depends explicitly on the specific human that has been put in the loop to find the most suitable trade-off for the specific scenario under examination. Among the different options, we follow the approach provided by Doshi-Velez and Kim (2017) to summarize the different categories of explainability based on the role played by the human in the evaluation (Fig. 2.3):

- **Application-grounded evaluation** (real humans—real tasks): This approach regards humans trying to complete real tasks relying on the produced explanations coming from the application of XAI to the ML system. In this case the assumption is that the human is an expert in the domain of the task (e.g. a doctor doing a diagnosis for a specific disease). The evaluation is conducted based on the results achieved by the human (better quality, less errors) relying on the explanations.
- **Human-grounded evaluation** (real humans—simple tasks): In this case we don't have domain experts but just lay humans to judge the explanations. The evaluation approach depends on the quality of explanation independently from the associated prediction. For example, the human may be provided an explanation and an input and is asked to simulate the model's output (regardless of the true output). We can think of a credit approval system for example, where a loan

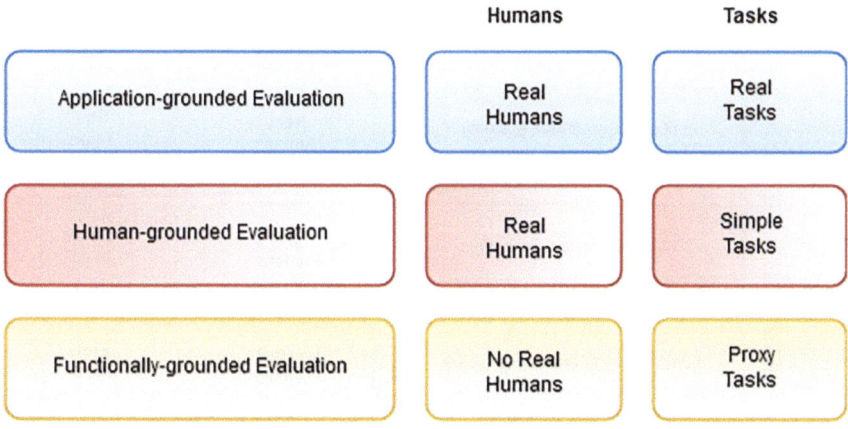

Fig. 2.3 Categories of explainability based on the role played by the human in the evaluation

officer must explain to a bank customer why their loan has been refused. In this case the system can show to the costumer some minimal set of pertinent negative features (e.g. number of accounts or age) that once altered would change their eligibility into a different state.

- **Functionally grounded evaluation** (no real humans—proxy tasks): This kind of evaluation is usually adopted when there are already a set of models that have been validated by human-grounded experiments as explainable, which are then used as a proxy for the real model. They are used when there is no access to humans to perform evaluation. So back to the example of the credit approval system, in this case we don't have humans to judge the explanations but we use an interpretable model that should act as a proxy to evaluate the quality of the explanations. A common proxy is a decision tree as a highly interpretable model, but careful trade-offs need to be achieved between choosing a full interpretable proxy and a lees interpretable proxy method that better represents the model behavior. The relation between interpretability and completeness helps also in better answering the question about the real need of explainability for ML models as we will discuss in the last section of this chapter.

Looking at the overall path we have followed so far, there is not an easy way to assess the level of explainability of a model quantitatively. And the main reason for this, is that this level has to be assessed by a human with different knowledge of the domain and different needs from case to case (up to the point that the evaluation may not include humans at all as in the functionally grounded approach).

We set explainability as something more than interpretability but at the same time we discussed how going toward explainability may mean increasing completeness (accurate description of the system to anticipate his behavior) but trading off interpretability (descriptions that are simple enough for a human with his specific domain knowledge).

The main takeaway of this digression is to use the factors we analyzed to properly set the priorities for XAI depending on the context. Looking again at Fig. 2.4,

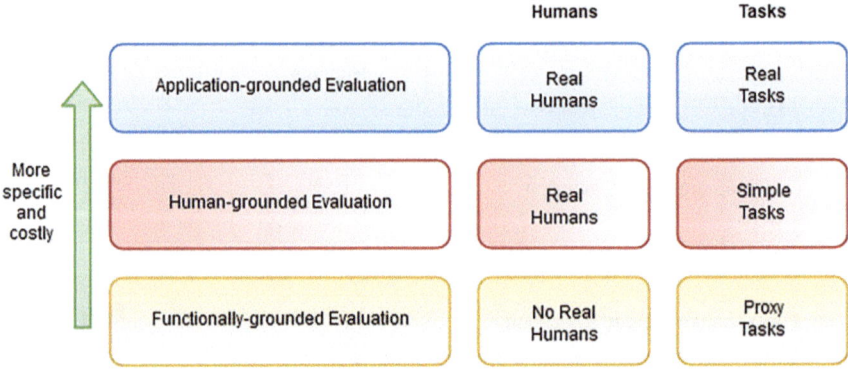

Fig. 2.4 Categories of explainability based on the role played by the human in the evaluation, arrow indicates increasing costs and complexity

we see that it is the same one we used to set the different types of evaluation but with an additional arrow on the left side: going up from functionally grounded to application-grounded evaluation usually means to increase cost and complexity. And we are now aware about the fact that increasing completeness of explanations can reduce interpretability. With these coordinates in mind, we are in the position to make decisions about the right evaluation for XAI based on the goals and constraints (Fig. 2.4).

2.2 How to Make Machine Learning Models Explainable

There is not a unique approach either to have a quantitative assessment of explainability or to define a taxonomy of XAI methods. Independently from the categories that we will use, it is important to have a real feeling of the factors that will orient the choices on the methods to make ML models explainable. Because of this, we will walk through a real-case scenario that, albeit very simple, provides a tangible representation of these concepts. Suppose that you work in a marketing department and you are asked to provide a rough model to forecast the sales of smartphones depending of the age of the costumers. The rough data are shown in Fig. 2.5.

Looking at the diagram, it is evident how the purchases are scattered in the plane without too much "regularity" going up and down. What we are saying here as "up and down" can be translated into a more formal knowledge of the learned function characteristics. Basically, we can distinguish three main behaviors of the learned function that give us indications about the level of explainability of the model:

- **Linear functions**: These are the most transparent class of machine learning models. Having linear functions means that every change in an input feature produces a change in the learned function at a defined rate and in one specific direction with a magnitude that can be read directly in an available coefficient of

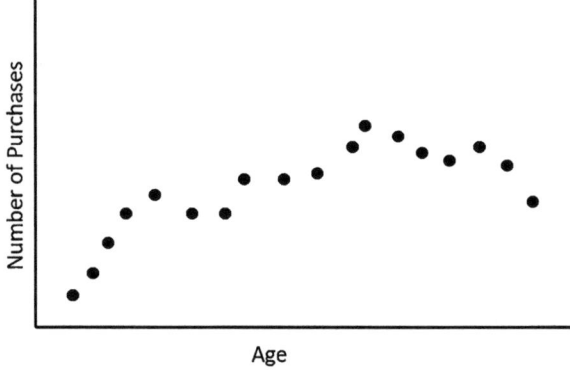

Fig. 2.5 Purchases of smartphones depending on age of the customer

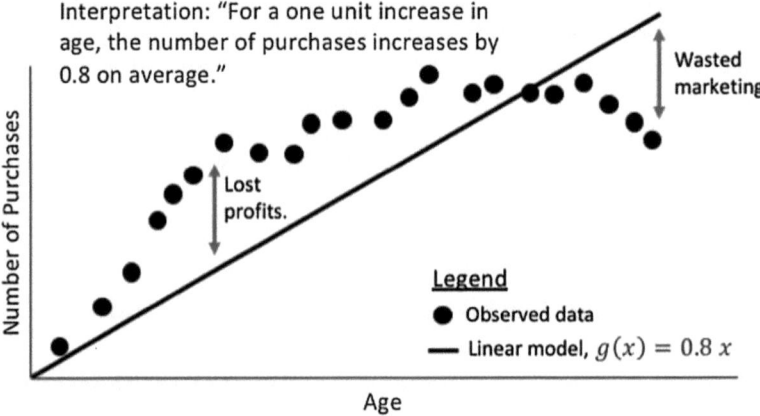

Fig. 2.6 A linear monotonic function provides straightforward, ready-to-use explanations, capturing the global characteristic of how purchases vary with each additional unit of age

the model. Of course, all linear functions are also monotonic. Just to recall, a monotonic function is function that is increasing on its entire domain or decreasing on its entire domain.
- **Nonlinear, monotonic functions:** in this case we don't have a single coefficient as direct evidence of the magnitude of the change in the output given a change in a feature input but in any case, the change remains in one direction only for a given input feature variation.
- **Nonlinear, non-monotonic functions:** these are hard to interpret because the response function output changes in different directions and at different rates for changes in an input feature.

Let's get back now to our example but now trying to predict the sales. Fig. 2.6 shows the case of a model trying to predict the number of purchases of smartphones based on the age of the consumer with a linear monotonic function.

The explanation is ready to go, the slope of the line tells us how much the number of purchases is expected to increase for a variation of one unit of age. Now look at the same problem in Fig. 2.7 but with the adoption of a nonlinear, non-monotonic function.

As you may easily realize, we lost the possibility of getting a single slope to explain the relation between the number of purchases and the age. The first linear model doesn't capture enough information of how purchases are related to age of the customers. This means that assuming that older customers are expected to buy more smartphones is not true across all the ages, there are intervals in which the trend changes. And to capture this trend we need to add complexity to our ML model adopting a nonlinear non-monotonic function.

But the comparison between these two figures tells us another important thing: the linear model doesn't fit so well the observed data, so you are getting very good explanations at the price of losing accuracy on predictions.

2.2 How to Make Machine Learning Models Explainable

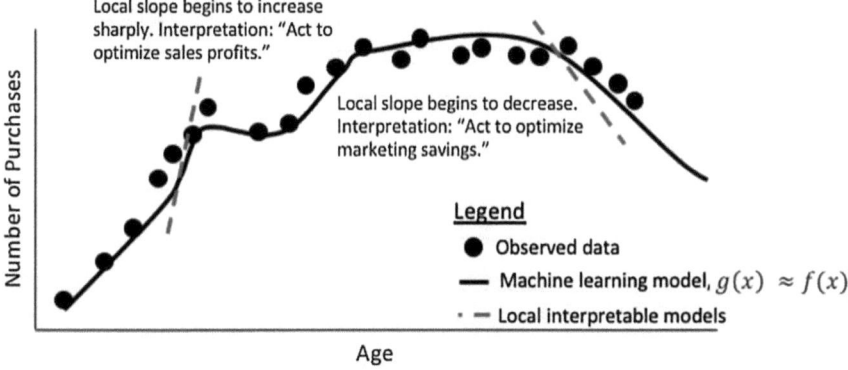

Fig. 2.7 With nonlinear non-monotonic function we lose a global easily explainable model in favor of accuracy improvement

And yes… for smart readers like you this is a confirmation of the common belief we promised to demystify in Chap. 1: there is a trade-off between performance and explainability, getting more performance may reduce explainability but keep trusting us and we will show how it is not always the case. We may guess a potential objection to our conclusion related to possible overfitting. We may argue that the second better performing model is overfitting the data and working on reducing overfitting would keep the performance and improve the explainability. At this stage, we say that this could be the case but we are not yet able to detail this argument. Assuming no overfitting for the moment (the model here is just qualitative without any quantitative details), the idea behind this example is just to show how improving performance "**in general**" may challenge explainability but we will see in Chap. 5 that the best practices for building better models (in terms of avoiding overfitting and bias among the others) may also help explainability and vice versa: explainability my guide ML model building.

Later in this chapter, we will look again at this figure to say something more about the dashed lines and what they mean in terms of explainability. Dashed lines represent local approximations of the model, it is like to use a linear model and get explanations that are valid only for a small interval to fit the local behavior of the more complex overall function. It is the time now, to progress further toward a taxonomy of the different XAI methods keeping in mind the ideas about the functions.

The first distinction is about **intrinsic explainability** and **post hoc explainability** (Fig. 2.8). In the first case, the model is built already as "intrinsically" explainable (think about the linear model of the previous example above) while in the second case the explainability is achieved at a later time after the model creation (post hoc refers to this fact, that the explainability is achieved in a second phase with an existing and running model in place).

Figure 2.8 shows the categories that we will use to go through the different XAI techniques following the approach of Du et al. (2019).

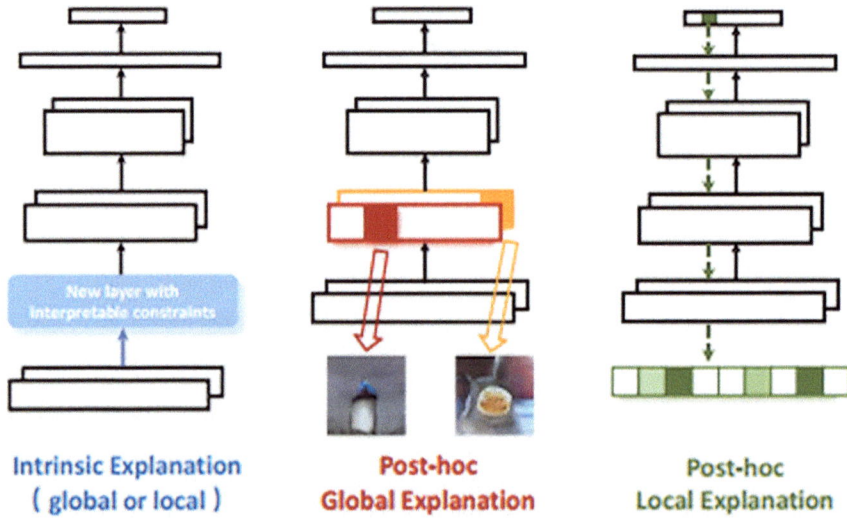

Fig. 2.8 Categories of different XAI techniques, Du et al. (2019)

All three schemes are about analyzing a Deep Neural Network (but the type of model is immaterial), and the first distinction is in fact between providing an intrinsic explanation and to give post hoc ones.

The second level of distinction in the taxonomy that we envision is more about scope; whatever intrinsic or post hoc explanations, we may further split between global (user can understand how the model works across all the range of data) and local explanations (specific explanations are provided for individual predictions). Let's go to the next sections to better details these general categories.

2.2.1 Intrinsic Explanations

There are two main classes of models that can be defined as intrinsically explainable (synonyms that can be used are white-box or interpretable models). In the first class, the subset of algorithms that produces the ML models can be directly interpreted. We take as representatives of this class the **linear regression, logistic regression, and decision tree models**. We will detail all of them in the next chapter, but it is important to start understanding why we call them intrinsically explainable.

Linear regression is used to model the dependence of a target (Y) from a set of features $(x_1 \ldots x_k)$ with a linear relation.

$$Y = m_0 + m_1 x_1 + m_2 x_2 + .. + m_k x_k \tag{2.1}$$

Equation (2.1) Standard linear relation between a target Y and a set of features $(x_1 \ldots x_k)$.

2.2 How to Make Machine Learning Models Explainable

The obvious advantage in terms of explainability is that we have the weights $(m_1...m_k)$ to compare to understand the relative importance of the different features. This is a more general case of the example we did to forecast purchases. In that example, we have just one feature (age) to model the relation with purchases. Here we have more than one feature as purchases would depend not only on age but also on other factors (e.g. gendersex, salary). Whatever the situation, the weights directly show the importance of the feature in predicting the outcome.

Logistic regression is a variation on linear regression to handle classification problems. We will talk again about it in Chap. 3 with a real working case, but it is important to start familiarizing yourself with it.

Do you remember the example of dogs and wolves we described in Chap. 1? It is a classical classification problem that is not addressed well by linear regression. Assuming you have these two classes (Dog vs. Wolf) and images to classify, if you heuristically try to use a linear regression model, it will fit and split the data between the two classes. But it just finds the best line (or hyperplane for more than two dimensions) to interpolate and split the set. That could be a problem.

In case of logistic regression, what we are searching for is something that gives us as output the probability that a specific item is a dog or a wolf and the probability by definition runs between 0 and 1. So in our case, we would set 0 for dogs and 1 for wolves. But using linear regression results with a fitting to the data that will produce numbers below 0 and above 1 which are not good for classification. So, you don't have a direct interpretation of the output as the probability for a given item. This is fixed in logistic regression but with a bit more complicated function than Eq. (2.1) above:

$$P(Y=1) = 1/\left(1 + \exp\left(-\left(m_0 + m_1 x_1 + m_2 x_2 + .. + m_k x_k\right)\right)\right) \qquad (2.2)$$

> To deepen the concepts Eq. (2.2) solves two theoretical problems in Machine Learning: it is the simplest equation that maximizes the Likelihood function of probability theory and it gives a fair robust convex loss function. This makes it is resilient to noise and very easy to train. For such reasons, it has a very wide application even as part of complex models.

In the case of logistic regression, we don't have a direct mapping between the weights and the effect on the outcome (no linear relation) but it is still possible to make sense of them to produce explanations as we will see in the next chapter. Linear Regression and Logistic Regression have in common the linear structure, such that if you draw in a bi-dimensional graph the samples you are trying to predict or classify, both Linear Regression and Logistic Regression split the different classes with a straight line. In this case we say that they both have linear decision boundaries.

The third main representative for intrinsic explainable models is **decision tree**. It is strongly different from logistic regression and linear regression. It doesn't have a simple linear decision boundary, can be used both for classification and regression, but it is not based on a function to fit the data. It works by partitioning the information until the right subsets are identified. The splits are performed on putting cutoff values on the features and the procedure can address also the case of nonlinear relations between features and outcome but keeping a strong level of explainability. We can look at Fig. 2.9 to have a first intuition on what is happening.

The tree shows the probability of surviving to Titanic disaster based on some features as sex, cabin class, and age. There are two labels: "yes" is for survival and "no" for not survival. In each box you can see the percentual of that label. So in the first box we can read "no" 67.7%, the proportion of people that have not survived, which also is impliedly saying that the complementary 32.3% has survived.

Traversing the tree, it is easy to get explanations of how the features are playing to determine the probability of survival. In the box below, we provide some further details about how decision tree works but consider that we will talk again and deeper about decision trees in Chap. 3.

We want to be very clear here and avoid any misunderstanding with our readers. In this context we said just enough about linear regression, logistic regression and decision tree to pass the idea of explainability as an "intrinsic" characteristic of the models without the need of any further technique to interpret them. Each of these three models will be explored in depth later. For now, it is important to get comfortable with the idea of "intrinsic explainability" to position it in the general taxonomy of the available methods.

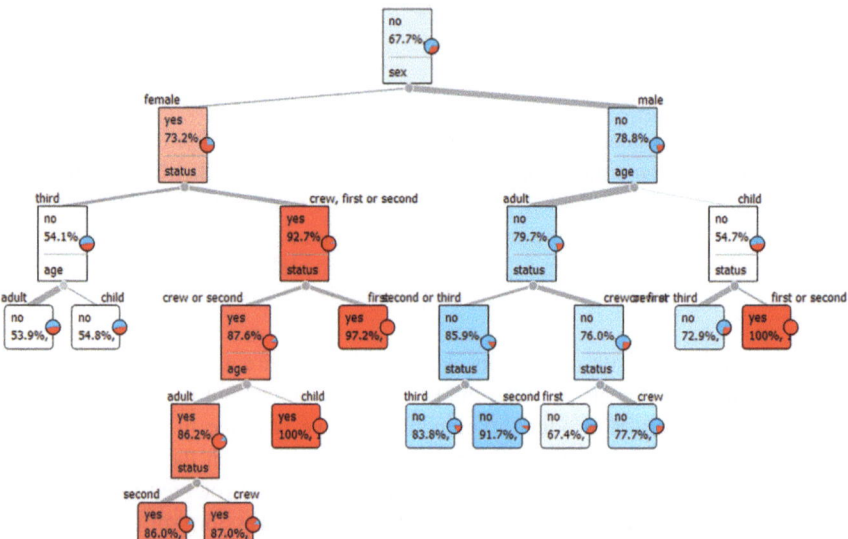

Fig. 2.9 A decision tree applied to the Titanic Dataset. It is easy to explain how the model predicts the probabilities of survival

2.2 How to Make Machine Learning Models Explainable

> To the reader who is acquainted with machine learning we remember Decision Trees do their splitting to achieve the **maximum purity** in the target variable, which means they maximize how homogeneous the groups are.
>
> If you have ten bicycles and you are splitting by color, if you end with ten red bicycles and 0 blue the group is 100% pure while if you end with five red bicycles and five blue bicycles you achieved a 100% impure group. Both Gini index and entropy are quantitative and general measures of this idea of purity. **Gini index** is directly linked with the example we made, and we would have zero for the purest case (ten red bicycles, 0 blue) and 0.5 for the worst case (five red bicycles, five blue ones). **Entropy** is a bit more complex in terms of formula and shaped to have a measure of disorder in a group (that is another aspect of pureness), by the way the boundaries would be the same with 0 entropy with ten red bicycles and 0 blue (that is also a very ordered group because you can achieve it in just one configuration and a full split on the target variable that is color) and entropy equals to one in the case of five red bicycles and five blues ones (disordered state in respect to color).

Beside the models that provide explainability by design, as mentioned, there is another class of intrinsic explainable models we'll call **tempered models**. In this case we may start from a model that is non-explainable but assuming we have the ability of modifying it (as with the specific case of a model we are building from scratch or we have access to its internals), we change it by adding interpretability constraints.

To get the idea of this, you may want to force the nonlinear model depicted in Fig. 2.7 (predicting purchases of smartphones depending from age) but force the relation to be monotonic, to guarantee that the direction of change in the outcome is always the same and simplify explainability (if a feature grows it always produces the same effect on the outcome). Yes, you may have already guessed that this would mean to get back to linear model of Fig. 2.6 and this is exactly what we would do to have a tempered model in this simple case. What is the risk here? The risk here is that forcing the constraints from outside the model may reduce the performance of the model albeit improving explainability as we already discussed from another perspective in Sect. 2.2 discussing Figs. 2.6 and 2.7.

(Yes, another time the myth seems to be confirmed about the trade-off between performance and explainability, but we will see an explicit example about XGBoost models in which adding explainability constraints will enhance the performance, so stay tuned and go with the flow. The spirit of this journey is to provide the information when needed not as general reference "all at once".)

2.2.2 Post Hoc Explanations

What if we are not in the case of "intrinsic explanations"? We can still rely on **post hoc explanations with two main variants: model-agnostic explanations and model-specific explanations**. As we said "post hoc" here means that the explanations are generated with an existing model already in place that is not always interpretable.

The techniques that belong to the ***model-agnostic category*** treat the model as a black box without accessing to the model's internal parameters. The strength of the agnostic methods is that they may be applied to any ML model to generate explanations. A typical example to get the idea before details is the "Permutation Feature Importance."

In this method the relative importance of a feature compared to the others is evaluated by looking at how the predictions are impacted by a permutation of the values of that specific feature in the dataset. Indirectly we may build explanations leveraging the features that contribute more on building the output.

Only as note here to keep in mind and to be discussed in depth in Chap. 4, another promising attempt is to train an intrinsically explainable model from the output of the *Blackbox* one. This approach is known as "knowledge distillation." We use the powerful and complex black-box model to search the complete dataset for the solutions and then use the results to guide a simpler and intrinsically explainable ML model to replicate the behavior but only on the narrowed-down solution space. Basically, the black-box model acts as a teacher for the explainable model to replicate the same results but with the possibility of providing explanations.

> We cannot stress enough the importance of model-agnostic explanations from a practical point of view. Suppose you are called to explain a model created by a data scientist, but you don't know anything of the methodology used to train the model and you have to test the weakness of the model and explain its answers to someone else. In this case, you may use an agnostic method, maybe a local one like LIME or SHAP (wait for future chapters for details on these methods). Agnostic methods have the advantage of being extremely friendly to the user and easy to use so they can be used even from someone who doesn't know anything about Machine Learning or Computer Science. This is exactly the reason that is causing so much interest and investments on these methods in industry.

The other possibility is to rely on ***model-specific explanations*** that are built specifically for each model through the examination of the model internal structure and parameters. Model specific explanations are usually complex because they need to deal with the inner structure of the ML model. An example is to use back-propagation (widely known in ML) from the other way around: trace back the model from output to input following the gradients (direction of MAX change) to identify the feature that most contributes on building the output. Another example comes from decision trees.

There is a simple yet more powerful version of decision tree called random forest which is not explainable, but do you remember how decision tree partitions information? A decision tree partitions information using an indicator like Gini impurity or entropy (see the note above), so we can empirically say that sudden variation in this indicator corresponds to important decisions and an XAI system may weigh the importance of features leveraging the related variation of this indicator that is a measurement of information gain at each split until the final label is assigned.

Again, with the risk of being boring, we are not assuming that you may have a full understanding of how these methods works with these few lines, but it is important at this stage just to start distinguishing between the different approaches and getting an intuition of the main concepts.

At this point of our journey we are still missing the practical skills to make these methods working (we will tackle that in Chap. 3) but we are able to distinguish between the two main families of XAI methods: the intrinsic explanations and the post hoc ones and see the reasons behind the choice. In one case, for intrinsic explanations the model can be interpreted as is, looking at the parameters, while for post hoc situations the model is not directly interpretable, and we need to work on it as an agnostic black box or by playing with model internals to get features' relative importance. Also, we placed these methods into the wider context of "Man in the Loop" paradigm to show how the humans that asks for explanations for XAI are also active players to enhance the overall performance of the ML system.

2.2.3 Global or Local Explainability

As we mentioned, the distinction between global and local explainability is in terms of scope. With global explanations, we are mostly interested in how the model works from a global point of view, the mechanism that generates the outputs. In the case of local explanations, we are searching for an explanation of a single prediction or outcome.

The idea here is to approximate the ML model locally near a given input with an intrinsic explainable model that can be directly interpreted (as above for linear regression, logistic regression or decision tree). The local surrogate is used to make sense of specific predictions, but these explanations are valid only for the neighborhood of the specific input.

Let's look again at our model to predict purchases (Fig. 2.10):

The nonlinear, non-monotonic function fits very well with the data. It is hard to provide explanations, but the dashed lines suggest an approach: get local linear approximations of the function to provide explanations around some specific area. We won't have a single slope but different slopes in different regions to have

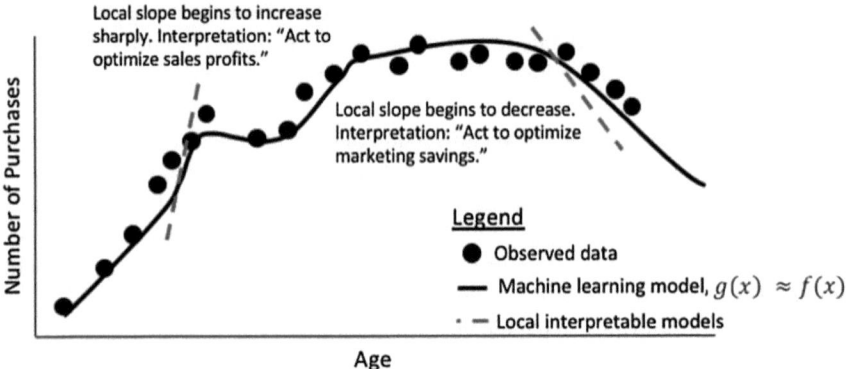

Fig. 2.10 Model to predict purchases

interpretations of what's going on. The dashed lines are exactly the local approximation of the model we are talking about. As you may see, we need a different line for different inputs to get an easy explanation in terms of a linear regression model that can locally approximate the nonlinear, non-monotonic one.

This example is also good to understand what we meant with the trade-off between interpretability and completeness. In case of linear regression, we had one single coefficient to explain the overall behavior of number of purchases with the age. But we saw how the fit with the data was not so good. With a more complex function we have better predictions, but we cannot provide explanations relying on a single parameter as before. Of course, we don't want to have function too complex to overfit data. To understand what's going on and anticipate the behavior of the system, we need to increase completeness of explanations. We rely on local approximations to interpret the results, so we produce different weights that have a validity limited to some set of data, losing a simple explanation for the overall. Providing explanations to humans in such a way could be difficult.

Let's try to be very clear about this point with our example. As we said, the scenario is that our marketing department has been asked to forecast the sales of a selected item depending on the age of the customers. We come back with a complex function that is very accurate in predicting the sales depending on age but up to now this is just Machine Learning without XAI. XAI starts to play its role to explain to the people that assigned us the work, "how" our model works in order to trust it. We have two options:

- **Achieve full interpretability**: We use the linear regression model instead of the complex function and say that the purchases are expected to increase with age.
- **Achieve completeness**: We cannot present the relation between purchases and age as global result but discuss the data with local linear surrogates. In some areas purchases increase with age, in others purchases decrease with increasing age. You may easily understand that this would make the scenario a bit more complicated to be explained and to be trusted. We are in the position of anticipat-

2.2 How to Make Machine Learning Models Explainable

ing very well model's output with our local linear surrogates but losing some level of interpretability.

Another point needs to be considered to avoid over-simplification. In this first example, we supposed that the model relies on one feature, age. This is not like that in the real world in which we would have used a lot of features to model the case (e.g. salary and sex for example, similar to the case of linear regression but with multiple variables). And producing explanations with multiple features becomes more complex because you lose the possibility of a quick look at the diagram to see what's going on as is the case with age. We used this simplification just to make more visible the relation between interpretability and completeness, but we will have the rest of the book to see how to deal in terms of XAI with huge number of features.

To recap with a different visual picture of what we have learned so far, we can have a look at Fig. 2.11.

The two big families of *intrinsic explanations* and *black box methods* can be grouped in terms of what we are using to get explanations. In case of intrinsic explanations, we are mainly relying on "internals" because, as we saw, the parameters (or weights) may already provide the right level of explainability.

In case of black box methods, we start from the model "predictions" and probe the model to understand behavior (how the predictions change) or we get local approximations valid for subset of the predictions; in alternative we use model

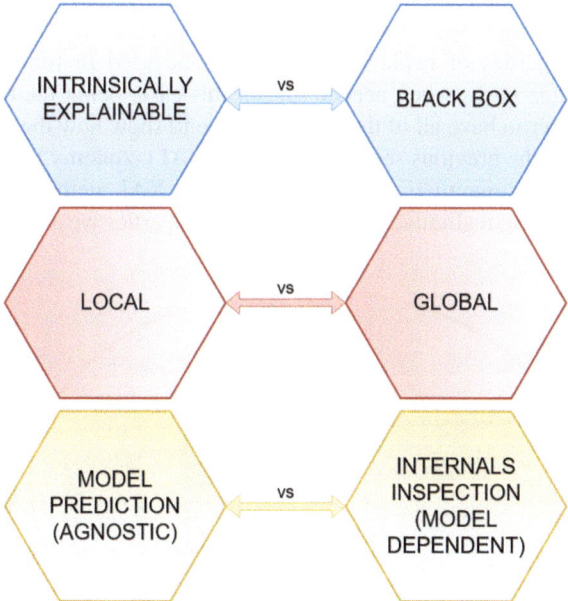

Fig. 2.11 A recap of explanations. We first divide models in intrinsically explainable and black box models. We then add the scope of explanations: local or global ones. Finally we split the methodologies in agnostic and model dependent

specific explanations *opening the box* but also in this case we don't have "ready-to-go" explanations provided by the internals as in the case of intrinsic explanations.

Let's get back to our XAI flow we presented in Chap. 1 as per Fig. 2.12 below, you might note that the classification of Du et al. (2019) used the term post hoc to put more emphasis on the fact that in this type of methods explainability is achieved at later time, after the ML model creation. The concepts do not change but we prefer to use our flow (Fig. 2.6) in the rest of the book in which the first split is made between the ML models that are explained from the outside as black boxes (agnostic approach) and the ones that are explained looking at the internals (including the case of Intrinsic explainable models). The other main difference is related to intrinsic explainable models: we consider them as a case of model-dependent approach to produce explanations (while Du et al. put them in a dedicated category); albeit explainable, we rely on specific model internals to interpret them. We don't use the term post hoc but it is important to keep it in mind as it is also present in XAI papers. We also think that it is better for you as reader to get used from the very beginning to different angles of looking at XAI considering the variety of views that you will be exposed in XAI literature as you dive deeper (Fig. 2.12).

2.3 Properties of Explanations

We need a last step before starting hands-on work with Python code to do XAI starting in the next chapter. We want to group here the main terms that have a specific meaning as properties of explanations that will be used in following chapters. Usually we define terms when needed, but in this case, being the terms logically grouped we prefer to have all of them in one place to show how they are related.

As we said in the previous sections regarding XAI taxonomy, it is not currently possible to have a quantitative assessment of the XAI methods and generated explanations but just qualitative evaluations. The properties we are going to discuss

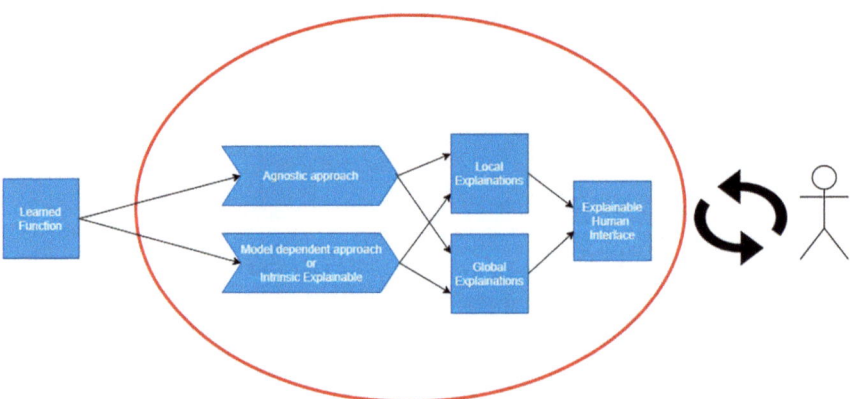

Fig. 2.12 XAI Main flow

2.3 Properties of Explanations

in this section follow the same path. We want to be as precise as possible in characterizing explanations and XAI methods although we cannot assign a "score" for the different properties. We may envision a real case in which our "human in the loop" collaborating with ML model may provide an XAI report in which the methods and produced explanations are tagged with the properties we are going to discuss in the list below, following the work of Robnik-Šikonja and Bohanec (2018).

The first group is used to characterize the explanations and the methods from a global perspective (not the individual explanations):

Completeness: The accuracy of the description of the systems up to the possibility of anticipating the results of the model.

Expressive Power: This regards the language adopted for the explanations. As we are seeing, each explanation could be expressed in different ways, with different techniques and for different "humans" (different needs, scope, and knowledge). There are different options for the explanations with different expressive power, they can be in the form of propositional logic (i.e. if-then-else), histograms, decision tress, or natural language to mention the main approaches.

Translucency: It describes how much the explanation is based on the investigation of the ML model internals. The two boundaries cases we may take as an example are the interpretable models like linear regression, in which the weights are directly used to provide explanations, and then the methods in which inputs are changed to see the variation in output (common in agnostic methods) with zero translucency.

Portability: This assesses the span of machine learning models covered by the specific XAI method. Agnostic methods will have a high portability in general while model specific explanations will have the lowest portability.

Algorithmic Complexity This is related to the computation complexity of the methods to generate explanations. It is very important in terms of being a potential bottleneck to provide explanations in case of huge complexity.

The group below refers to properties of individual explanations (subset of the ones in Robnik-Šikonja & Bohanec, 2018):

Accuracy: This is related to the usual definition of accuracy that comes out from ML but from an XAI point of view. In ML, accuracy is a performance metric defined as number of correct predictions to the total number of input samples. For our purposes, accuracy is an indication of how well an explanation may predict unseen data. This is related to the argument of generating knowledge as an application of XAI. Explanation may be used for prediction instead of the ML model and it should match at least the level of accuracy achieved by the ML system.

Consistency: This property describes the similarity between explanations generated from different models but trained on the same task. If the models generate similar predictions, the expectation is that related explanations are similar. We will see how consistency will be critical in the selection of XAI methodologies.

Stability: This property compares explanations between similar instances for a specific model. It differs from consistency which compares different models. If a small variation in a feature produces a huge change in the explanations (assuming that the same change has not produce a huge effect on the prediction) the explanations are not stable, and the explanation method has a high variance that is not good in terms of trust.

Comprehensibility: This is related to the arguments exposed in the "Human in the loop" section of this chapter. The attempt is to have an idea of how well humans may understand the generated explanations.

2.4 Summary

- Evaluate explanations from a human point of view. Recognize the role of humans and tailor explanations for specific audience.
- Distinguish between interpretability and completeness for explanations and set the proper trade-off between the two depending on the main goal: achieve a detailed description of the system or privilege easy explanations for the audience.
- Place XAI methods in a proper taxonomy
  - Recognize Intrinsic explanations vs. Post hoc explanations
  - Use explanations in the right scope: Global vs. Local
- Locate XAI methods inside main XAI flow that will be used in the next chapters to address real-case scenarios.
- Envision a report to have an assessment of explanations and adopted XAI methods, using the right properties.
- Learn how to think critically about XAI and challenge the real need of XAI to set the right level of explainability required.

In the next chapter, we will start hands-on work on interpretable models with specific examples leveraging Python. The goal will be to have practical cases to work on and provide explanations on the ML models predictions, relying on the properties of these models that are intrinsically explainable.

References

Doshi-Velez, F., & Kim, B. (2017). Towards a rigorous science of interpretable machine learning. *arXiv preprint arXiv:1702.08608*.
Du, M., Liu, N., & Hu, X. (2019). Techniques for interpretable machine learning. *Communications of the ACM, 63*(1), 68–77.
Gilpin, L. H., Bau, D., Yuan, B. Z., Bajwa, A., Specter, M., & Kagal, L. (2018). Explaining explanations: An overview of interpretability of machine learning. In *2018 IEEE 5th International Conference on data science and advanced analytics (DSAA)* (pp. 80–89). IEEE.
Robnik-Šikonja, M., & Bohanec, M. (2018). Perturbation-based explanations of prediction models. In *Human and machine learning* (pp. 159–175). Springer.

Chapter 3
Intrinsic Explainable Models

What I cannot create, I do not understand

Richard Feynman

> This Chapter Covers:
> - XAI methods for Intrinsic Explainable models:
> - Linear and Logistic Regression
> - Decision Tree
> - K-Nearest Neighbors (KNN)

The main objective of this chapter is to show how to provide explanations for intrinsic explainable models. As we said, for this category of ML models, XAI can be achieved by looking at the internals with the proper interpretations of the weights and parameters that build the model. We will make practical examples (using Python code) that will deal with the quality of wine, survival properties in a *Titanic*-like disaster, and for the ML-addicted the ever-green categorization of Iris flowers.

Do you remember the XAI flow (Fig. 3.1) we envisioned in Chap. 1?

As you may see from Fig. 3.1 and as already explained, we consider intrinsic explainable models in the same path of model-dependent approach for XAI: we can provide explanations because these models are not opaque, we can look at their internals, but each model is different from the other in terms of what to look at to provide explanations. Being intrinsically explainable, we won't' have any issue on providing global (how the model works overall) or local explanations (give rationales for a single prediction).

We will focus on the concepts so that people may translate the same flows from Python to other programming languages or tools (e.g. R).

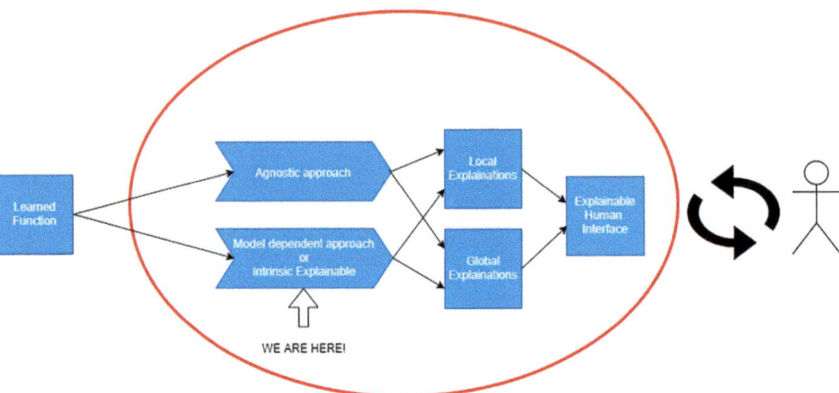

Fig. 3.1 XAI flow: intrinsic explainable models

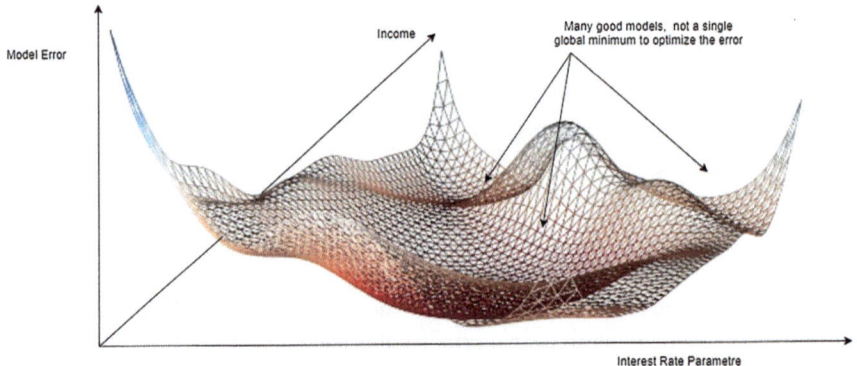

Fig. 3.2 An Illustration of the error surface of machine learning model

3.1 Loss Function

Before going into details of our real-life scenarios, we need to revisit some concepts related to the loss function we already mentioned in Chap. 1. Do you remember the case we talked about before, about the risk associated to grant a loan to a customer (Fig. 3.2)?

As we said, if we model the error as a landscape, each choice of parameters makes a different model and we prefer the model that minimizes the loss function. Every local minimum could be a good choice for the model. Each different model would generate a different set of interpretations. So, the loss function is fundamental from a purely ML perspective but also from a more specific XAI angle. (The loss function in the case of Linear Regression is also called *Empirical Risk.*)

3.1 Loss Function

A sufficiently powerful model such as Neural Networks can have many minima in the loss function so the following situations produce a very complex loss function landscape:

- We have a terrible choice of features, some of which are irrelevant;
- Between samples there are strong outliers, samples that are very different from the majority of the samples;
- The problem itself is very difficult or badly posed (or, in fact, insolvable)

The connection we are making with XAI is that the regularization methods that are well known in Machine Learning and that are applied to handle such kind of complex situations also help in finding the most relevant features in a model.

Regularization makes the loss function smoother and provides indications about the relative importance of features in producing the output which helps provide explanations of the model.

> Just to clarify: we already used the terms "features importance" with different flavors (such as "relative feature importance"). There is no quantitative definition of this importance. The meaning is how much that feature contributes to building the output. Just as an example, we expect that the color of your favorite shoes won't contribute on predicting your health (low feature importance) while your weight might be an important feature in such a scenario.

As we know, we always express the training of a machine learning model via the minimization of a loss function. For example, we can express the loss as the mean of quadratic deviations from the expected value of the model as

$$\text{Loss}(w) = \frac{1}{2N} \sum_{i=1}^{N} (y_i - h(x_i; w))^2 \tag{3.1}$$

Think of it as the total sum of errors you commit taking the outputs of a hypothesis function instead of the true values of the examples. A choice of parameters (weights) for the hypothesis function uniquely defines the model. If we plot the loss value of the model in the parameter space, we would expect something like (Fig. 3.3):

Finding the minimum of loss in the parameters is equivalent to choosing the best model that is satisfying the prescribed loss function.

The most used method for finding such a minimum is gradient descent (GD). In gradient descent, we simply recursively update the weights in the reverse direction of the gradient and proportionally at the gradient's magnitude.

$$w_i = w_i - \eta \frac{\partial \text{Loss}}{\partial w_i} \tag{3.2}$$

For example, if locally the loss is increasing in space parameter we decrease the parameter value (Fig. 3.4).

From the courses on Machine Learning we know GD is very sensitive to the initial position and the roughness of the loss function landscape. This affect the convergence of the training process, for example we can be attracted and be stuck in a local minimum.

Technically, we can smooth the loss function partially by solving the roughness problem with **regularization**. So regularization gives us a more stable convergence of the gradient descent method, and it even accelerates it.

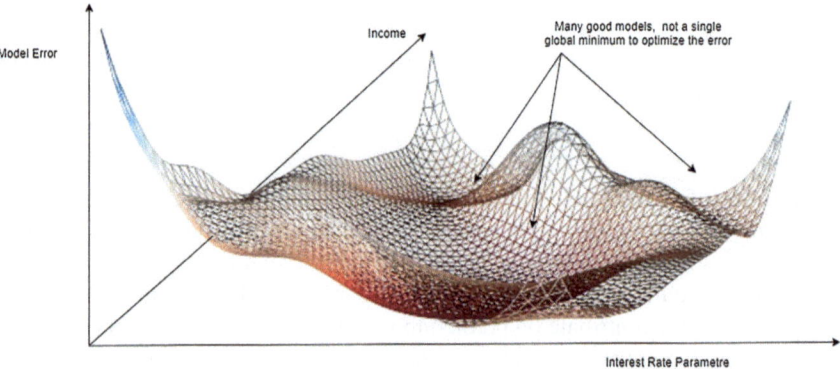

Fig. 3.3 (Smooth Loss Function in a generic parameter space)

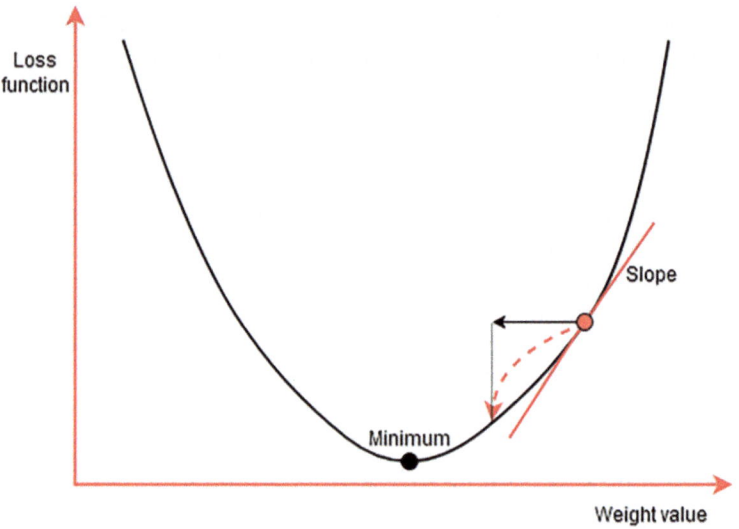

Fig. 3.4 In gradient descent we jump to the opposite direction to the gradient proportionally to the value of the gradient

There are many forms of regularization especially in the domain of neural networks but the most used ones are the *Tikhonov regularization (also called ridge or l_2 regularization)*

$$\text{Loss}_{\text{ridge}}(w) = \frac{1}{2N}\sum_{i=1}^{N}(y_i - h(x_i; w))^2 + \lambda \sum_{i=1}^{M} w_j^2 \quad (3.3)$$

and the Lasso or l_1 regularization

$$\text{Loss}_{\text{lasso}}(w) = \frac{1}{2N}\sum_{i=1}^{N}(y_i - h(x_i; w))^2 + \alpha \sum_{i=1}^{M} |w_j| \quad (3.4)$$

Adding a positive quadratic term in the parameters (or the sum of absolute values of weights) to the standard loss function transforms the loss into a smoother and more convex one. Moreover, the new positive term forces the weights to stay small, and this property gives less sensitivity to noise. In the absence of noise, and outliers ridge regularization usually gives more accurate results than Lasso regularization. It is even differentiable, but Lasso is more resilient to outliers because it gives less importance to greater deviations.

A remarkable property of Lasso regression is that it gives more weights to minor deviations. That helps us consider the true relevance of a feature and forces the weight relative to irrelevant features down to zero.

As an example, think of adding as a feature in a bank model for loans the zodiac sign of the loaner. With Tikhonov regularization you will have a small weight relative to the sign. With Lasso regularization that weight will be exactly zero.

So Lasso regularization will be one of our tools of choice to identifying the most relevant features in a model. We can even rank the features by feature importance.

We are now in the position of exploiting these concepts in the practical scenarios we will detail in the following.

3.2 Linear Regression

We start our journey through intrinsic explainable models with the application of linear regression to the problem of predicting the quality of red wine on a qualitative scale of 0–10, depending on some specific set of features. Suppose a wine producer heard about the "miracles" that machine learning may perform to improve business nowadays. The producer knows almost nothing about math, but he thinks that a good data scientist (that is you as the reader) may create insight into the chemical analysis of wine.

The idea of the wine producer is to leverage the results to increase the wine price and/or reposition the product on the market.

Here we are assuming that you as a data scientist decided to use a linear regression ML model to make a prediction about chemical analysis and wine quality. You want to provide explanations and interpretation of the results so to answer questions like: "Which characteristics of wine have more impact on quality?"

The tricky point will be that the answers to such questions are meant for the wine producer, so it won't be enough to show numbers. XAI must do its job of providing "human understandable" explanations.

Remember that linear regression belongs to the category of intrinsic explainable models. This means that we are in a position to get interpretations directly from the model weights. But let's start by getting familiar with the wine quality data that we will use from UCI Machine Learning Repository (UCI, 2009).

```
Wines.head() #A
#The Wine-set as our dataset
```

As we see (Table 3.1), we rely on the following 11 features to predict quality:
"fixed acidity," "volatile acidity," "citric acid," "residual sugar," "chlorides," "free sulfur dioxide," "total sulfur dioxide," "density," "pH," "sulfates," "alcohol."

It helps to take a first look at the general description of the feature statistics to see variation around max and min values.

```
Wines.describe()
```

A quick look at the table that is generated as output (we don't report it here because it is pretty large) shows the different scales for the various features besides the info about the min, max values and percentiles.

As we said, we will use linear regression to build our ML model and get predictions on wine quality. Let's repeat very quickly how linear regression would appear in two dimensions, as in the case in which quality would depend only on one feature (e.g. acidity).

The equation of the line that fits the data well (Fig. 3.5) is

$$Y = m_0 + m_1 x_1 \tag{3.5}$$

We use this two-dimensional simplification to have a visual explanation of the two weights m_0 and m_1:

m_0 represents the quality value in case of acidity = 0 ($x|1 = 0$)
m_1 represents the increase (decrease if negative) of quality for a unit increase of acidity

This needs to be generalized to the case of multiple features as below (11 features in our case, $k = 1…11$):

$$Y = m_0 + m_1 x_1 + m_2 x_2 + .. + m_k x_k \tag{3.6}$$

We can now write a few lines in Python to build the linear regression model for predictions. We will show below only the most meaningful lines (full code available with the book).

3.2 Linear Regression

Table 3.1 The wine DataFrame

	Fixed acidity	Volatile acidity	Citric acid	Residual sugar	Chlorides	Free sulfur dioxide	Total sulfur dioxide	Density	pH	Sulfates	Alcohol	Quality
0	7.4	0.7	0	1.9	0.076	11	34	0.9978	3.51	0.56	9.4	5
1	7.8	0.88	0	2.6	0.098	25	67	0.9968	3.2	0.68	9.8	5
2	7.8	0.76	0.04	2.3	0.092	15	54	0.997	3.26	0.65	9.8	5
3	11.2	0.28	0.56	1.9	0.075	17	60	0.998	3.16	0.58	9.8	6
4	7.4	0.7	0	1.9	0.076	11	34	0.9978	3.51	0.56	9.4	5

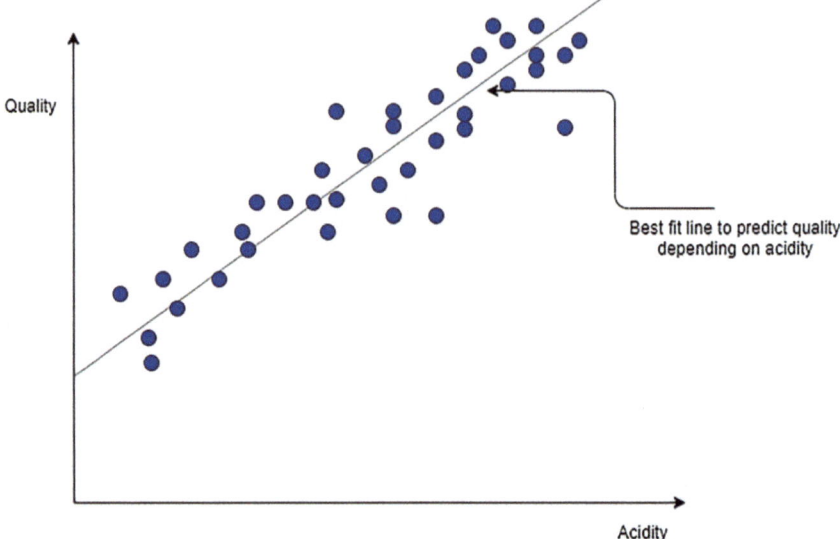

Fig. 3.5 Example of a linear regression in only two dimensions: Acidity and the target (Quality)

The code is based on the Scikit-learn free machine learning library for Python that everyone with some basic experience in Machine Learning knows and uses in everyday life.

You may see below the standard code to get Linear Regression on wine data.

```
df = pd.read_csv('winequality-red.csv')
X=df.iloc[:,:-1].values
Y=df.iloc[:,-1].values
x_train,x_test,y_train,y_test=train_test_split(X,Y,random_
state=3) #A

regressor = LinearRegression()
regressor.fit(x_train, y_train) #B

coefficients=pd.DataFrame(regressor.coef_,col_names)
coefficients.columns=['Coefficient']

#A usual splitting of data between train and test
#B Fitting to produce the coefficients
```

With the few lines above we found the coefficients for the linear regression that provide us directly the explanations we are searching for in this specific cases (Table 3.2).

3.2 Linear Regression

Table 3.2 Sorted coefficients of the linear regression

	Coefficient
Sulfates	0.823543
Alcohol	0.294189
Fixed acidity	0.023246
Residual sugar	0.008099
Free sulfur dioxide	0.005519
Total sulfur dioxide	−0.003546
Citric acid	−0.141105
pH	−0.406550
Volatile acidity	−0.991400
Chlorides	−1.592219
Density	−6.047890

```
print(coefficients.sort_values(by='Coefficient', ascending=False))
```

We see the ones that are negatively correlated with quality with a minus sign and the ones that positively impact the quality with a plus sign. The top three contributors are the negative values: **density, chlorides, and volatile acidity**; while the top three positive ones are **sulfates, alcohol, and fixed acidity**. So, without relying on complex tools or artifacts we have shown how we can make sense of a linear regression in terms of getting explanations about how the prediction about quality is produced.

With minimum effort and amount of code you as the data scientist are already in the position of providing feedback to the wine producer. He wanted to know how to improve the quality of his wine. The direct answer is that he should work on the levels of sulfates and alcohol. Would that be enough? Would the wine producer trust our recommendation without any further explanation? We don't think so. We may try to bring him the considerations about the linear regression and the weights, but it would be hard relying only on these arguments to convince him about what to do to improve the quality.

The wine producer is not comfortable with mathematics and functions so talking about the linear function that minimizes the loss function and the related coefficients would not help. We need to back-up our results with some further artifacts to provide an effective explanation.

Do you remember that we talked about correlation in the previous chapters? Correlation is exactly what we need here to provide better explanations to our wine producer.

Correlation is a measure of the degree of the linear relation between two variables and can vary from −1 (full negative correlation, one variable's increase makes the other to decrease) to 1 (positive correlation, the two variables increase together). Every variable has obviously correlation = 1 with itself.

Let's start looking at the Table 3.3 of correlation coefficients, of the 11 features with the output, below

Table 3.3 Correlations with our target "Quality" of the features of the model

Correlations with target	
Alcohol	0.476166
Volatile acidity	−0.390558
Sulfates	0.251397
Citric acid	0.226373
Total sulfur dioxide	−0.185112
Density	−0.174919
Chlorides	−0.128907
Fixed acidity	0.124052
pH	−0.057731
Free sulfur dioxide	−0.050554
Residual sugar	0.013732

```
correlations = df.corr()['quality'].drop('quality')
correlations.iloc[ (-correlations.abs()).argsort()]
```

Looking at the Table 3.3, values we see how volatile acidity and alcohol are the features that are more correlated with quality (negatively for volatile acidity and positively for alcohol).

As we discussed about the difference between correlation and causation, this does not necessarily mean that alcohol and volatile acidity are "the main causes" of quality. For example, there could be a third unknown feature that control both alcohol and volatile acidity). But up to the boundaries of the current scenario with a linear model, volatile acidity and alcohol are the features that best "explain" the changes in quality.

One useful part about correlation is that it supports the possibility of a visual representation of what's going through a visualization called a heatmap. Obviously, we can calculate the correlation of any two variables in the dataset. We have 11 features and one output (wine's quality) therefore we can calculate 12 × 12 = 144 correlations but due to the symmetry of correlations $\frac{12(12-1)}{2} = 66$ of them are unique.

With a few lines of code, we can generate the heatmap we mentioned:

```
import seaborn as sns
sns.heatmap(df.corr(), annot=True, linewidths=.5, ax=ax,
cmap="twilight")
plt.show()
```

Looking at Fig. 3.6, we may understand the name "heatmap" for the ones that are not used to it. It has to be read like a visual table in which each feature is correlated with another through a coefficient that corresponds to a different color with the scale shown on the right.

3.2 Linear Regression

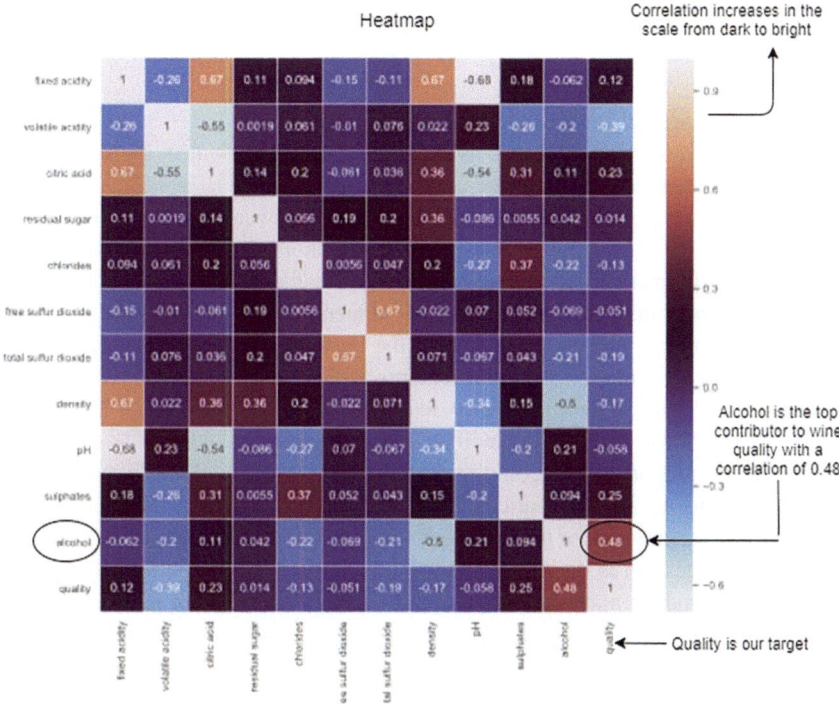

Fig. 3.6 Heatmap shows the correlation coefficients among the different features, focus is on correlation with our target: Quality

We see as expected that the main diagonal of the square is white (max correlation) because the diagonal shows the correlation of each feature with itself (= 1).

Apart from that, we may see the results that were numerically expressed in the previous table: we have the confirmation that **alcohol is the most important feature to improve quality**.

We have to stress that the present naïve linear model is not the state of the art for the solution for we have set aside three theoretical questions.

From a theoretical point of view, using only the weights gives a model that is scale dependent. That is, if we measure density with a physical unit ten times larger, we will have a weight ten times smaller, so we can't directly compare the relative strengths between two weights, but we can do it with correlations which are adimensional. This point is usually addressed by standardization i.e. subtracting from the features their mean and scaling dividing by the standard deviation of the features. The standardization gives us features with mean 0 and variance of 1.

We can easily see that the correlation of a coefficient with the target can have another interpretation. If we are in the case of a one dimensional linear regression we have the following formula between the m_1 coefficient of the regression and the correlation ρ between X and Y.

$$m_1 = \rho \frac{\sigma_Y}{\sigma_X} \tag{3.7}$$

Here σ_Y and σ_X are the standard deviations of Y and X.

So we can think the correlation as the regression coefficient in the special case $\sigma_X = \sigma_Y = 1$ of standardized quantities.

The second important question is the correlation between features: a heavy correlation between features gives a phenomenon called multicollinearity. Multicollinearity can yield solutions with error of weight is extremely large i.e. the weights are badly determined and possibly numerically unstable. Think of it, let's have two features x_1 and $x_2 = 2 * x_1$ here the correlation between features is 1. A general linear combination between features will be $w_1 x_1 + w_2 x_2 = x_1(w_1 + 2w_2)$. Using gradient descent we will exactly determine the weight $w_3 = (w_1 + 2w_2)$ but the relative contribution of w_1 and w_2 will be undetermined.

The usual method to overcome multicollinearity is to exclude features highly correlated or to *whiten* the features via Principal Component Analysis. We don't resume these techniques for they are well found in the literature.

The third question is on badly determined weight even in absence of multicollinearity. A standard approach is to calculate the ratio of the coefficient and its uncertainty.

For a standard linear regression with one feature we have the proportion

$$\frac{m}{\sigma_m} \sim \frac{m}{\left(\frac{\sigma_\varepsilon}{\sigma_X}\right)} \tag{3.8}$$

Where σ_ε is the standard deviation of the regression and σ_X is the standard deviation of the feature. If we have a weight m smaller than its uncertainty σ_m we can't determine even the sign of the weight, so low values of the ratio are undesirable. From the equation, we also see that the variance of a feature can't be too small.

Now we are roughly in the position of providing a better explanation to our wine producer: we may start showing the results of linear regression. In case he feels lost in the numbers we may rely on the visual heatmap representation to support our explanation: our model produced a prediction that alcohol is strongly coupled with quality; changing alcohol level improves quality.

We are now ready to summarize the explanations we collected to answer the questions about the reasons of our predictions and recommendation on wine quality based on the outcome of the ML linear regression model.

Let's have a look at the Table 3.4 below:

The idea here is that in order to answer the question "Which are the main wine characteristics to work on to improve quality?" it is not enough to look at the weights and selecting those who have biggest absolute value.

We also need to look at the correlation with the target and to the variance of the feature. Density for example has the largest weight value but a very small correlation coefficient.

3.2 Linear Regression

Table 3.4 Wine features' weights and correlation with quality

Feature	Weight	Correlation with quality
Alcohol	0.29	0.48
Sulfates	0.82	0.25
Density	−6.05	−0.17
Chlorides	−1.59	−0.13

Table 3.5 Coefficients after the feature selection with Lasso. The features have been standardized and the α has been chosen to have six non zero weights

Coefficient of Lasso Regression	
Alcohol	0.292478
Volatile acidity	−0.170318
Sulfates	0.079738
Total sulfur dioxide	−0.036544
Fixed acidity	0.020537
Chlorides	−0.002670
Citric acid	0.000000
Residual sugar	0.000000
Free sulfur dioxide	0.000000
Density	−0.000000
pH	−0.000000

We have already seen that correlation is indeed the m of a linear regression when the target and the feature have been standardized. So the weight of the Density *looks* big for the scale choice but in effect, it is small.

Also Eq. (3.5) reminds us that the uncertainty on weight could be even larger than the weight itself. We can resolve this problem restricting only to the more meaningful features.

The idea is simple but powerful. In Lasso regularization, we add a positive term in l_1 norm to force the weights associated with less significant features to be precisely zero.

If we gradually increase the Lasso constant each feature will be zeroed one a time. The first to disappear will be the least important feature and last surviving feature will be the most important one. In this manner we can construct a more robust ranking of features by feature importance.

This procedure can be done only at training time. More useful techniques are done post hoc (using intrinsic or agnostic methods) i.e. after the model's training.

We standardize the features to unit variance and train a Lasso model with different alpha (Table 3.5)

```
x_train_scaled = preprocessing.StandardScaler().fit_
transform(x_train)
# scaling features

from sklearn import linear_model
regressor = linear_model.Lasso(alpha=0.045)
# selecting a Lasso regressor model
```

```
regressor.fit(x_train_scaled,y_train)
# training the Lasso regressor

coefficients=pd.DataFrame(regressor.coef_,col_names)
coefficients.columns=['Coefficient']
print(coefficients.iloc[ (-coefficients.Coefficient.abs()).argsort()])
```

For example with a choice of alpha = 0.045 we find show what are the first six feature by importance and in fact we "density" has been found as one of the less relevant features.

Do you remember the properties of explanations we explained in Chap. 2? Let's see how they fit with this real case-scenario looking at Table 3.6.

3.3 Logistic Regression

In the previous section, we used linear regression to deal with wine quality and then getting explanation with XAI. In this section, we face a scenario of classification instead of prediction.

Suppose that a scholar in biology that is using a ML classification system to distinguish between specific type of flowers. The focus here is not just on the classification but, given the classification with a certain accuracy, to provide explanations about the criteria that the ML model is adopting to assign the flower category. For this purpose, we will use the evergreen Iris flower dataset that is well known in ML community, shifting the focus from classification to explanations.

Let's state better our problem: we have a flower dataset containing instances of Iris species. We will build a ML Logistic Regression model to classify Iris flower instances as virginica, setosa, or versicolor based on four features: pedal length, pedal height, sepal length, and sepal height. But our focus is not on the ML model that does the classification but on the explanations that we need to provide in this Logistic Regression case.

Table 3.6 Properties of explanations

Property	Assessment
Completeness	Full completeness achieved without the need of trading-off with interpretability being an intrinsic explainable model.
Expressive power	Correlation coefficients provide a direct interpretation of the linear regression weights
Translucency	High, we can look directly at the internals to provide explanations
Portability	Low, explanations rely specifically on linear regression machinery
Algorithmic complexity	Low, no need of complex methods to generate explanations
Comprehensibility	Good level of human understandable explanations to build as much confidence as possible in the wine producer

3.3 Logistic Regression

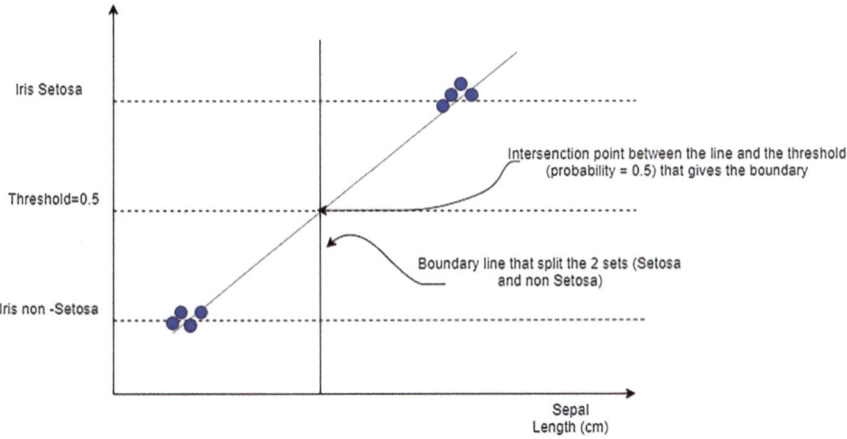

Fig. 3.7 Linear regression to classify Iris flowers type based on Sepal Length only)

Assuming that the ML model does a good job on classifying the flowers, our goal is to provide methods to answer questions on how this classification is performed: which are the most important features among the four to split the plants into categories of *Iris virginica*, setosa, or versicolor?

From an XAI perspective it is important to start from recalling why we adopt a Logistic Regression model for classification instead of an easier Linear Regression. Suppose for the moment that there is only one feature (the Sepal Length just for example) that controls if a flower in the dataset is an *Iris setosa* or not and suppose to have the situation depicted in Fig. 3.7.

The threshold is the limit that we set to 0.5 as to get the probability that a specific Iris instance is Setosa or not. As we see from the figure, a linear regression would properly classify the two sets. Then suppose that you have a further datapoint like in the following graph (Fig. 3.8).

The additional point totally changes the regression line we use for predictions. In this new scenario, the previous set of flowers would be now categorized as non-Setosa, which is not the case. The way that one point changes the regression line shows how linear regression is not suitable for classification problems. This is the reason why we need to switch to logistic regression we already mentioned in the previous chapter.

Before jumping to the details of our flower classification problem, we need to get back to theory to understand what changes in terms of XAI, moving from Linear Regression to Logistic Regression. Mainly, we will lose the possibility of a straightforward interpretation of the model coefficients to provide explanations.

Let's get back to Linear Regression formula 3.6 to model target (Y) with a set of features ($x_1 \ldots x_k$), this is how it changes for Logistic Regression

$$P(Y=1) = 1/\left(1 + \left(\exp-\left(m_0 + m_1 x_1 + m_2 x_2 + \ldots + m_k x_k\right)\right)\right) \quad (3.9)$$

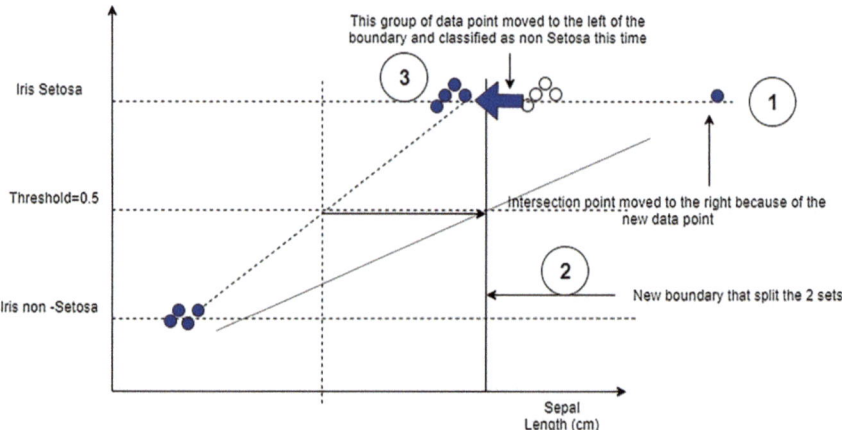

Fig. 3.8 Linear regression is broken by the additional data point on the top right, the numbers provide the flow)

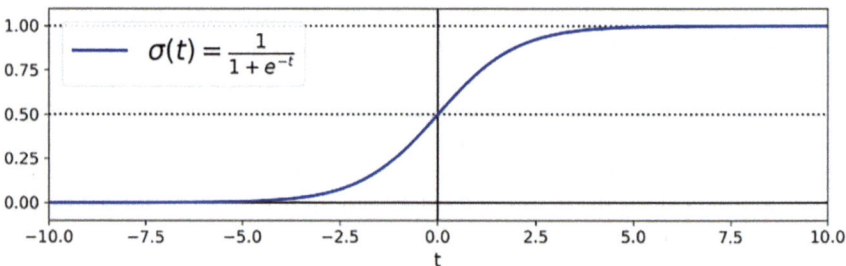

Fig. 3.9 Logistic function is used to fix the classification problems with linear regression

It is evident how we cannot rely directly on $m_1 \ldots m_k$ weights to get feature importance because they appear inside the exponential function. You may find in the box below an idea of why we need to use the exponential function, something to recall but that is generally known to people familiar with basics in ML (Fig. 3.9).

Let's review the need to use exponential functions. Basically, the problem we saw, where the addition of one data point may break the classification through linear regression, is fixed using the so-called logistic function: $\sigma(t) = 1/(1 + (\exp - (y)))$ that appears like this:
The effect that we obtain putting our linear regression formula instead of t is exactly the one we want: put a boundary on probabilities between 0 and 1 as it must be and avoid the problem of the changing line with the additional data point.

3.3 Logistic Regression

So, what we can do to interpret the $m_1...m_k$ in terms of relative importance of features to produce explanations, as we did for linear regression case?

In (3.9), we are dealing with probabilities. In our specific scenario, we want to assign a probability for a flower classification. The first step is to tweak (3.9) to have a probability ratio. We can understand the meaning of $P(Y=1)$ as the probability for a flower being an *Iris setosa* and $1 - P(Y=1)$ as the complementary one: the probability for the same flower does not belong in that category.

The first step is to take out the linear combination inside the exponential using the log function through some basic math:

$$\log \frac{P(Y=1)}{1-P(Y=1)} = \log \frac{P(Y=1)}{P(Y=0)} = m_0 + m_1 x_1 + m_2 x_2 + .. + m_k x_k \quad (3.10)$$

As we see the formula is back to linear but with log terms.

The term $\log \frac{P(Y=1)}{P(Y=0)}$ I is named log-odds or *logit* where the odds mean the ratio of probability of the event and probability of no event.

This lets us express in a different way the relative weight of a feature compared to the others:

$$\frac{\text{odds}(x_k+1)}{\text{odds}} = \exp(m_k) \quad (3.11)$$

We skipped some intermediate steps to get formula (3.11) that are not fundamental for main concept and may hide the result.

While in the case of linear regression m_k is directly the weight (the relative importance) of feature k, in the case of logistic regression we get something similar but a bit more complicated. With linear regression, the change of 1 unit in the feature k causes the target to change by the weight m_k. In logistic regression, the same change of 1 unit changes the odds by a multiplicative factor $\exp(m_k)$ all other things being equal. It may seem a bit abstract but we will clearly need these concepts in our case of flower classification.

To anticipate a bit what will follows and to fix what we said with something more tangible, first assume that m_k (just for example) to be the petal length. We may say that increasing it by one unit would enhance the probability for the flower instance of being an *Iris setosa* by a factor $\exp(m_k)$. But to get to this point, we now need to have some Python code with real datasets and numbers. Let's have a first look at the data set after the usual code to import libraries and load the Iris data (UCI, 1988; Table 3.7).

Table 3.7 Iris dataset sample

	Id	SepalLengthCm	SepalWidthCm	PetalLengthCm	PetalWidthCm	Species
0	1	5.1	3.5	1.4	0.2	Iris-setosa
1	2	4.9	3	1.4	0.2	Iris-setosa
2	3	4.7	3.2	1.3	0.2	Iris-setosa
3	4	4.6	3.1	1.5	0.2	Iris-setosa
4	5	5	3.6	1.4	0.2	Iris-setosa

```
X = iris.data[:, :2]   # we only take the first two features.
y = iris.target
df=pd.
DataFrame(X,
columns
= ['Sepal_Length','Sepal_Width','Petal_Length','Petal_Width'])
df['species_id']=y
species_map={0:'Setosa',1:'Versicolor',2:'Virginica'}
df['species_names']=df['species_id'].map(species_map)
df.head()
```

Nothing special so far, just standard Python code to get the dataset; the sample shows some flower instances with the features and the species name. Next step is to train the model on the training data set and then go for the classification on the test dataset as usual.

```
# Split the data into a train and a test set
perm = np.random.permutation(len(X))
f= df.loc[perm]
x_train, x_test = X[perm][30:], X[perm][:30]
y_train, y_test = y[perm][30:], y[perm][:30]

# Train the model
from sklearn.linear_model import LogisticRegression
log_reg = LogisticRegression()
log_reg.fit(x_train,y_train)
```

And after that we test the performance of the model

```
# Test the model
predictions = log_reg.predict(x_test)
print(predictions)# printing predictions

print()# Printing new line
```

3.3 Logistic Regression

```
#Check precision, recall, f1-score
from sklearn.metrics import classification_report,accuracy_score
print( classification_report(y_test, predictions) )
print( accuracy_score(y_test, predictions))
```

The output is the following (Table 3.8)

['Versicolor' 'Setosa' 'Virginica' 'Versicolor' 'Versicolor' 'Setosa' 'Versicolor' 'Virginica' 'Versicolor' 'Versicolor' 'Virginica' 'Setosa' 'Setosa' 'Setosa' 'Setosa' 'Versicolor' 'Virginica' 'Versicolor' 'Versicolor' 'Virginica' 'Setosa' 'Virginica' 'Setosa' 'Virginica' 'Virginica' 'Virginica' 'Virginica' 'Setosa' 'Setosa']

The precision is the ratio tp/(tp + fp) where tp is the number of true positives and fp the number of false positives. We can think of precision as the ability of the classifier not to label as positive a sample that is negative.

The recall is the ratio tp/(tp + fn) where fn is the number of false negatives. We can think recall as the ability of the classifier to find all the positive samples.

The F-beta is simply the harmonic mean of recall and precision.

The support is the number of samples of y_test in each class.

All these concepts may be pretty clear to people that are familiar with ML. And this is the point where ML usually stops if we don't invoke XAI. Let's try to be very clear here: the ML model does a wonderful job on classifying Iris flowers but from these metrics we don't have any evidence of the main features that are used to do this. How can our scholar in biology present the results? Which is more important for the classification, the sepal length or width? The petal length or width? These are the basic and natural questions that come from XAI and that we need to answer.

So let's do one step back and before going for the classification in three types let's start from splitting Setosa from non-Setosa flowers to have a purely binary classification.

Do you remember Eq. (3.10) that we used to have something similar to linear regression to interpret coefficients?

Let's write the same equation but for our specific case in which we have our 4 features: sepal length, sepal width, petal length, petal width for $m_1...m_4$. We saw how to use (3.11) to get the impact of each coefficient on the odds ratio; here we show how $m_0 + m_1x_1 + m_2x_2 + ... + m_kx_k$ can also be interpreted as the decision boundary for the classification.

Let's do a scatter plot of the flowers dataset to understand this point better.

Table 3.8 Scores of Iris Classification using Logistic Regression

	Precision	Recall	F1-score	support
Setosa	1.00	1.00	1.00	10
Versicolor	1.00	1.00	1.00	9
Virginica	1.00	1.00	1.00	11
Avg/total	1.00	1.00	1.00	30
1.0				

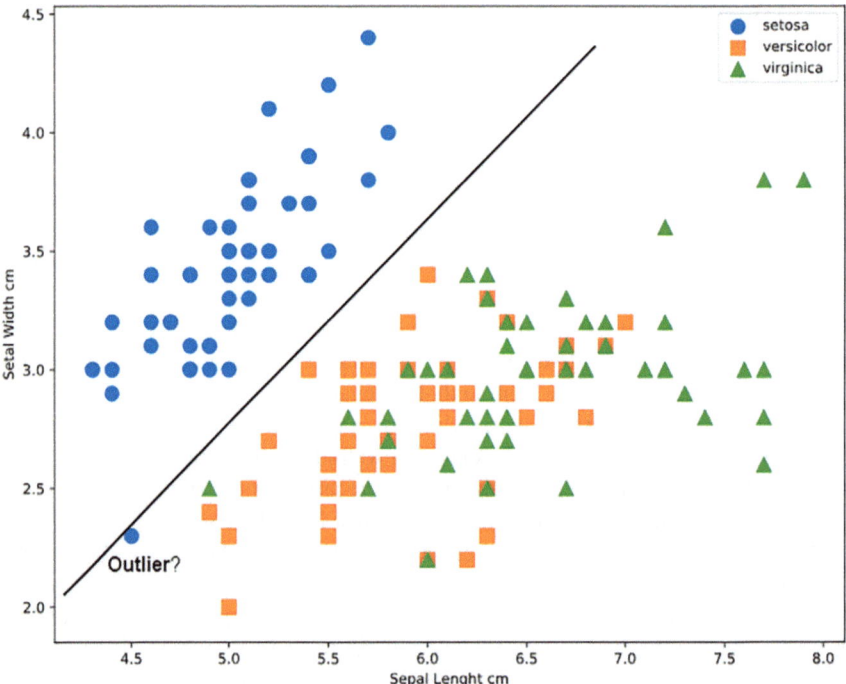

Fig. 3.10 Iris classification based on Sepal_Length and Sepal_Width features

```
marker_map = ['o', 's', '^']
unique = np.unique(df['species_id'])

for marker, val in zip(marker_map, unique):
    toUse = (df['species_id'] == val)
    plt.scatter(X[toUse,0], X[toUse,1], marker=marker,
cmap="twilight", label=species_map[val], s=100)

plt.xlabel('Sepal Lenght cm')
plt.ylabel('Setal Width cm')
plt.legend()
plt.show()
```

Setosa are the point marked as circles, the two sets are quite separated using representing the scatter plot on features Sepal Length and Sepal Width (Fig. 3.10).

Let's see what happens if we do the same but using Sepal Length and Petal Width (Fig. 3.11).

3.3 Logistic Regression

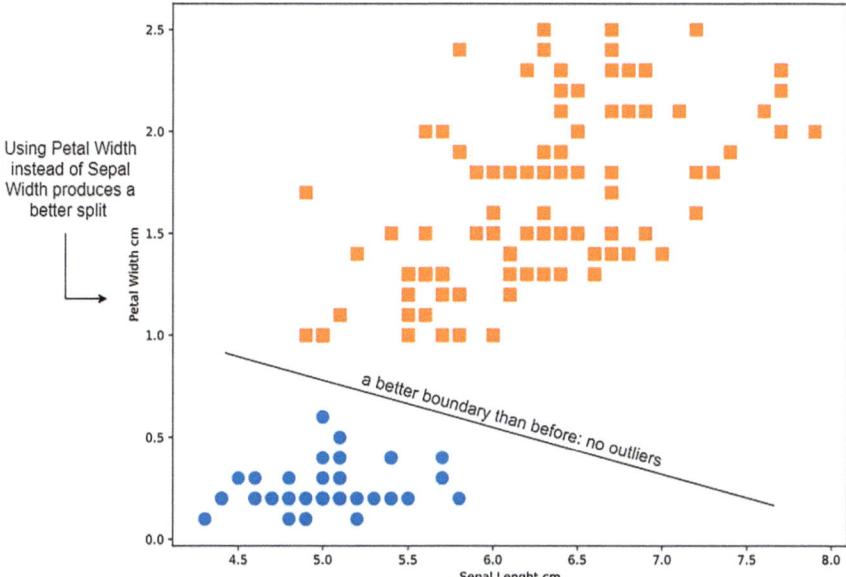

Fig. 3.11 Iris classification with Sepal Length and Petal Width instead of Sepal Width produce a better split

```
y = np.array(y)
marker_map = ['o', 's', 's']  # here we use same symbol for
versicolor and virginica
unique = np.unique(y)
for marker, val in zip(marker_map, unique):
    toUse = (y == val)
    plt.scatter(X[toUse,0], X[toUse,1], marker=marker,
cmap="twilight", s=100)
plt.xlabel('Sepal Lenght cm')
plt.ylabel('Petal Width cm')
plt.show()
```

The sets of *Iris setosa* and non-setosa are better separated. (We manually added the line that represents the boundary between the two groups.)

From this simple case on binary classification, we see how to identify the couple of features that are most important to classify *Iris setosa* vs. Iris non-setosa flowers. Plotting the dataset on the plane of these features (Sepal Length and Petal Width), we have a clear linear boundary to delimit the two sets. But how to find the equation of this linear boundary to have a quantitative answer? It is easy to access the coefficients of the logistic regression model to get the linear boundary equation directly.

```
W, b = log_reg.coef_, log_reg.intercept_
W,b
```

Output: (array([[1.3983599 , 3.91315269]]),
array([-10.48150545]))

What do these numbers mean? They are the coefficient of the linear boundary that separates the setosa from non-Setosa flowers (Fig. 3.12):

$$\log\left(\frac{P(Y = \text{Setosa})}{P(Y \neq \text{Setosa})}\right) = m_0 + m_1 x_1 + m_2 x_2 \tag{3.12}$$

where x_1 = Sepal Length, x_2 = Petal Width

And the same coefficients are the ones that express the odds ratio:

$$\text{odds}(x_k + 1) / \text{odds} = \exp(m_k) \tag{4}$$

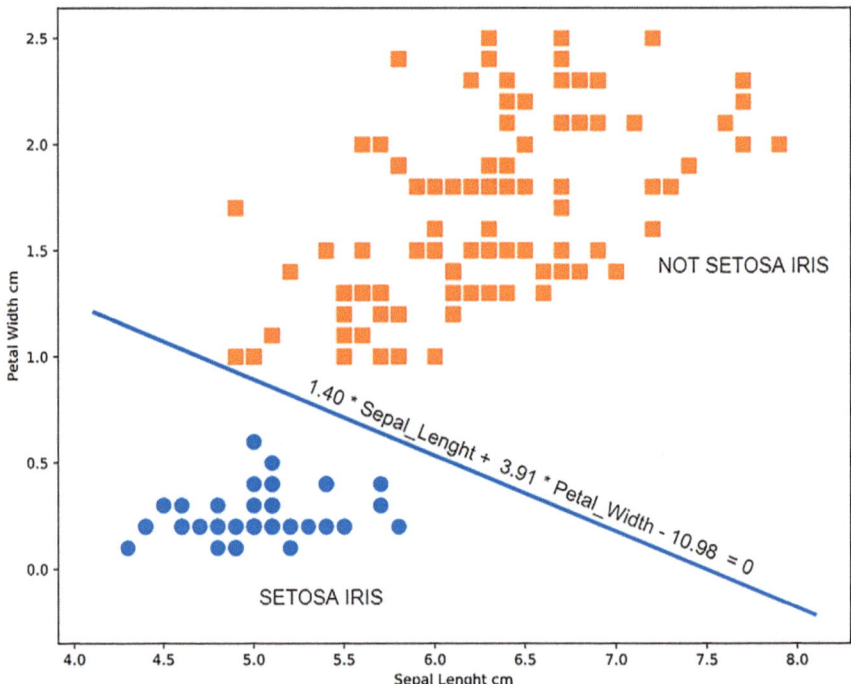

Fig. 3.12 Iris classification that shows the equation of the boundary line

3.3 Logistic Regression

For our specific case:

$$\frac{\text{odds}(x_1+1)}{\text{odds}} = \exp(m_0) \quad x_1 = \text{Sepal Length} \quad (3.13)$$

$$\frac{\text{odds}(x_2+1)}{\text{odds}} = \exp(m_1) \quad x_2 = \text{Petal Width} \quad (3.14)$$

Equations (3.13) and (3.14) express how the probability of iris being setosa or not changes for the increment of 1 cm in sepal length and petal width. And because of the scatter plots we know that sepal length and petal length are the features that produce the best separation of the two sets. How would these arguments change in case of having three categories: Setosa, versicolor, and virginica? Not very much. We would do the same steps but with more coefficients and scatter plots to examine. The W and b of the previous equations would become:

```
W, b = log_reg.coef_, log_reg.intercept_
W,b
(array([[ 0.3711229 ,  1.409712  , -2.15210117, -0.95474179],
        [ 0.49400451, -1.58897112,  0.43717015, -1.11187838],
        [-1.55895271, -1.58893375,  2.39874554,  2.15556209]]),
 array([ 0.2478905 ,  0.86408083, -1.00411267]))
```

Having three categories would produce a 4*3 matrix of coefficients (four features, three categories) and three intercepts for the three categories. We would use these numbers to repeat what we did before for the binary classification and get the odds ratio to obtain the relative weights of the different features on explaining the classification (Table 3.9)

So what is the strategy for our scholar in biology to share his results from an XAI perspective?

Remember that one thing is just to build a ML model to classify flowers, another thing is to provide explanations of how this ML model is achieving the classification or, said in other terms, which are the most important features to recognize the Iris type.

Our scholar in biology would start explaining the results for the binary case. In particular, he could give a high level overview of how odds are related to the probability of an iris being a setosa or not depending on the features. The assumption is

Table 3.9 Odds ratio

	Setosa	Versicolor	Virginica
Sepal length	1.44936119	1.61875962	0.21035554
Sepal width	4.09477593	0.20667587	0.20414071
Petal length	0.11623966	1.55210045	11.00944064
Petal width	0.38491152	0.33653155	8.63283183

that the audience could be comfortable with basics in mathematics. From an XAI perspective, the scholar could take the recommended further step to show how the coefficients are related to the liner boundaries that split the flowers dataset into different types.

As shown in Fig. 3.10, we may point to sepal length and petal width as the features that are the most important ones for our classification problem. Then to further describe the explanations, the scholar may also explain how to obtain the odds ratio that express the feature weights, in the complete set of three flower species.

Let's repeat the assessment of explanations' properties we already did for linear regression, using the Table 3.10.

3.4 Decision Trees

In this section we will be asked by an insurance company to explain our model of risk. Making this request more practical we deal with the scenario of a marine insurance company and we use a Decision Tree ML model to predict the survival probabilities. The insurance company is asked to provide explanations and criteria to back up the survival rates and we will show how to get them.

We use a decision tree for, as we will see, they are the more logical choice in case of categorical tabular data. We download a reduced dataset of the famous disaster and use it to train and test a model that provides information on the fate of passengers on the Titanic, according to sex, age and passenger's class (pclass). The target of the model is to predict survival as a Yes/No option. The insurance company has asked for explanations and criteria to bring up the survival rates and we will show how to get them.

We have chosen this particular easy example, well known in the ML field, to revise Decision Trees for those who already know machine learning basics and maybe want to deepen the concepts. We will focus on the aspects that we need from an XAI perspective.

Table 3.10 Properties of explanations

Property	Assessment
Completeness	Full completeness achieved without the need of trading-off with interpretability being an intrinsic explainable model.
Expressive power	Less than linear regression case. Interpretations of coefficients is not so straightforward.
Translucency	As any intrinsic explainable model, we can look at the internals. Weights are used to provide explanations but not so directly as in linear regression case.
Portability	Method is not portable, specific for logistic regression.
Algorithmic complexity	Low but not trivial as in linear regression case.
Comprehensibility	Explanations are human understandable also for not technical people.

3.4 Decision Trees

Before going to the scenario and XAI, we need to review some concepts related to decision trees. Feel free to skip ahead if you feel comfortable with decision tree theory.

There are various implementations of decision trees. For simplicity, we will refer to the Classification and Regression Tree (CART) algorithm as implemented in Scikit Learn library. CART was introduced by Breiman in 1984 and is the first "universal" algorithm in the sense that it can accomplish both Classification and Regression Tasks.

CART constructs a binary tree using some logic for splitting the initial dataset into branches so that the splits increasingly adapt data to the target labels.

The Decision Tree partitions the feature space into rectangles approximating the possible relations between features. It similar to how a doctor may say, "If your weight is over 80 kg, you risk diabetes."

So Decision Trees model human reasoning (Fig. 3.13). With such a clear visual representation of the decision process, we can promptly answer counterfactual questions such as "What if?" and make explanations via contrastive arguments such as "your loan would be accepted if you were five years older" by simply looking at the representation.

Let's resume the theory behind Decision Trees.

We call impurity our primary indicator of how to do the splits.

For classification tasks like the dataset of Titanic, we can find in literature different types of impurity:

- Gini Impurity
- Shannon Entropy
- Classification Error

We call p_i the class proportion of occurrences in class i in respect to all the classes. The class proportion depends on the choice of the feature X_i and a predefined threshold value t_i for the split. In binary Decision Trees, every class is set of samples such that $X_i < t_i$ or $X_i > t_i$.

Now we can write the equations for the three impurity types:

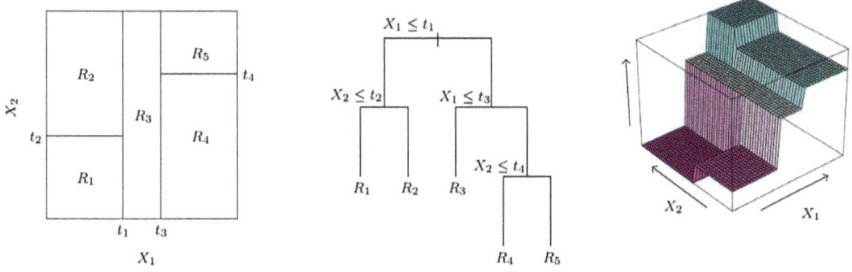

Fig. 3.13 Decision Trees partition feature space approximating the true functional relation

Gini equation $\text{Gini} = 1 - \Sigma_{i=1}^{C} (p_i)^2$
Shannon Entropy $\text{Entropy} = \Sigma_{i=1}^{C} - p_i \log_2 (p_i)$
Classification Error $CE = 1 - \max(p_i)$

here C is the number of classes.

We know that for a perfectly classified item impurity would be zero. As we already said in the previous chapter, a node is 100% impure when it is split evenly 50/50 and 100% pure when all node data belongs to a single class.

So after we have selected the type of Impurity to use in training a Decision Tree, we "simply" minimize the total impurity on each node finding the appropriate number of nodes and the corresponding couples of feature x_i and splitting threshold t_i.

We pick a loss function J as

$$J(i, t_i) = p_{1\,(x_i < t_i)} + p_{2(x_i > t_i)}$$

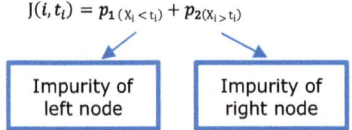

and minimize it finding the appropriate X_i and t_i in each node.

This Choice of Loss function is different from the Loss function we have already seen. It is a *local* training loss function used for each node and not a global one. It defines the steps of the model training not the complete error on the training set.

We want to stress that the general problem of finding the best Decision Tree is too computationally expensive, in fact non polynomial in time (NP-Complete). So in practice, we employ a greedy strategy on the splits employing a top-down, greedy search to test each feature at every node of the tree.

For speed reasons, Gini Impurity is the default choice of CART algorithm in Scikit Learn. As you can see in the following figure, the results don't change so much if you use it with respect to Entropy (Fig. 3.14).

In the case of regression, we build Regression Trees and the impurity is simply variance. To train a Regression Tree is to find in each node the appropriate couples X_i and t_i that minimize prediction variance with respect to training samples. But we don't show further details. Now we will briefly talk about the advantages and disadvantages of Decision Trees with respect to Linear Regression and Logistic Regression we have already seen.

Decision Trees can natively model nonlinear relations in the feature where, for example, Linear Regression cannot do it automatically. Also, Linear Regression does not consider the interaction between features such as how specific values of a feature constrain the variability of another feature. Decision Trees can easily do that. We can add that using a greedy algorithm gives fast computation time for the tree and is only needed to test enough attributes until all data is classified.

Decision Trees models have a high capacity in describing (and memorizing) data as opposed to linear regression which has low capacity; in fact, they can have zero

3.4 Decision Trees

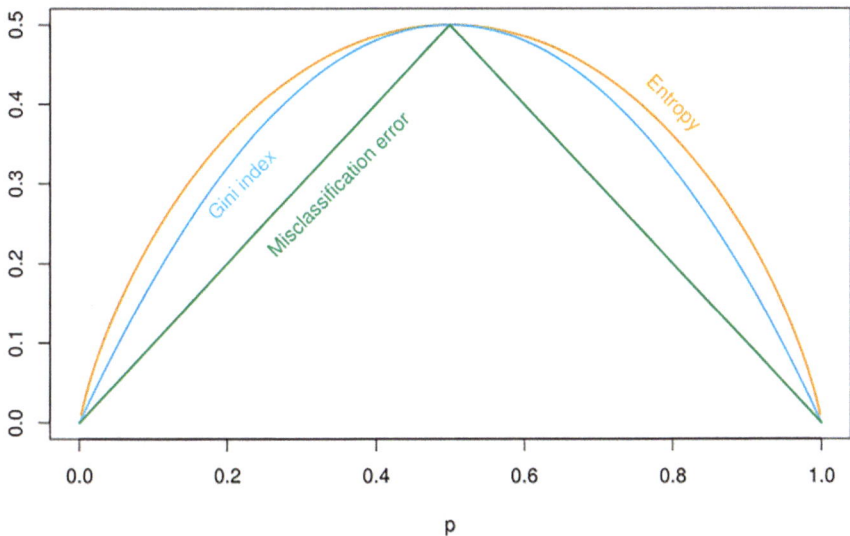

Fig. 3.14 Impurity as a function of class proportion p_i

regression error on training data. But a high capacity comes with an increased risk of overfitting. As you know, overfitting means the Model is so powerful that it "memorizes" data and noise alike but becomes incapable of predicting the outcome of some independent test samples. First variants of Decision Trees like ID3 addressed this problem following Occam's razor principle in the attempt to create the smallest possible decision tree. They selected the splits starting from those with higher purity gain.

But it was soon discovered that little gain in purity in some early node could give a substantial gain in the latter, so a criterion to make the DT simple would impact its performance. A more modern approach goes for deep trees, and after the training start a pruning process in which the internal nodes of the tree not really needed to explain the total accuracy of the model are deleted, simplifying the tree. A simpler model is also more robust to outliers.

Decision Trees give to us understandable prediction rules that are created from the training data in a directly interpretable manner.

The limit of such intrinsic interpretability lies in the depth of trees. In fact, the same feature can be reused in a different part of the same tree and as we have seen little variation in the purity of a node in early splits can give us large effects later. So for interpreting feature importance in Decision Trees, we have to track for each feature the TOTAL PURITY VARIATION by simply adding all variations of that feature in every apparition of that feature.

With these concepts in mind, we can go back to the scenario we presented at the beginning of this section: an insurance company needs to provide explanations about its ML model that predicts survival rates for marine accidents. We use a model based on the famous Titanic dataset and learn how to explain its forecasts. Having categorical data, we decide to use a Decision Tree model. Using Scikit-Learn we will train the model to explain its predictions and, as support, will calculate Permutation Importance as a measure of feature importance.

Behind of the permutation importance methods is the following: assuming to have a trained model, we shuffle the values in a specific column and do the predictions again using the shuffled dataset. The predictions are expected to worsen because of the shuffling (dataset has been hacked!). We repeat the shuffling for each column (one column per time) and see which columns have more impact on the prediction (deteriorating more the performance). The columns that cause more deterioration are the most important features. Said in other terms, if shuffling the values of a feature screws-up the model, then the ML model is heavily relying on this feature to do the job. That feature is expected to be very important.

Feature Importance is not specific of decision trees. It is an agnostic method that works not knowing anything the model inner workings.

We stress that Permutation Importance in its simplicity is a post hoc procedure. We check the model workings against some test dataset AFTER the training process of the model. We can calculate Permutation Importance with just a line of code in Eli5. We have introduced a Post hoc technique at this early stage of the book for its simplicity and to compare its results with the other methods we have already introduced.

We can now go to our code to work on our scenario; as usual let's start by having a look at the dataset (Waskom, 2014).

```
#Load the data from Seaborn library
titanic = sns.load_dataset('titanic')

#Print the first 10 rows of data
titanic.head(10)
```

	Survived	Pclass	Sex	Age
0	0	3	Male	22.0
1	1	1	Female	38.0
2	1	3	Female	26.0
3	1	1	Female	35.0
4	0	3	Male	35.0
5	0	3	Male	NaN
6	0	1	Male	54.0
7	0	3	Male	2.0
8	1	3	Female	27.0
9	1	2	Female	14.0

We select convert the categorical data in dummy numerical ones.

3.4 Decision Trees

```
from sklearn.preprocessing import LabelEncoder
labelencoder = LabelEncoder()
##Encode sex column
titanic.iloc[:,2]= labelencoder.fit_transform(titanic.
iloc[:,2].values)
```

And split features and target columns

```
#Split the data into independent 'X' and dependent 'Y' variables
X_train = titanic.iloc[:, 1:4].values
Y_train = titanic.iloc[:, 0].values
```

We train the model

```
from sklearn.tree import DecisionTreeClassifier
tree = DecisionTreeClassifier(max_depth=3)
tree.fit(X_train, Y_train)

#output of training
DecisionTreeClassifier(class_weight=None, criterion='gini', max_
depth=3,
                    max_features=None, max_leaf_nodes=None,
min_impurity_decrease=0.0, min_impurity_split=None,
                    min_samples_leaf=1, min_samples_split=2,
                    min_weight_fraction_leaf=0.0,
                    presort=False,
                    random_state=None, splitter='best')
```

and with a little of effort graph, the corresponding learned Tree

```
from IPython.display import Image
from sklearn.externals.six import StringIO
from sklearn.tree import export_graphviz
import pydot

dot_data = StringIO()
export_graphviz(tree, out_file=dot_data,feature_names=features,fill
ed=True,rounded=True)

graph = pydot.graph_from_dot_data(dot_data.getvalue())
Image(graph[0].create_png())
```

We can now explain entirely the ratio behind each prediction of the model (Fig. 3.15). This is exactly what we demand from an intrinsically explainable model,

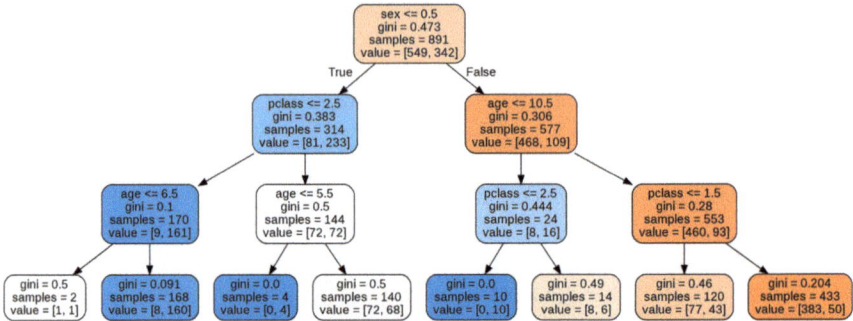

Fig. 3.15 Decision tree to predict survival rates based on the different features

but which features are more relevant? This is the fundamental question from an XAI perspective. We are not just asked to do predictions but also to provide explanations.

We have two ways of assessing feature importance in DT. One is ranking the features by the most relevant decrease in Gini index but as we already said such a method is neither robust or consistent, so we merely calculate permutation importance in this case (for the reasons we explained above).

So far, we have skipped stating the obvious about Scikit-learn installation and main usage, but it is important to note that Scikit-learn provides dependencies for the ELI5 package. ELI5 stands for "Explain it like 5 (years old)" and is a well-established explainability library. As with any common ML Python package, it is enough to import ELI5 to access its features:

```
import eli5    #A
eli5.show_weights() #B

#A import eli5 package
#B Use show_weights() API to display classifier weigths
```

In the lines above we import ELI5 and call one of its APIs just to show the syntax. We can now calculate Permutation Importance with just a line of code in Eli5.

```
import eli5
from eli5.sklearn import PermutationImportance
perm = PermutationImportance(tree, random_state=1).fit(X_train, Y_train)
eli5.show_weights(perm)
```

Weight	Feature
0.1481 ± 0.0167	Sex
0.1003 ± 0.0152	Age
0.0301 ± 0.0105	Pclass

3.5 K-Nearest Neighbors (KNN)

So Sex and Age are much more critical than Passengers' class.

Look at the decision tree graph. Each node (rectangle) contains both survived and not survived passengers. The **value** array contains the number of survived and not-survived passengers. So in the first rectangle of 891 people the majority, 549 passengers, will not survive the accident. The fist split puts women on the left and men on the right.

Now in the two subsets, the women subset has a majority who has survived. The second split is on age and we see that in the male subset on the right splitting by age tells us that male children have a higher chance to survive.

We can argue that, greedily, features that we use first are more relevant but it is not valid in general. Permutation importance reassures us that this is precisely the case in this example.

We close the section with summarising some disadvantages of pure Decision Trees.

The regions in the feature space will always be rectangular (we use only a feature at a time) and the transition from a region to another will not be smooth. In fact, Decision Trees struggle to describe even linear relations between features. A variation of DT called MARS is an attempt to add smoothness and natively nonlinear and nonlinear relations between features retaining intrinsic explainability.

As for properties (Table 3.11).

3.5 K-Nearest Neighbors (KNN)

We now return to the task of wine quality prediction but using K-Nearest Neighbors (KNN), another useful and intrinsically explainable methodology.

KNN was introduced in an unpublished report US Air Force School of Aviation Medicine in 1951 by Fix and Hodges and it is one of the more established Machine Learning Model of all Artificial Intelligence. Remember that we want to understand

Table 3.11 Properties of explainations

Property	Assessment
Completeness	Full completeness achieved without the need of trading-off with interpretability being an intrinsic explainable model.
Expressive power	High expressive power in fact DTs mimic to some extent human reasoning
Transluncency	Intrinsic explainable easy to guess results
Portability	In fact many models are derived from decision trees such random Forest and boosted trees so DT results can be incorporated in such models
Algorithmic complexity	Decision tree are NP-complete but we resort to heuristic for fast evaluation
Comprehensibility	Easy explanations to humans

which features are the most important ones to increase the quality and KNN will provide deeper insight on the explanations.

A KNN is easily explainable both using counterfactual examples that give visual explanations like "What if wine's acidity would increase?" and with a little modification even contrastive examples. With contrastive explanations, we create descriptions based on the missing abnormalities.

We can look to classification results with similar features. For each near sample with a different classification, we look for abnormalities in the corresponding features.

The core idea of KNN is to train the model merely memorizing all the samples and making predictions by an average or a majority voting process involving the results of some memorized examples (in fact, k of them) that have features most similar to the features of the item we want to predict.

In the example of the Titanic dataset, a passenger is predicted to survive if at least four out of seven passengers with similar features (same age, same boarding class, and same-sex) have survived.

Technically, KNN differs from other learning algorithms both in training complexity, which is merely $O(1)$ and in inference complexity that is much slower than other methods having to sort the samples continuously by the nearness. Without using some heuristic inference complexity is $O(N^2)$.

So let's go back to the wine producer who asked us how to improve the quality of wine and the reasons behind such answers. Just for visualization, we pick a large k to reduce the noise that affects the data and after the splitting of data in a training set and a test set we use only two features for the model.

```
# Importing the dataset
dataset = pd.read_csv('wine_data.csv')
X = dataset.iloc[:, 1:13].values
y = dataset.iloc[:, 0].values

# Splitting the dataset into the Training set and Test set
from sklearn.model_selection import train_test_split
X_train, X_test, y_train, y_test = train_test_split(X, y, test_size = 0.10)

# Fitting KNN to the Training set
from sklearn.neighbors import KNeighborsClassifier

classifier=KNeighborsClassifier(n_neighbors=15, metric="euclidean")
trained_model=classifier.fit(X_train[:,0:2],y_train)
```

We draw the boundaries using a mesh grid. For each node in the grid, the model predicts the corresponding class

3.5 K-Nearest Neighbors (KNN)

```
X=X_train
h=0.05
x_min, x_max = X[:, 0].min() - 1, X[:, 0].max() + 1
y_min, y_max = X[:, 1].min() - 1, X[:, 1].max() + 1
xx, yy = np.meshgrid(np.arange(x_min, x_max, h),
                     np.arange(y_min, y_max, h))
Z = trained_model.predict(np.c_[xx.ravel(), yy.ravel()])
kk=np.c_[xx.ravel(), yy.ravel()]

# Put the result into a color plot
Z = Z.reshape(xx.shape)
plt.figure(figsize=(14, 8))
plt.pcolormesh(xx, yy, Z)
plt.scatter(X[:, 0], X[:, 1], c=y_train)
plt.title("Wine KNN classification (k = 15)")
plt.show()
```

Circles correspond to Top Quality wines and triangles to poor quality ones. The x axis represents alcohol content and the y axis volatile acidity (Fig. 3.16).

Now you can clearly visualize using the figure which examples in the training set have a similar composition to a new wine for which you want to predict the quality.

And you can even see which wines are "abnormal" in the sense that they are members of other classes of wine quality and they follow the regularities of a different class.

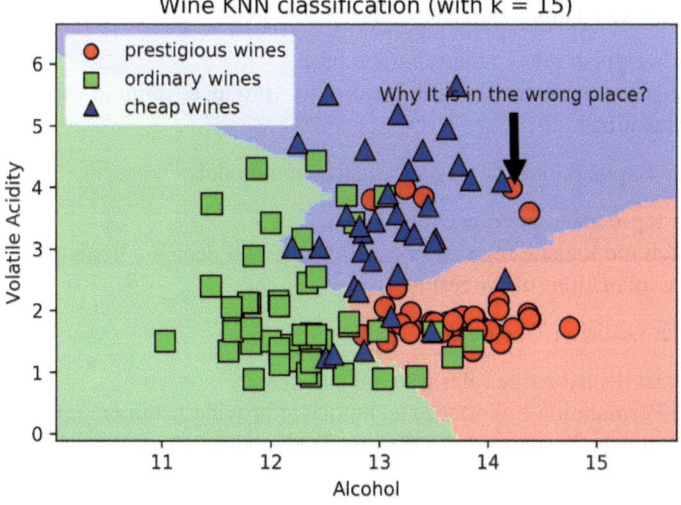

Fig. 3.16 Wine quality classification with KNN

Table 3.12 Explanations' properties

Property	Assessment
Completeness	Full completeness achieved without the need of trading-off with interpretability being an intrinsic explainable model.
Expressive power	High expressive power in terms of counterfactual and contrastive explanations
Translucency	Intrinsic explainable easy to guess results
Portability	KNN has a unique class in its own not portable.
Algorithmic complexity	Simple training, complex inference step
Comprehensibility	Easy explanations to humans

So you can give counterfactual explanations ideally changing the position of wine in the picture. For example, "If you decrease acidity you will go in the red zone of more prestigious wines."

And you can also give contrastive explanations pointing at the regular structure of blue points in the blue "poor wines" zone. For example, "All cheap wines tend to have a volatile acidity <2"

We close with the usual properties (Table 3.12):

3.6 Summary

We have seen how intrinsic interpretable models can be interpreted and how to produce "human understandable" explanations:

- Use l_1 regularization for XAI to get feature importance
- Produce Explanations for Linear Regression models
    - Use weights to rank feature importance
    - Interpret correlation coefficients to produce human understandable explanations
- Provide Explanations for Logistic Regression models
    - Use log-odds to provide explanations
    - Match the logistic regression coefficients with decision boundaries to enrich the explanations of the results
- Interpret Decision Tree models
    - Extract decision tree rules for explanations
    - Use Permutation Importance technique to provide features importance
    - Mitigate the limitations of Decision Tree models
- Provide counterfactual explanations using KNN models

In the next chapter, we will start our journey through model agnostic methods for XAI. The main difference is that we will lose the possibility of "easy" explanations as for the case of intrinsic explainable models, but we will learn how to get explanations through powerful methods that can be applied "agnostically" to different ML models

References

UCI. (1988). *Iris data set*. Retrieved from http://archive.ics.uci.edu/ml/datasets/Iris/
UCI. (2009). *Wine quality data set*. Retrieved from https://archive.ics.uci.edu/ml/datasets/wine+quality
Waskom, M. (2014). *Seaborn dataset*. Retrieved fom https://github.com/mwaskom/seaborn-data/blob/master/titanic.csv

Chapter 4
Model-Agnostic Methods for XAI

> *Then why do you want to know?*
> *Because learning does not consist only of knowing what we must or we can do, but also of knowing what we could do and perhaps should not do.*
>
> Umberto Eco, The Name of the Rose

This Chapter Covers:

- Permutation Importance
- Partial Dependence Plot
- Accumulated Local Effects (ALE)
- Shapley Additive exPlanations (SHAP)
- Shapley Values Theory

In this chapter, we start our journey through XAI model agnostic methods that are, as we said, potent techniques to produce explanations without relying on ML models internals that are "opaque." Additionally, we will explore Accumulated Local Effects (ALE), a method that addresses limitations of Partial Dependence Plots when dealing with correlated features—a common scenario in real-world datasets. ALE provides more accurate interpretations by focusing on local changes in the feature space rather than averaging across potentially unrealistic feature combinations.

The main strength of model-agnostic methods is that they can be applied to whatever ML model, including intrinsic explainable models. The ML model is considered as a black box, and these methods provide explanations without any prerequisite knowledge of ML model internals. We want to be very clear on this last statement: as you may remember, in Chap. 3 we used permutation importance to produce explanations on the Titanic decision tree.

In that case, we relied on the "intrinsic explainability" of the decision tree but we saw how permutation importance produced enhanced interpretability. Here we start

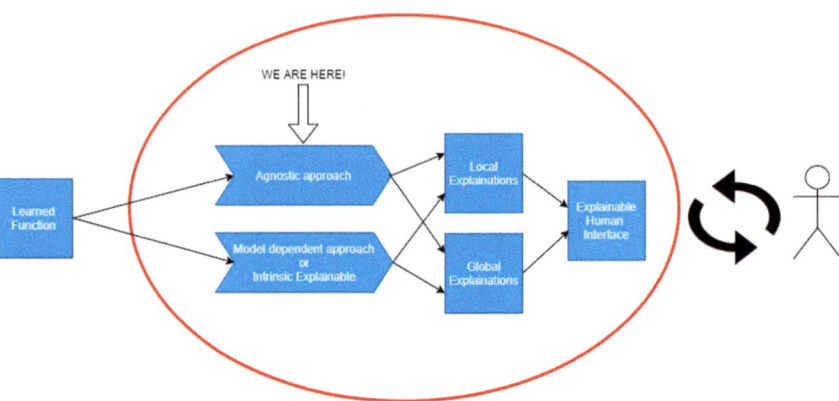

Fig. 4.1 XAI flow: agnostic approach

again from permutation importance methods but applied to a case in which we don't have the possibility of producing intrinsic explanations.

Returning to our flow (Fig. 4.1), we follow an agnostic approach and, for each method discussed, we will identify whether the explanations provided are of a local or global nature.

We will use two main real-case scenarios to explain how these methods work. The first one is based on a scenario presented in XAI Kaggle learning section that we recommend (Becker, 2020): basically, it deals with a Machine Learning division of a company that is in betting business for sporting events.

You will learn how to answer "What" and "How" questions on the scenario using XAI agnostic methods. Permutation Importance will allow the identification of the most important features while Partial Dependence Plot will provide details on "How" the features are impacting the predictions. Moreover, you will be able to use SHAP to produce explanations on specific instances instead of just global explanations.

Having in mind how to use these methods, then we will provide theoretical foundations of SHAP before switching to the scenario of a taxi cab company to provide to customers explanations on the fares (predicted by a boosted treed ML model) they will pay in real time.

We will use the cab scenario to revisit SHAP in comparison to LIME and see how it performs with boosted trees. Also you will learn SHAP's limitations and how to tackle them.

4.1 Global Explanations: Permutation Importance and Partial Dependence Plot

We are in the Machine Learning division of a company that is in betting business for sporting events. We rely on a complex machine learning model that takes as input team statistics, and it predicts, as output, whether or not a team will have the "Player

4.1 Global Explanations: Permutation Importance and Partial Dependence Plot

of the Match" prize (Becker, 2020). This prediction is then passed to bookmakers in real time. Our stakeholders are concerned about how the ML model works and which are the most important criteria used to predict the team having or not "Player of the Match" award. As XAI experts, the first question we are asked to answer is: **What are the more significant features of your model?**

4.1.1 Ranking Features by Permutation Importance

The permutation Importance method tries to answer this specific question. We recall what already explained in Chap. 3: we assume to have an "opaque" trained model to interpret and understand the relative importance of the features leveraged by the model to make predictions.

What we do is to shuffle the values in a column (feature) and make the prediction again but with the shuffled values.

The expectation is that the error associated with the prediction will increase depending on the importance of the shuffled feature: the more important is the specific feature, the more the predictions will worsen because of the shuffling.

Assuming that the model is not relying on a feature to make the predictions, the shuffling of the values of that feature won't impact the performance.

So let's go to our scenario with the first snip of code here:

```
import numpy as np
import pandas as pd
from sklearn.model_selection import train_test_split
from sklearn.ensemble import RandomForestClassifier

data = pd.read_csv('../input/fifa-2018-match-statistics/FIFA 2018
Statistics.csv')
y = (data['Player of the Match'] == "Yes")  # Convert from string
"Yes"/"No" to binary
feature_names = [i for i in data.columns if data[i].dtype in
[np.int64]]
X = data[feature_names]
train_X, val_X, train_y, val_y = train_test_split(X, y,
random_state=1)
my_model = RandomForestClassifier(n_estimators=100,
                                  random_state=0).fit(train_X,
train_y)
```

We use the statistics available from FIFA, after importing the file and cleaning the data, we just do the usual ML split between training and test data before building our model that, in this case, is a Random Forest (Becker, 2020).

Let's have a look at the data to make sense of what's going on. The Table 4.1 is extracted from the original CSV file.

As you may see, we have the teams, the date, a lot of features, and our target feature that is Player of the Match (Y/N) saying if Team had the player of the match prize or no for that specific match against the "Opponent."

After our data manipulation aimed at building our Random Forest model, the data look like this (Table 4.2):

```
X.head()
```

The relevant thing to note is that we just kept the numerical data, transformed the Player of the Match (Y/N) in 1 or 0 values; the first column left to "Goal Scored," is a numeric id for the match.

As we said, we are using Random Forest to do "Player of the Match" predictions, but we are not interested in the ML details of Random Forest; we are focused on how to generate explanations around the predictions provided by our ML model. Random Forest is built as a huge ensemble of decision trees; each decision tree contributes to the final prediction that will result as the "most voted" one, a kind of

Table 4.1 Fifa 2018 matches, target feature is Man of The Match column

Date	Team	Opponent	Goal scored	Ball possession %	Attempts	On-target	Off-target	Man of the match
14-06-2018	Russia	Saudi Arabia	5	40	13	7	3	Yes
14-06-2018	Saudi Arabia	Russia	0	60	6	0	3	No
15-06-2018	Egypt	Uruguay	0	43	8	3	3	No
15-06-2018	Uruguay	Egypt	1	57	14	4	6	Yes
15-06-2018	Morocco	Iran	0	64	13	3	6	No
15-06-2018	Iran	Morocco	1	36	8	2	5	Yes

Table 4.2 Statistics after a first data manipulation needed to build the ML Model

	Goal scored	Ball possession %	Attempts	On-target	Off-target	Blocked	Corners	Offsides	Free kicks
0	5	40	13	7	3	3	6	3	11
1	0	60	6	0	3	3	2	1	25
2	0	43	8	3	3	2	0	1	7
3	1	57	14	4	6	4	5	1	13
4	0	64	13	3	6	4	5	0	14
5	1	36	8	2	5	1	2	0	22

4.1 Global Explanations: Permutation Importance and Partial Dependence Plot

wisdom of the crowd. It performs better than a decision tree at the cost of losing the intrinsic interpretability of a single decision tree. We will answer the question, "What are the most important features of your model?" with permutation importance method, and the same flow can be adopted for any other opaque model, different from Random Forest.

```
import eli5    #A
from eli5.sklearn import PermutationImportance
perm = PermutationImportance(my_model, random_state=1).fit(val_X,
val_y) #B

eli5.show_weights(perm, feature_names = val_X.columns.
tolist())              #C
#A Here we Import eli5 library
#B Train The permutation importance model on the validation set
#C Show the feature importance
```

Few lines of code are enough to generate our explanations with Permutation Importance (Fig. 4.2):

Let's explain the table. The features are ranked by their relative importance, so the first and most important result is that we may directly answer our question: "What are the most important features of your model?": Goals Scored is the most important feature that our Random Forest ML model uses to predict if the team will have or not the Player of the Match prize.

The importances are calculated by shuffling all the values of a feature and observing how the model's performance changes. Numerically the importances are the value of the Loss function *minus* the value of the Loss function after the shuffling.

Out[14]:

Weight	Feature
0.1750 ± 0.0848	Goal Scored
0.0500 ± 0.0637	Distance Covered (Kms)
0.0437 ± 0.0637	Yellow Card
0.0187 ± 0.0500	Off-Target
0.0187 ± 0.0637	Free Kicks
0.0187 ± 0.0637	Fouls Committed
0.0125 ± 0.0637	Pass Accuracy %
0.0125 ± 0.0306	Blocked
0.0063 ± 0.0612	Saves
0.0063 ± 0.0250	Ball Possession %
0 ± 0.0000	Red
0 ± 0.0000	Yellow & Red
0.0000 ± 0.0559	On-Target
-0.0063 ± 0.0729	Offsides
-0.0063 ± 0.0919	Corners
-0.0063 ± 0.0250	Goals in PSO
-0.0187 ± 0.0306	Attempts
-0.0500 ± 0.0637	Passes

Fig. 4.2 Permutation importance output. Every weight is shown with its uncertainty. (Becker, 2020)

The related error is estimated statistically through repetitions of the shuffling.

Notice the smart idea behind this procedure. If you shuffle the values of a set of variables the statistical properties like mean, variance and so on are retained, but we have destroyed the causal dependence between the target.

Here the numbers are variations of the Loss function. They show by how much the Loss function will increase so they provide to us merely a ranking. They don't show to us a relative importance variation: the "Goal Scored" feature is not three times more important than the "Distance covered" one. It gives three times more variation of the Loss function that is not the same thing as to contribute three times more to the answer of the model.

In Machine Learning courses, we have seen how dropping less important features (under a specified value you have to guess) can improve model performance. The reason is usually to drop irrelevant features.

But we must have special attention for the case of two or more correlated features. Think of it, if two features are highly correlated with a though experiment think of two features that are copies of each other. The model can indifferently use one feature or the other. So when you shuffle one of the two features, the model will use the other to retain some performance so the importance of the shuffled features will be underestimated. **Permutation Importance will underestimate highly correlated features**.

4.1.2 Permutation Importance on the Train Set

At the bottom of the ranking that we have some negative values. A negative value may sound strange but it simply states that *the model without those features has an increased accuracy*. Such phenomenon is in fact normal in the training of a model and excluding those features and then retraining the model anew increases the overall performance.

Wait! A model trained on a dataset shouldn't be using those bad features at all! In fact that's precisely true on the training set not on the validation/test set that poses to the model a novel task. The negatives values are a case of bad generalization so more in detail, the negative values are a form of overfitting.

To double check this hypothesis we apply permutation importance to the train set and the expectation is that we won't have these negative values. Let's do this quick exercise for confirmation.

```
import eli5
from eli5.sklearn import PermutationImportance
perm = PermutationImportance(my_model, random_state=1).
fit(train_X,train_y) #A
eli5.show_weights(perm, feature_names = val_X.columns.tolist())

#A Here we train the Permutation Importance model on the
train.set
```

4.1 Global Explanations: Permutation Importance and Partial Dependence Plot

It is the same code as before but we changed the test set with the training set.
Here the output:
The table confirms our idea (Fig. 4.3): this time we don't have any negative value on the train set and "goals scored" is confirmed as the most important feature that affects the output.

This confirms our guess of overfitting as further shown by the changes in the rest of the ranking. In particular, "Attempts" is now at second place while it is at the bottom of the ranking performed with the test set (also in red). This is a further evidence of the fact that the ML model is badly using features that are not important to overfit the results. **As general recommendation the feature importance methods should be always applied on the test set.** This exercise has been performed just to deal with negative values and check the overfitting.

4.1.3 Partial Dependence Plot

The main strength of this XAI method is to provide a simple and direct answer about the most important feature. The output is a nice and easy table that can be directly passed to our stakeholders. But it doesn't help no answering the "How": we may be interested or asked to answer how goals scored may change the predictions. Is there any threshold on goals scored to increase the probability of having the Player of the Match prize? Permutation Importance cannot help on this and in a while we will see how to tackle this further point.

Getting back to the point of answering the "How" instead of "What," we introduce the partial dependence plot method (PDP) to deal with this.

Fig. 4.3 Permutation Importance output on the training set, no anomalous negative values here

```
Out[8]:
```

Weight	Feature
0.1375 ± 0.0243	Goal Scored
0.0187 ± 0.0156	Attempts
0.0104 ± 0.0132	Free Kicks
0.0104 ± 0.0000	Blocked
0.0083 ± 0.0083	Distance Covered (Kms)
0.0062 ± 0.0102	Pass Accuracy %
0.0062 ± 0.0102	On-Target
0.0042 ± 0.0102	Ball Possession %
0.0021 ± 0.0083	Fouls Committed
0.0021 ± 0.0083	Passes
0.0021 ± 0.0083	Corners
0 ± 0.0000	Yellow Card
0 ± 0.0000	Saves
0 ± 0.0000	Red
0 ± 0.0000	Offsides
0 ± 0.0000	Off-Target
0 ± 0.0000	Yellow & Red
0 ± 0.0000	Goals in PSO

PDP sketches the functional form of the relationship between an input feature and the target, as we will see it can also be extended to more that one input feature. As Permutation Importance PDP is used on a model that has already been fit and we will use it to see "how" the predictions are changed by changes in the number of goals (our most important feature as from Permutation Importance). What is performed under the covers by PDP method is to evaluate the effect of changes in a feature over multiple rows to get an average behavior and provide the related functional relationship. It is important to note that averaging may hide a subtlety, the fact that the functional relation may be increasing or decreasing for different rows and this won't appear in the final result that will show just the "average" behavior. Also in the simplest case, interactions between features is not taken into consideration but we will see the case of a two-dimensional PDP later.

Let's go to the real code to touch with hands what we are talking about.

Here the main flow in which we consider the PDP for goals scored that according to our previous analysis is the most important feature.

```
from pdpbox import pdp, get_dataset, info_plots   #A


feature_to_plot = 'Goal Scored'   #B
pdp_dist = pdp.pdp_isolate(model=my_model, dataset=val_X, model_features=feature_names, feature=feature_to_plot)

pdp.pdp_plot(pdp_dist, feature_to_plot)
plt.show()

#A Import from the pdpbox library
#B We select the 'Goal scored' feature
```

With this bunch of lines, we just select the feature we want to analyze (Goal Scored) and pass the info to PDP library to do the job, here the results.

Nice diagram (Fig. 4.4) provided by PDP library, right? On x axis we have the goals scored while on y axis we have the estimated change in the prediction respect to the baseline value that is set to 0. The shaded area is an indication of the confidence level.

We see a first interesting outcome: we already got from Permutation Importance that Goals Scored is the most important feature, here we see that we have a strong and positive increase in prediction with 1 goal scored but after that the trend is almost flat, scoring a lot of goals doesn't change to match the overall prediction of having the "Player of the match" prize.

Let's do the same exercise but with a different feature, the one that was in second position in the Permutation Importance ranking that is the covered distance: "Distance Covered (kms)".

4.1 Global Explanations: Permutation Importance and Partial Dependence Plot

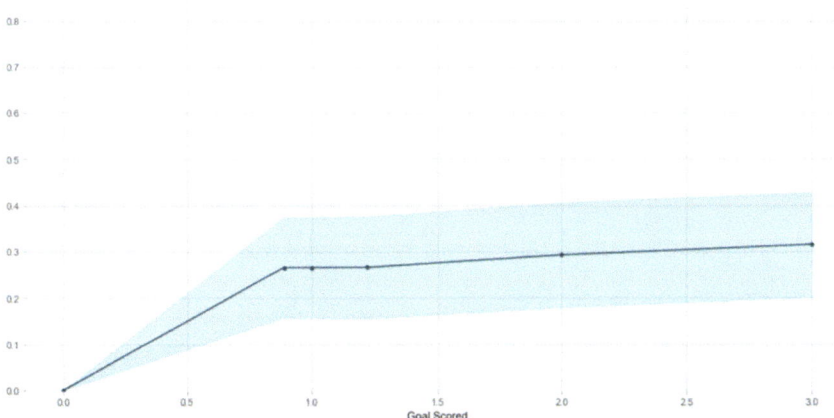

Fig. 4.4 Partial Depence Plot diagram that shows how "Goals Scored" influences the prediction (Becker, 2020)

```
feature_to_plot = 'Distance Covered (kms)' #A


pdp_dist = pdp.pdp_isolate(model=my_model, dataset=val_X, model_features=feature_names, feature=feature_to_plot)


pdp.pdp_plot(pdp_dist, feature_to_plot)
plt.show()
```

#A We select the 'Distance Covered (kms)' feature

Note that the scale on y axis is now between 0 and 0.20 with a max around 0.08 while for "Goal Scored" we have a max around 0.27. This confirms the fact that "Goals Scored" is the most important feature for our Random Forest model (Fig. 4.5).

Also here we have an interesting scenario: the increase of the distance covered has a positive impact on the probability of having the "Player of the Match" but if the distance covered by the team is more than 100 Km or so, the trend goes in the opposite direction: running too much decreases the probability for the "Player of the Match" award that was not evident from Permutation Importance analysis only.

Remember that, here we have just the average effect of one feature on the target, we are not able to see how the values are playing for each predicted result. On average goals are slightly increasing the prediction but we could have cases (matches)

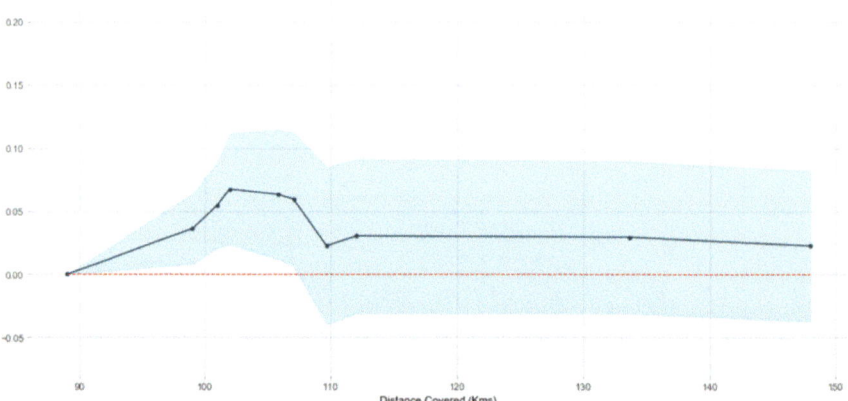

Fig. 4.5 Partial Depence Plot diagram that shows how "Distance Covered" influences the prediction

in which the trend is the opposite and others in which the number of goals strongly increased the overall probability, the overall result is the average we see. This depends from the fact that we are not taking into consideration the effect and interplay of the others features but isolating just one. Here we want to exploit another nice feature of PDP library that allows to look at the effects of 2 features at the same time, so to narrow down the mutual interactions.

The few lines of code will do the trick:

```
features_to_plot = ['Goal Scored', 'Distance Covered (Kms)']
inter1 = pdp.pdp_interact(model=my_model, dataset=val_X,
model_features=feature_names, features=features_to_plot)   #A

pdp.
pdp_interact_plot(pdp_interact_out=inter1, feature_
names=features_to_plot)
plt.show()

#A  PDP for feature interaction
```

Some new behavior emerges from this diagram if compared with the previous 2 in which we have just one feature pe time (Fig. 4.6). Maximum increase of probability for "Player of the Match" prize is with goals between 2 and 3 and a distance covered about 100 Kms.

Looking at the single diagram of goals scored it seems there is just a slight variation above 1 goal, here we see that there is a clear area of best values (yellow area).

4.1 Global Explanations: Permutation Importance and Partial Dependence Plot

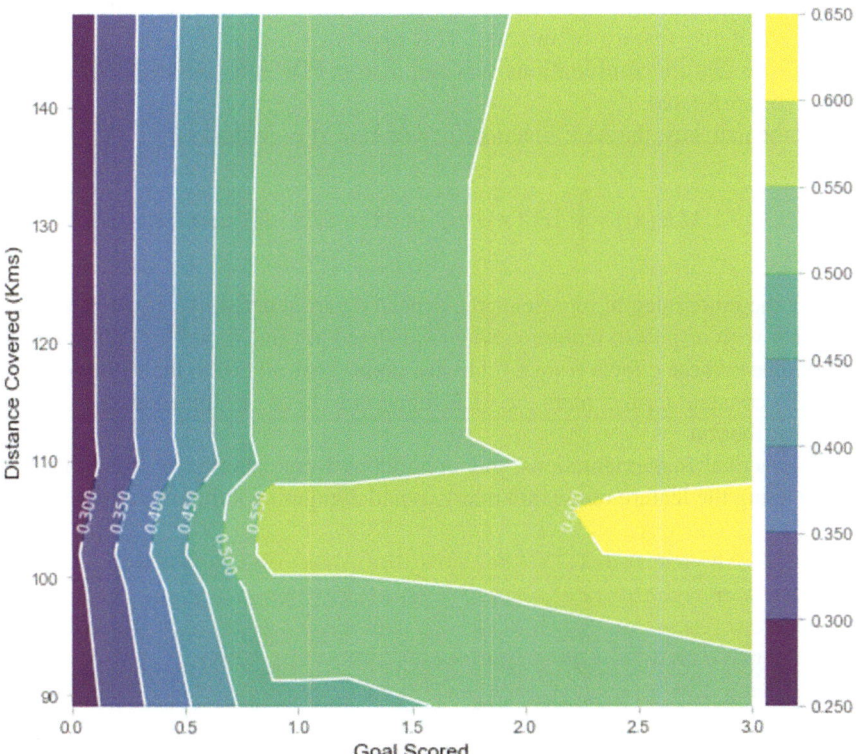

Fig. 4.6 PDP diagram that shows the iteraction of the two main features and their impact on the prediction

Also, it is confirmed that maximum effect from distance covered is achieved around 100 Kms but with more goals also longer distances produce the same overall effect. We did this exercise just with the two most important features but you can easily guess that it could be worth to explore also other combinations to deep dive the scenario and produce detailed explanations.

4.1.4 Accumulated Local Effects (ALE)

Traditional methods for visualizing feature effects in machine learning models, such as Partial Dependence Plots (PDPs), can produce misleading results when features are correlated. Accumulated Local Effects (ALE) as addresses this limitation by providing an unbiased approach to visualizing how features influence predictions, even in the presence of strong feature correlations.

The fundamental insight behind ALE is to focus on local changes in the feature space rather than marginal effects across the entire distribution. While PDPs estimate the marginal effect of a feature by averaging predictions across the distribution of other features, ALE accumulates local effects within narrow intervals of the feature of interest. This approach avoids the extrapolation to unrealistic feature combinations that can distort PDP interpretations when features are correlated.

Mathematically, the ALE function for a feature xj is defined as:

$$ALE(\text{xj}) = \int_{0}^{\text{xj}} \left[\int \partial f(x) / \partial xj \ p(x|\text{xj}=z) dx \right] dz - \text{constant}$$

Where the inner integral represents the expected partial derivative of the prediction function with respect to feature xj when xj is fixed at value z, and the outer integral accumulates these effects from a reference point (typically 0) to the value of interest. The constant term centers the ALE function to have zero mean across the feature distribution.

In practical implementations, this continuous formulation is approximated by discretizing the feature range into intervals and computing finite differences:

```python
def compute_ale(model, X, feature_idx, num_intervals=50):
    """Compute ALE for a specific feature"""
    # Sort data by the feature of interest
    sorted_indices = np.argsort(X[:, feature_idx])
    sorted_X = X[sorted_indices]

    # Create intervals along the feature range
    unique_values = np.unique(sorted_X[:, feature_idx])
    if len(unique_values) <= num_intervals:
        # For categorical or discrete features
        intervals = unique_values
    else:
        # For continuous features
        intervals = np.quantile(sorted_X[:, feature_idx],
                                np.linspace(0, 1, num_intervals+1))

    # Initialize ALE values
    ale_values = np.zeros(len(intervals)-1)

    # Compute local effects for each interval
    for i in range(len(intervals)-1):
```

4.1 Global Explanations: Permutation Importance and Partial Dependence Plot

```
        # Find points in this interval
        mask = ((sorted_X[:, feature_idx] >= intervals[i]) &
                (sorted_X[:, feature_idx] < intervals[i+1]))
        points_in_interval = sorted_X[mask]

        if len(points_in_interval) == 0:
            continue

        # For each point, compute effect of changing feature value
        effects = []
        for point in points_in_interval:
            x_lower = point.copy()
            x_upper = point.copy()
            x_lower[feature_idx] = intervals[i]
            x_upper[feature_idx] = intervals[i+1]

            effect = model.predict([x_upper])[0] - model.predict([x_lower])[0]
            effects.append(effect)

        # Average effect within this interval
        ale_values[i] = np.mean(effects)

    # Accumulate effects
    ale_cumulative = np.cumsum(ale_values)

    # Center the ALE function (zero mean)
    ale_centered = ale_cumulative - np.mean(ale_cumulative)

    return intervals, ale_centered
```

The real-world impact of this approach is particularly evident in domains with naturally correlated features. In real estate valuation, for instance, square footage and neighborhood location are strongly correlated—larger homes tend to be concentrated in certain areas. A PDP analysis might suggest that adding 1000 square feet would increase a home's value by the same amount regardless of location, a clearly unrealistic conclusion.

An ALE analysis of the same data would show that the effect of increased square footage varies substantially by location bracket. In a study of the Boston housing market, researchers found that the value increase per square foot in central neighborhoods was 2.3 times higher than in peripheral areas—a critically important distinction that PDPs obscured due to correlation effects.

Similarly, in credit risk modeling, income and debt levels typically show strong correlation. PDPs might suggest that increasing income always reduces default risk by the same amount, regardless of debt level. ALE analysis reveals the more realistic pattern: income increases reduce default risk most significantly at moderate debt levels, with diminishing effects for individuals with either very low or very high debt-to-income ratios.

The practical implementation of ALE has been significantly simplified by integration into interpretability libraries:

```
from alibi.explainers import ALE


# Initialize the ALE explainer
ale_explainer = ALE(predictor=model.predict,
                    feature_names=feature_names)

# Compute ALE for all features
ale_explanation = ale_explainer.explain(X_test)

# Visualize ALE for specific features
fig, ax = plt.subplots(1, 2, figsize=(12, 5))
plot_ale(ale_explanation, features=[0, 1], ax=ax)
```

Beyond avoiding distortions from correlated features, ALE offers several additional advantages. The method is model-agnostic, applicable to any prediction function regardless of its internal structure. It focuses on actual data regions rather than hypothetical combinations, enhancing the reliability of the interpretations. And it can be computed efficiently even for complex models, making it practical for regular use in model development and validation.

ALE also provides a foundation for extending feature effect analysis to interactions. By computing second-order ALE functions, analysts can identify and visualize how pairs of features jointly influence predictions beyond their individual effects. This capability is particularly valuable in domains where interaction effects are theoretically expected but difficult to quantify precisely.

The development of ALE exemplifies how advances in statistical methodology can directly address practical challenges in model interpretation. By recognizing the limitations of existing approaches and developing theoretically sound alternatives, researchers have provided practitioners with more reliable tools for understanding and communicating model behavior, particularly in domains where feature correlations are unavoidable.

4.1.5 Properties of Explanations

Let's summarize as usual the properties of the explanations we provided (Table 4.3):

Please, compare this Table 4.4 with the analogous one we got for intrinsic explainable models in the previous chapter.

4.1 Global Explanations: Permutation Importance and Partial Dependence Plot

Table 4.3 Explanations Properties assessment for Permutation Importance and PDP methods, and ALE methods

Property	Assessment
Completeness	Interpretability achieved with agnostic method, completeness is low, limited possibility of anticipating model predictions (we can just look at goals scored as rough indicator)
Expressive power	Good in terms of getting evidence of the most important feature but on average and without details of features interactions (or limited)
Translucency	Low, we don't have insight into model internals
Portability	High, the method doesn't rely on the ML model specs.
Algorithmic complexity	Low, no need of complex methods to generate explanations
Comprehensibility	Good level of human understandable explanations

Table 4.4 Properties of explanations for intrinsic explainable models

Property	Assessment
Completeness	Full completeness achieved without the need of trading-off with interpretability being an intrinsic explainable model.
Expressive power	Less than linear regression case. Interpretations of coefficients is not so straightforward.
Translucency	As any intrinsic explainable model, we can look at the internals. Weights are used to provide explanations but not so directly as in linear regression case.
Portability	Method is not portable, specific for logistic regression.
Algorithmic complexity	Low but not trivial as in linear regression case.
Comprehensibility	Explanations are human understandable also for not technical people.

It is evident that for agnostic methods we are losing completeness (we don't have a full understanding of the model) but gaining portability because agnostic methods are not model dependent.

It is useful to remark the scope of the explanations we provided, they are global explanations. As we saw we leveraged Permutation Importance and PDP that relies on averages to show functional relationships and explanations. For ALE specifically, Translucency remains low, but Expressive Power is improved compared to PDP when dealing with correlated features, providing more realistic interpretations of feature effects. We are not in the position, with these methods, of answering specific questions on a prediction for specific data points. The objective of the next section is to use SHAP to switch to local scope and answer questions on a particular data point prediction.

4.2 Local Explanations: XAI with Shapley Additive Explanations

Do you remember the beginning of our journey through XAI started in Chap. 1?

The classical example that is always presented to introduce XAI is that someone, say Helen, goes to the bank to ask for a loan but it is refused. The obvious question that follows is "Why?" and the bank might be in troubles if an opaque ML model has generated the answer without XAI. As global methods, permutation importance and partial dependence plots don't help in this case. Helen is not interested in getting "global" explanations about how the ML model works but wants an answer on her specific case.

If we move to our working scenario of "Player of the Match" prize, so far we provided explanations about the most important features and the functional relationship of these features with the prediction but we are not able to answer to the direct question:

Considering the features in the figure how much the specific prediction for this match has been driven by the number of goals scored by Uruguay?

This is the same problem from Helen: she doesn't want to know how the bank ML model generally uses the features but know her case, why her loan has been refused.

The same for "Player of the Match": this time, we don't want to know the most important features of the model but get explanations on a specific match, say Uruguay-Russia.

We are transitioning from global explanations to local explanations using SHAP library. Here SHAP stands for Shapley Additive explanations (Table 4.5).

From Permutation Importance and PDP we know that "on average" number of goals is main driver of the prediction, but we also know (remember the 2D PDP plot) that the features may have mutual interactions that can change the situation. Moreover, here we want to know how much the prediction has "likely" been increased, compared to a baseline, by the fact that Uruguay scored exactly three goals. SHAP helps in these specific cases, where we need an answer on a single prediction and we are less interested in an understanding of the "average" behavior of the model (Becker, 2020).

Table 4.5 Specific match that will be analyzed with Shap

Date	Team	Opponent	Goal scored	Ball possession %	Attempts	On-target	Off-target	Man of the match
25-06-2018	Uruguay	Russia	3	56	17	7	6	Yes

4.2.1 Shapley Values: A Game-Theoretical Approach

SHAP method relies on Shapley value, named by Lord Shapley in 1951 that introduced this concept to find solutions in cooperative games. To set the stage, game theory is a theoretical framework to deal with situations in which we have competing players and we search for the optimal decisions that depend from the strategy adopted by the other players. The creators of modern game theory were mathematicians John von Neumann, John Nash and the economist, Oskar Morgenstern.

Cooperative game theory is a specific case of game theory in which the assumption is to have a group of players that make decisions as coalitions building cooperative behavior. Shapley values deal with this specific scenario: we have a coalition of players in a game that, with a specific strategy, achieve a collective payoff, we want to know the fairest way to split the payoff among these players according to the contribution that each of them provided.

How to estimate this marginal contribution of each player?

Assuming to have three players, Bob, Luke and Anna that join the game one after the other, the most straightforward answer would be to consider the payoff achieved by each of them:

Bob joined and got a payoff of 7, then Luke joined to bring a payoff of 3, and last Anna added a payoff of 2; so the sequence would per (7,3,2).

But here we are not taking into consideration the fact that is changing the sequence in which the players join the game may change the respective payoff (because of the different game background conditions they may find at the time they enter). Also, we need to consider the case in which all the players join the game simultaneously.

Shapley Values answer this question by doing an average over all possible sequences to find each player's marginal contribution. We'll see this adding some technical details in Sect. 4.3.

How is all this stuff related to machine learning an XAI? The analogy that is adopted is powerful, and it is an excellent example of strong ideas that pass the barriers from one domain of science to another.

We can replace "players" with "features" that are now playing to build the prediction that is our payoff. Shapley values will tell us how the payoff is fairly distributed among the features, that is which features contributes more for a specific prediction that is the outcome of a game.

If you prefer, you can also think of Shapley values as a fair repartition of wages for workers.

Each worker's wage will be proportional to his contribution, and his contribution is calculated precisely via Shapley values.

The XAI method is called SHAP that is an acronym from Shapley Additive exPlanations and provides explanations of a single prediction through a linear combination (additive model) of the underlying Shapley values. Let's see how it works in practice.

4.2.2 The First Use of SHAP

Back to our specific question, we have a match "Uruguay vs. Russia," and we want to know how much the prediction for "Player of the Match" prize that resulted by the ML model has been driven by the number of goals scored by Uruguay. We compare the prediction to a baseline value that is defined as the average value for all the predictions of all the matches.

```
row_to_show = 19
data_for_prediction = val_X.iloc[row_to_show]   #A
data_for_prediction_array = data_for_prediction.values.
reshape(1, -1)
my_model.predict_proba(data_for_prediction_array)

#A use 1 row of data here. Could use multiple rows if desired
```

These lines of code are just to select the right match (Uruguay—Russia) and see the prediction from our ML Random Forest model of having Uruguay assigned the "Player of the Match" award (52%):

```
array([[0.48, 0.52]])
```

Following the snip of code to import SHAP library and use it:

```
import shap   #A
k_explainer = shap.KernelExplainer(my_model.predict_proba,
train_X) #B

k_shap_values = k_explainer.shap_values(data_for_prediction)


#A package used to calculate Shap values
#B # use KernelSHAP to explain test set predictions
```

With these three lines of code, we have already produced the shap values that can be used to explain the specific prediction for the match, but they would be just numbers. The most exciting feature is the built-in graphic library that allows for a beautiful and interpretable plot of the results.

```
shap.force_plot(k_explainer.expected_value[1], k_shap_values[1],
data_for_prediction)
```

Let's look at the output and how to interpret it:

There are two kinds of main indicators in the diagram (Fig. 4.7). Features on the left are the ones that increase the predictions, and their relative length is and an indication of the importance of the features in determining the prediction.

4.2 Local Explanations: XAI with Shapley Additive Explanations

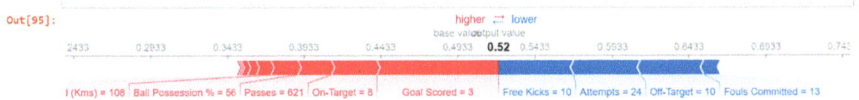

Fig. 4.7 SHAP diagram that shows how the features impact on the match Uruguay-Russia. A force diagram representing by how much the features change the value final value. For example we see that "Goal Scored = 3" has the most impact for it pushes the final value to the right with the biggest interval. (Becker, 2020)

Features on the right, same logic, are the ones that are expected to decrease the prediction value. For this specific match, we predicted a probability of 0.52 for Uruguay to have "Player of the Match" that is not so high considering that they scored three goals (remember that goals scored was identified as the most important feature).

The value 0.52 has to be compared with the baseline value of 0.50, which is the average of all the outputs and in this case, represents the maximal uncertainty. We see the explanation of this result: albeit the red driving features, we have the free kicks, attempts and off-target features that depress the overall probability. The shift from the baseline is the difference in length between the sum of red bars and the blue bars.

It is important to stress again the big difference with what we did before. With Permutation Importance and PDP we identified the most important features and provided an "average" functional relationship between these features and the prediction. We had no chance of getting into a specific prediction to answer a why question on a specif match. Herewith SHAP we are able to answer a question on a specific occurrence so to address problems in which a person wants to know what happened with his case (loan rejected) and is not interested in having general explanations about the relative importance of the features: we are interested in what happened to us ONLY! The number of goals scored in the Uruguay-Russia match would make us predict a higher value for the prediction, while for this specific match, SHAP told us that there are other factors to limit the prediction to 0.52.

But SHAP can do even more, and it is not limited to the deep-dive of a single prediction.

Do you remember one of the limitations of the Permutation Importance method? We can use it to know the relative importance of a feature, but we don't know if that feature contributed a lot for few predictions and almost nothing for the rest producing an average behavior. SHAP allows getting a summary plot in which we see the impact of each feature on each prediction.

```
shap_values = k_explainer.shap_values(val_X) #A
shap.summary_plot(shap_values[1], val_X)

#A We call the summary plot
```

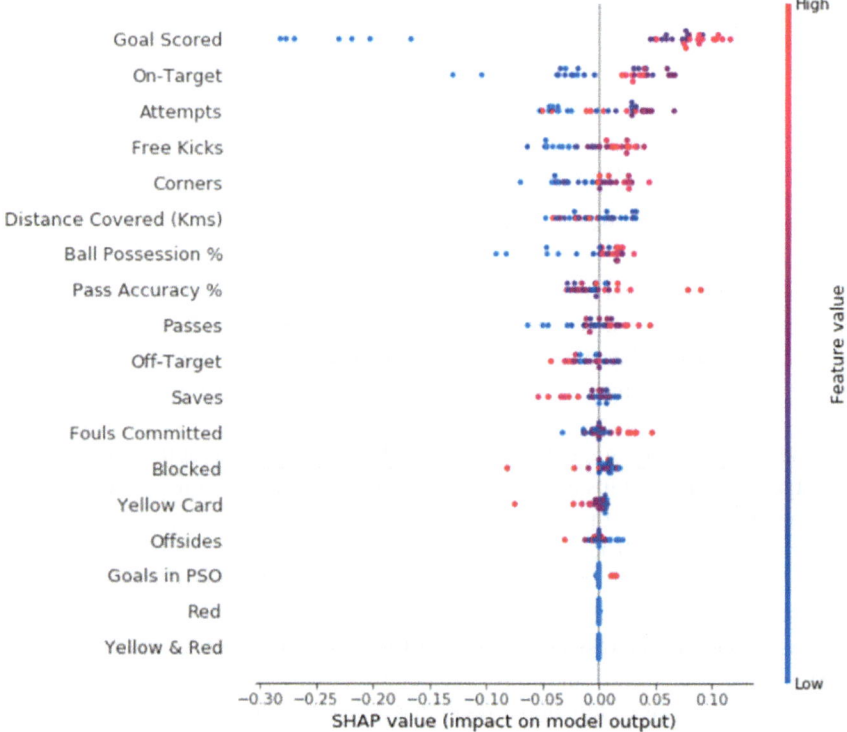

Fig. 4.8 SHAP diagram that shows the features ranking and the related impact on the march prediction. (Becker, 2020)

The code produces the following:

You see on the left the list of features and on x-axis the SHAP value (Fig. 4.8). The color of each dot represents if that feature is high or low for that specific row of data. The relative position of the dot on x-axis shows if that feature contributed positively or negatively to the prediction. In this way, you may quickly assess if, for each prediction, the feature is almost flat or impacting a lot some rows and almost nothing the others.

4.2.3 Properties of Explanations

Before closing this section let's do the usual assessment of the explanations we provided.

Let's summarize, as usual, the properties of the explanations we provided (Table 4.6):

Table 4.6 Explanations Properties assessment for SHAP

Property	Assessment
Completeness	We are focusing on explaining individual predictions, we don't have insight into the ML Model machinery
Expressive power	Strong expressive power for the single predictions that is what is asked usually to XAI
Translucency	Low, we don't have insight into model internals
Portability	High, the method doesn't rely on the ML model specs. to explain the predictions.
Algorithmic complexity	Medium, SHAP is easy to implement but it is not so easy to get a full understanding of the underlying concepts.
Comprehensibility	Good level of human understandable explanations, the diagram that is produced is very powerful.

After a fast, practical approach, we'll deepen our understanding of the mathematical foundations of the Shapley values and we meet SHAP library and of the various algorithms it provides to us.

4.3 The Road to KernelSHAP

To deepen the understanding, we briefly discuss the theoretical properties of Shapley values and why SHAP library is the de facto state of the art of post hoc explanations.

We will also look at LIME method in comparison with SHAP. LIME is an acronym for Local interpretable model-agnostic explanations.

4.3.1 The Shapley Formula

Let's say our built model is in the form $y = f(x)$.

An agnostic explainer works with models that are of the black-box type. So without any knowledge of the inner workings of the **model f** the explainer builds an **explaining model g** having access only to the outputs of model **f** and possibly some info on the training set or the domain.

As we have already said, Shapley values have a deep foundation in game theory a theoretical base that many other explanation methods lack. In the original foundational papers and subsequent work from Scott Lundberg and others (Lundberg & Lee, 2016; Lundberg & Lee, 2017) Shapley explanations have two important properties they are:

(s.1) they are **additive,** so we can share the quantity to explain between the features

$g(x) = \phi_0 + \sum_{i=1}^{M} \phi_i =$ a constant contribution + sum of each feature's importance

where ϕ_i is the contribution of each feature of the model and ϕ_0 is independent of features.

(s.2) they are **consistent,** or we can say *monotonic* in the sense that if a feature x_i has more influence on the model than another feature x_j then the $\phi_i > \phi_j$.

It has been demonstrated that Shapley values are unique, in the sense that they are the only possible explanations to have properties (s.1) and (s.2). This gives an enormous appeal to such a method.

Now, HOW can they be implemented? The formula for Shapley values is:

$$\phi_i = \sum_{S \subseteq N\{i\}} \frac{(M-|S|-1)!|S|!}{M!} \left[f_x(S \cup \{i\}) - f_x(S) \right] \quad (4.1)$$

Here we sum on every possible subset S of features not including the feature i we are investigating. So $f_x(S)$ is the expected output given the features subset S and $f_x(S \cup i) - f_x(S)$ is thus the contribution made by adding the feature i.

The combinatorial is a weighting factor that takes into account the multiple ways of creating subsets of features.

4.3.2 How to Calculate Shapley Values

Wait! How can we evaluate f not using some features? We have to use some background information to evaluate f with fictitious values instead of the features we are investigating. It is not a trivial matter, and usually, we use some data distribution, or we give to the method a background dataset to sample from randomly.

The number of possible subsets of features not including feature i is 2^{N-1}, where N is the total number of features so it increases very quickly with the number of features.

A standard calculation for large features is practically too time-consuming, so we can think only of an estimate approximating it via a Monte Carlo (random) approach.

The work of Lungren, implemented in the SHAP library, shows to us other attractive solutions (and speedups).

KernelShap we have used in the previous section is an agnostic **approximate** linear approximation and works for every possible model you may train.

TreeShap is not agnostic because it only works on tree-based models (even boosted trees), but it works in *linear* time, and it is even an **exact** calculation of Shapley values not an approximation.

There are also some specific methods for **DeepNeural Networks**, but we will see them in the following chapter.

4.3.3 Local Linear Surrogate Models (LIME)

The idea behind KernelShap is to construct a local surrogate model for the explanation model *g*. A surrogate model is an effective approximation of the model: a reconstruction of the model that can give approximately the same results of the model. A local surrogate model will approximate the real model for values near to a given sample.

As we have already seen, a local explanation model is more powerful than a global one. It can answer to the customer question "why my loan has been refused?" probing all the answers that the model will give for little variations of customer data. Technically it gives different explanations for each instance of the model.

The requirement of additivity (s.1) in the previous subsection forces us to use a locally linear model. Which, in fact, is an excellent idea for we know linear models are intrinsically explainable. So It turns out that KernelShap is not so different from another well know agnostic technique called LIME.

The idea of the original LIME by Ribeiro et al. (2016) (as we said LIME stands for Local Interpretable Model-Agnostic Explanations) is to find a surrogate linear model repeatedly calling the trained model *f*. Say *f* is a classifier: the value *f(x)* is the probability of a class for the instance *x* (Fig. 4.9).

To construct a linear model g we add some Gaussian noise to x to have some new perturbed points say $z_1, z_2, z_3, \ldots$. We call the model f on these new instances to have new class probabilities $f(z_1), f(z_2), f(z_3)\ldots$ finally we train a linear g on these new instances with the requirement that distant points are weighted less using a weight exponentially decreasing with the distance.

This process can be described with a Loss function of the parameters of g with two terms:

$$\text{Loss}(g) = L(f, \pi, g) + \Omega(g)$$

$L(f, \pi, g)$ is the usual sum of squared differences between the values instances $f(z_i)$ and the surrogate model g you would expect from a Loss function. But here it's

Fig. 4.9 Schematics for LIME. The class output of the model are circles or crosses and the dimensions reminds us of the weight so distant points are weighted less. (Ribeiro et al., 2016)

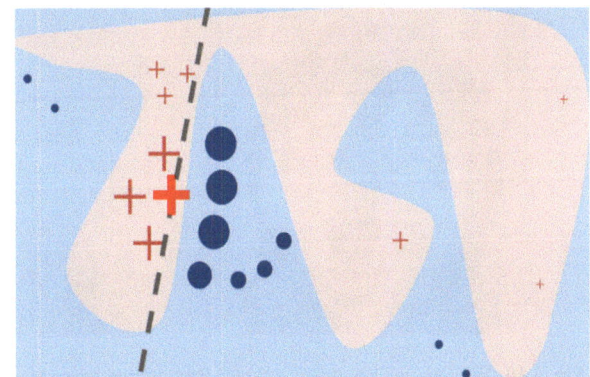

multiplied with a positive weight factor π decreasing with the distance from the original instance x.

In Lime, π is usually a decreasing exponential function.

The new term $\Omega(g)$ is a Lasso regularization to have a sparse representation. We can change $\Omega(g)$ to reduce the dimension of the explanation to say only K nonzero features.

When we restrain only to a few nonzero features (without losing the fidelity of our explanation) we increase simplicity and so we gain in interpretability.

In fact, in literature, usually, we differentiate the instance x (which like all the features of the model f) with its interpretable analog x' where only the meaningful characteristics are considered.

Think of the picture of the wolf we have already seen in the first chapter. In this case, x is a matrix with all the pixels values (in the three colors) while x' is only the selected region with snow (Fig. 4.10).

4.3.4 KernelSHAP Is a Unique Form of LIME

Now that we have a know of how a linear surrogate is constructed.

We can state the following striking property KernelShap is the only local linear surrogate explanation model g of the initial model f that gives us Shapley values.

In KernelShap, we use the LIME Loss but weighting by a "distance" π counting the number of possible subsets of features we use in the Shapley formula (4.1).

So, as a matter of fact, KernelShap is a special type of LIME.

We conclude by showing remarking one advantage of SHAP.

Shapley values are all of the same dimensions (say dollars) even if the corresponding features are not. So we could profitably use Shapley values themselves as new features for another model without using normalization. Or we can group

Fig. 4.10 Why do you say this is a Wolf? Because we have snow! (Ribeiro et al., 2016)

4.4 KernelSHAP and Interactions

4.4.1 The NewYork Cab Scenario

Let's start with our scenario that deals with Kaggle dataset for the NewYork Cab dataset (Kaggle, 2020). The objective is to predict the fare amount for cabs in New Yok based on pickup place and dropoff locations. The estimation is very basic and takes into consideration only the distance between the two positions.

We will revisit SHAP with greater attention to feature interactions. Additionally, we will assume a critical requirement for timely explanations, and to help meet this performance goal, we will record prediction times.

4.4.2 Train the Model with Preliminary Analysis

Let's open the file

```
import pandas as pd
import numpy as np
import matplotlib.pyplot as plt
from lightgbm import LGBMRegressor    #A
from sklearn.model_selection import train_test_split
from sklearn.metrics import r2_score

from sklearn.inspection import permutation_importance #A


# Data preprocessing.
data =  pd.read_csv("./smalltrain.csv", nrows=50000)

#A We will use Sklearn
```

now we filter outliers and train a gradient boosting model

```
data = data.query('pickup_latitude > 40.7 and pickup_latitude < 40.8 and ' +
```

```
                    'dropoff_latitude > 40.7 and dropoff_latitude <
40.8 and ' +
                    'pickup_longitude > -74 and pickup_longitude <
-73.9 and ' +
                    'dropoff_longitude > -74 and dropoff_longitude <
-73.9 and ' +
                    'fare_amount > 0'
                    )

y = data.fare_amount


base_features = ['pickup_longitude',
                'pickup_latitude',
                'dropoff_longitude',
                'dropoff_latitude',
                'passenger_count']

X = data[base_features]


X_train, X_test, y_train, y_test = train_test_split(X,y,test_
size=0.5, random_state=1111)


# Tain with LGBM Regressor
reg = LGBMRegressor( importance_type='split',  random_state=42,
num_leaves=120)   #B

reg.fit(X_train, y_train)
print(r2_score(y_train,reg.predict(X_train)))
print(r2_score(y_test,reg.predict(X_test)))

#A We will train a LGBM Regressor. Note that LGBM is a tree-
based model.
#B Here we train the LGBM Regressor
```

We get an accuracy of

```
0.6872889587581943   train
0.4777749929850299   test
```

Not a good model because the R2 score is pretty low and we can say it's even overfitting because the score on the train set is significantly larger than that on the test set.

4.4 KernelSHAP and Interactions

We can calculate the usual permutation importance

```
# Getting permutation importance.
result = permutation_importance(reg, X_test, y_test, n_
repeats=10, random_state=42)
perm_sorted_idx = result.importances_mean.argsort()

# Visualize two variable importance plots.
fig, ax1= plt.subplots(1, 1, figsize=(12, 5))

ax1.title.set_text('Permutation Importance')
ax1.boxplot(result.importances[perm_sorted_idx].T, vert=False,
            labels=X_test.columns[perm_sorted_idx])

fig.tight_layout()
plt.show()
```

As we have already seen, permutation importance shows to us the features that have a major impact on the loss function as in the y-axis of Fig. 4.11.

Now we make an "interaction PDP" plot between the two more important features (following the permutation importance)

```
print('Computing partial dependence plots...')
from sklearn.inspection import plot_partial_dependence
import time

tic = time.time()
fig, ax = plt.subplots(figsize=(5, 5))
plot_partial_dependence(reg, X_test, [(X_test.columns[0],X_test.
columns[3])],
```

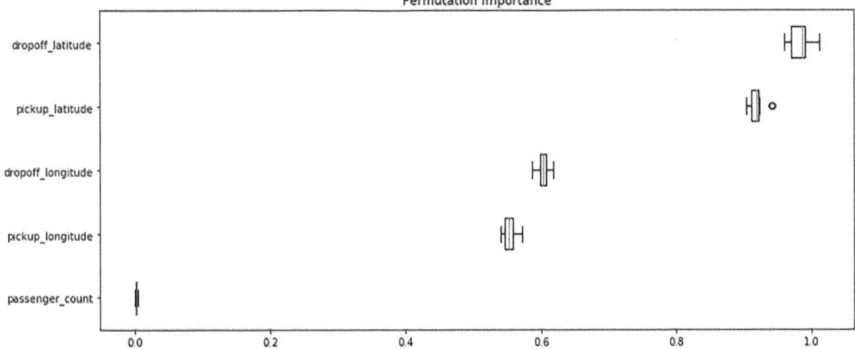

Fig. 4.11 We plot the Permutation Importance for the NewYork Cab dataset. We use a boxplot to express the uncertainty of the Importance estimated repeating n_repeats times the calculation

```
                        n_jobs=3, grid_resolution=20,ax=ax)
print("done in {:.3f}s".format(time.time() - tic))
ax.set_title('Partial dependence of NY taxi fare data - 2D')
plt.show()
```

With output

```
Computing partial dependence plots...
done in 21.406s
```

Again Fig. 4.12 can show us how the two features are playing together.

When two features do not interact, we can write $f(x) = f_j(x_j) + f_i(x_i)$, so each feature contributes independently from the other. And the picture suggests to us a complex interacting behavior like in the "Player of the match" scenario.

4.4.3 Making the Model Explainable with KernelShap

Now we search for local explainations using SHAP library

```
import shap
tic = time.time()
```

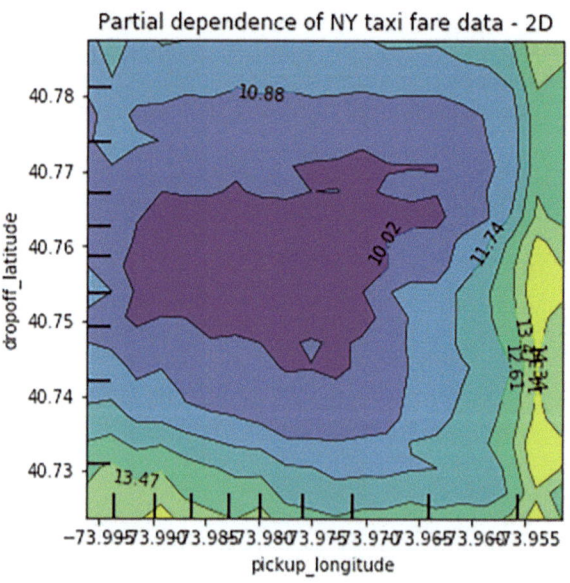

Fig. 4.12 Partial Dependence Plot for the two most important features

4.4 KernelSHAP and Interactions

```
background=shap.kmeans(X_train, 10)    #A


explainer = shap.KernelExplainer(reg.predict, background)    #B
shap_values = explainer.shap_values(X_test, nsamples=20)
print("done in {:.3f}s".format(time.time() - tic))

#A We need background information. For speed we'll summarize the
background as K = 10 samples
#B Using KernelSHAP
```

To speed up the creation of the model, we used a little trick instead of passing as a background, a big set of samples we used k-means to reduce the example set to only ten meaningful centroids. Nonetheless, model creation took **115.090s** on our machine.

As introduced at the beginning of the section, in this scenario, we are assuming that explanations should be available as fast as possible. KernelShap computation is too slow for tasks where speed is a must. We will see in a while how to improve performance.

4.4.4 Interactions of Features

Now for reference with our SHAP model trained we can do feature importance and a partial dependence plot using SHAP (Fig. 4.13).

```
# Variable importance-like plot.
shap.summary_plot(shap_values, X_test, plot_type="bar")

shap.dependence_plot("pickup_latitude", shap_values, X_test)    #A
```

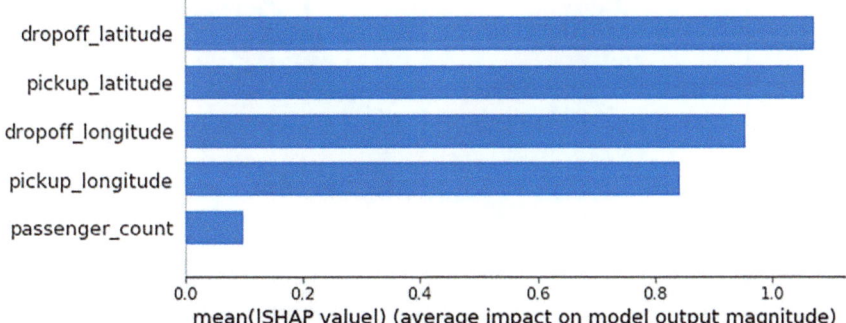

Fig. 4.13 Feature Importance diagram using SHAP

```
#A Let's make a PDP-like plot with SHAP
```

Here every point is a different sample (Fig. 4.14). On the left vertical axis, we have the SHAP value for that sample for the feature pickup_latitude, and on the horizontal axis the pickup_latitude_value. The model automatically finds the feature that is most likely interacting with "pickup_latitude" and uses these features to color the points. On the right vertical axis, we read that the interacting feature is dropoff_latitude.

If the two features were not interacting, the overall coloring would be uniformly distributed or the shades of a different color would be not intersecting.

We see instead a definite pattern with an intersection showing us a relevant interaction. What the picture roughly shows is that when dropoff_latitude is low (Lower Side of NYC) the contribution of pickup_latitude to prediction (i.e. the Shap value of pickup_latitude) is increasing the fee. Instead when dropoff_latitude is high it seems that the pickup_latitude contribution to the fee is decreasing with the increase of the pickup_latitude variable. As a matter of fact the features dropoff_latitude is changing how the pickup_latitude feature contributes to the fee.

4.5 A Faster SHAP for Boosted Trees

As mentioned, TreeSHAP includes a faster algorithm for tree-based models, including the LightGBM model used in the previous section. In addition to its speed, TreeSHAP provides exact Shapley value calculations.

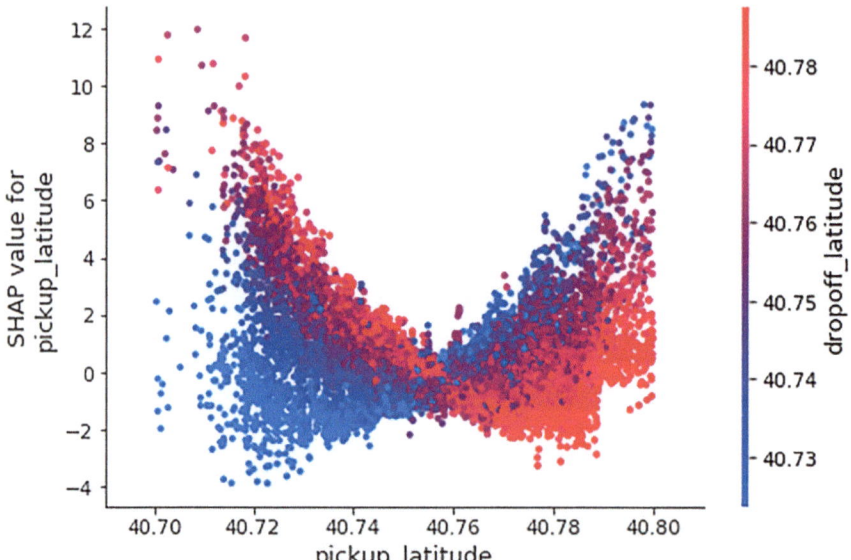

Fig. 4.14 Partial Dependence Plot using SHAP

4.5.1 Using TreeShap

The idea here is calculating the Shapley values as a weighted average at every node of the Shapley contribution of the branch. The algorithm, which does a clever reuse of previous values to collect results, can even estimate the background contribution from the tree structure.

Obviously, TreeShap is no more agnostic like KernelShap for it uses the model inner structure, but we have a wonderful tradeoff both in speed and precision of calculation.

Se let's retrain, on the same boosted model before, with TreeShap instead of KernelShap to meet the requirement of shortening the time to provide explanations.

```
import shap
print('Computing SHAP...')
tic = time.time()


explainer = shap.TreeExplainer(reg) #A


shap_values = explainer.shap_values(X_test)
print("done in {:.3f}s".format(time.time() - tic))
      pd.DataFrame(shap_values,columns=X_test.columns)

#A Using TreeSHAP, not KernelSHAP. Remember reg is a LGBM model
that is a tree-based model.
```

with output:

```
Computing SHAP...
Setting feature_perturbation = "tree_path_dependent" because no
background data was given.
done in 18.044s
```

It took only **18 s** to be compared with the previous result of **115 s**, and this is a huge improvement! Our stakeholders that were pushing to get timely explanations about the predicted fares will definitely be happy.

4.5.2 Providing Explanations

To complete the work with The NewYork Cab Company (or an Agency interested in the control/taxation of the fares), we have to show some examples and the SHAP at work with a force-plot (Fig. 4.15).

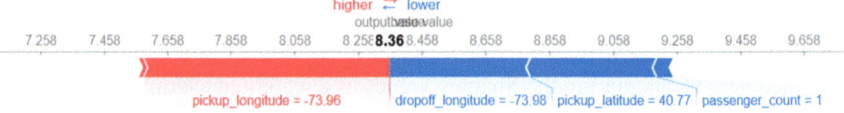

Fig. 4.15 SHAP diagram for Cab scenario

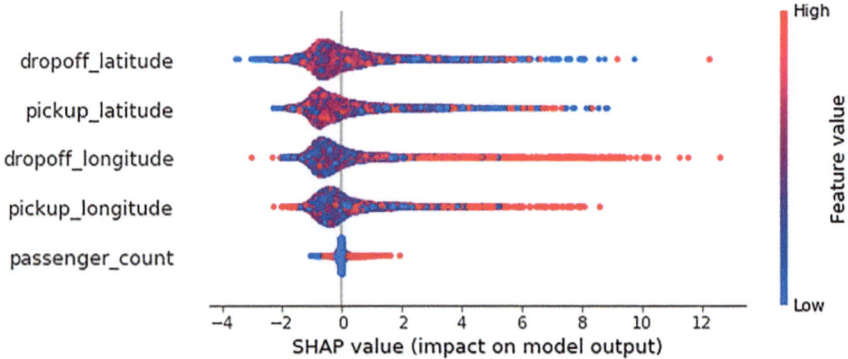

Fig. 4.16 SHAP summary plot for Cab scenario

```
shap.initjs() # print the JS visualization code to the notebook
# visualize the a prediction's explanation, decomposition between
average vs. row specific prediction.
shap.force_plot(explainer.expected_value, shap_values[50,:], X_
test.iloc[50,:])
```

We readily see the positive and negative contributions at the fare. Now, what if we want to see all the force plot for all the samples at once? We can have a summary of where each point is a different sample (Fig. 4.16).

```
# Each plot represents one data row, with SHAP value for each
variable,
# along with red-blue as the magnitude of the original data.
shap.summary_plot(shap_values, X_test)
```

Or we can rotate vertically the force plots and pack them horizontally to have one plot

```
# Pretty visualization of the SHAP values per data row. We limit
to the first 5000 samples
shap.force_plot(explainer.expected_value, shap_values[0:5000,:],
X_test)
```

4.6 A Naïve Criticism to SHAP

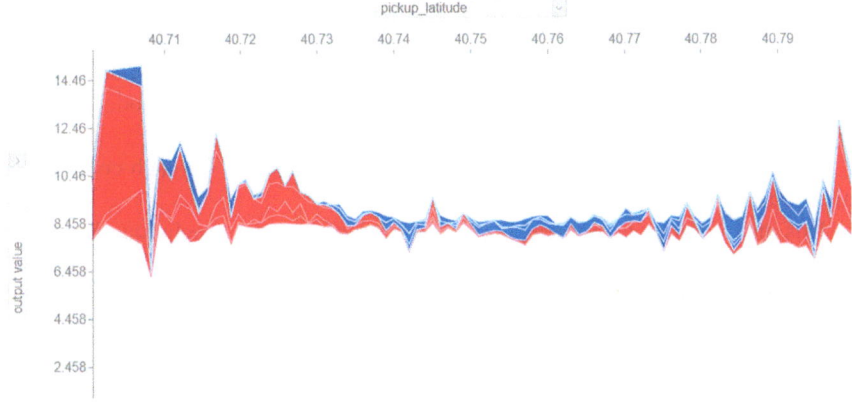

Fig. 4.17 Packed SHAP force plots

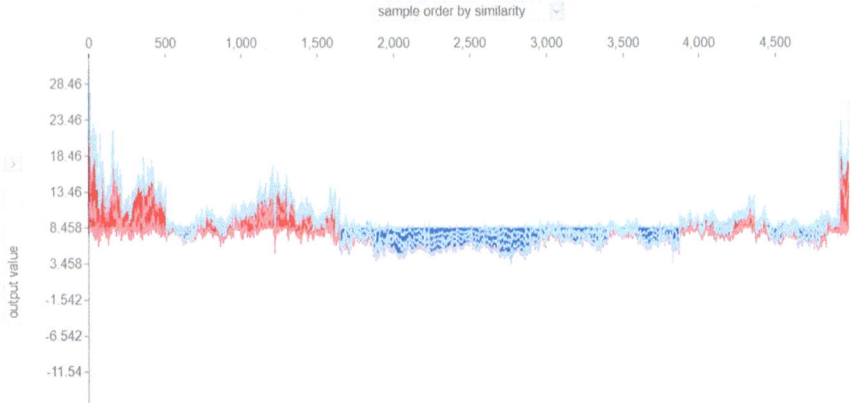

Fig. 4.18 SHAP diagram: similarity between explanations

sorting them by pickup_latitude (Fig. 4.17)

Or grouping SHAP explanation using the similarity between the explanations (Fig. 4.18)

4.6 A Naïve Criticism to SHAP

We conclude with a clever objection by Edden Gerber (2020) who proposes an explanation method inspired by SHAP named Naïve Shapley values.

The method of Gerber, albeit with its limits, clarify what we truly find in SHAP in contrast to what we would expect to find in SHAP.

If we revisit Formula 4.1, we see that a crucial aspect of calculating Shapley values is f_X, the evaluation of the model with certain features excluded.

As we know, SHAP method replaces the missing features with some background information (or statistics) about the missing features but what if the model f_X would be already independent of the missing features?

In fact, Gerber's Naïve Shapley values retrain a model f_X for each coalition of missing features and after that calculates (4.1).

This methodology is not new, it is just a form of bagging a technique already at the core of Random Forest and it is not an agnostic method anymore for it does not explain the original model f_X but retrains a bunch of new reference models.

But it can really be of use in the modeling phase, and its results will surprise us.

Now if we have to retrain the models we must have access to the same train set used to train the original model. We may inadvertently augment the model with predictive power it did not originally possess, and the training time increases exponentially with the number of features involved.

In fact Naïve Shap describes the properties of the data more than the model f_X we want to explain but the results are somewhat enlightening.

We start from training a model f on the adult census, a database already included in the SHAP library predicting the annual income of people.

In Fig. 4.19, we have both the explanation of TreeExplainer SHAP on the model *f* and the description of Naïve Shapley values of the same model plus the accessory trained models.

We see that the values are very similar, but the Naïve have more spread out values.

Look at the *Gender* feature; for instance, the original model can't clearly show the effect of Gender on the income, but Naïve Shapley values does it because there is gender disparity is in the data.

> We have already seen this effect in the section on permutation importance; the original model has learned to use other features than gender to predict income so it thinks gender is less important in the prediction than it is in the data.
>
> Instead "Naïve Shapley" retrains other models that are forced to use Gender when some other features are missing, so it is more representative of the real data and it doesn't explain well what the model does.
>
> As a matter of fact, we could train better models in terms of fairness comparing the trained model's SHAP values with the data analysis provided by the "Naïve Shapley values" approach.

In the next chapter we will introduce the explainable models for Deep Learning. And we will literally try to open the (Black) Box!

4.7 Summary

This chapter has been devoted to model agnostic methods starting with Permutation Importance and Partial Dependence plots before deep diving the more complex Shapley values and SHAP methods

4.7 Summary

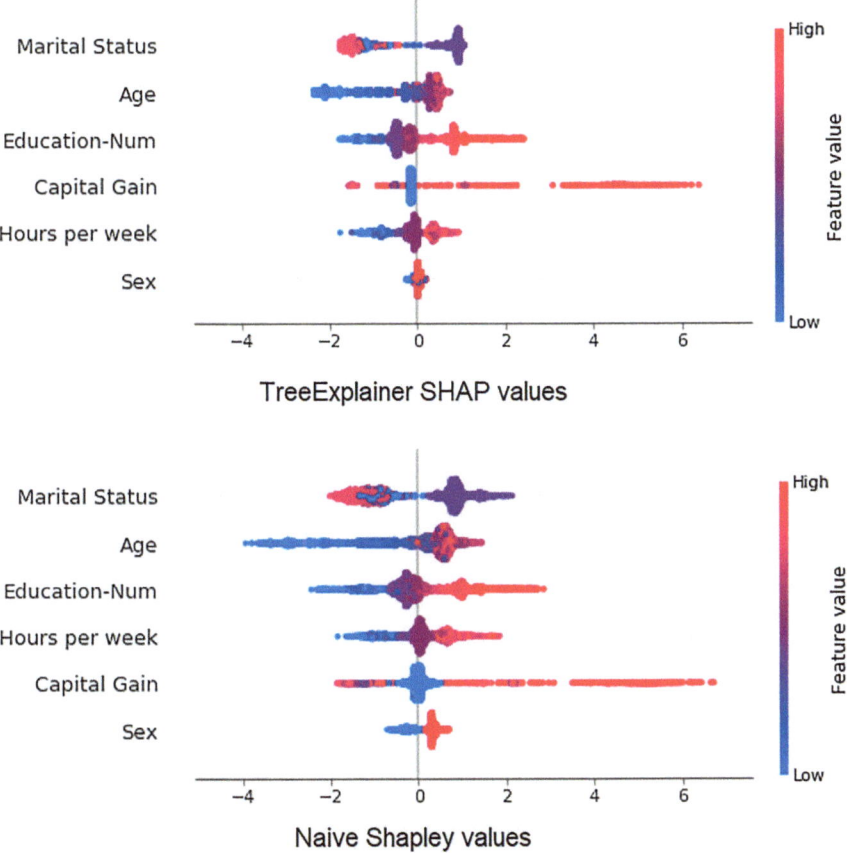

Fig. 4.19 The adult census database—UCI Machine Learning Repository (UCI, 1996)—SHAP documentation

- Use Permutation Importance to answer "What" questions on the most important features.
- Use Partial Dependence Plots to answer "How" questions to understand the impact of the features on the predictions.
- Use Accumulated Local Effects (ALE) instead of PDPs when dealing with correlated features to obtain more accurate interpretations of feature-prediction relationships.
- Provide Local Explanations using SHAP.
- Enrich our understanding of SHAP to be compared with LIME.
- Improve performance on generating explanations using TreeShap instead of KernelShap.
- Get awareness on the limits of SHAP to tailor adapt the best XAI strategy for the specific case.

References

Becker, D. (2020). *Machine learning explainability*. Retrieved from https://www.kaggle.com/learn/machine-learning-explainability

Gerber, E. (2020). *A new perspective on Shapley values, part II: The Naïve Shapley method*. Retrieved from https://edden-gerber.github.io/shapley-part-2/

Kaggle. (2020). *New York taxi fare prediction*. Retrieved from https://www.kaggle.com/c/new-york-city-taxi-fare-prediction

Lundberg, S., & Lee, S. I. (2016). An unexpected unity among methods for interpreting model predictions. https://doi.org/10.48550/arXiv.1611.07478

Lundberg, S. M., & Lee, S. I. (2017). A unified approach to interpreting model predictions. In *Advances in neural information processing systems* (pp. 4765–4774). Curran Associates.

Ribeiro, M. T., Singh, S., & Guestrin, C. (2016). "Why should I trust you?" Explaining the predictions of any classifier. In *Proceedings of the 22nd ACM SIGKDD international conference on knowledge discovery and data mining* (pp. 1135–1144). ACM.

UCI. (1996). *Adult data set*. Retrieved from https://archive.ics.uci.edu/ml/datasets/adult

Chapter 5
Explaining Deep Learning Models

The sculpture is already complete within the marble block, before I start my work. It is already there, I just have to chisel away the superfluous material.

Michelangelo

This Chapter Covers:

- Occlusion
- Gradient Models
- Activation-Based Models
- Unsupervised Activation Models
- Future Prospectives

    – Provide explanations for Deep Learning Models in Computer Vision.
    – Build an explainable agnostic model for a black box in computer vision.
    – Use Saliency Maps to provide explanations focusing on regions of major interest.
    – The reader will have a glimpse of the future of the field:

        Build interpretable CNN
        Use unsupervised learning to do exploratory analysis on a model.

In this chapter, we will talk about XAI methods for Deep Learning models.

The explanation of deep learning models is a matter of active research, so we will illustrate the criticalities and advantages of what are the methods of today and could be those of the future.

In this chapter, we want to stress a fundamental concept: having an explainable model is a way to create a robust and reliable model. So there is no need for trading between Explainability and robustness; on the contrary, Explainability naturally makes the model more robust.

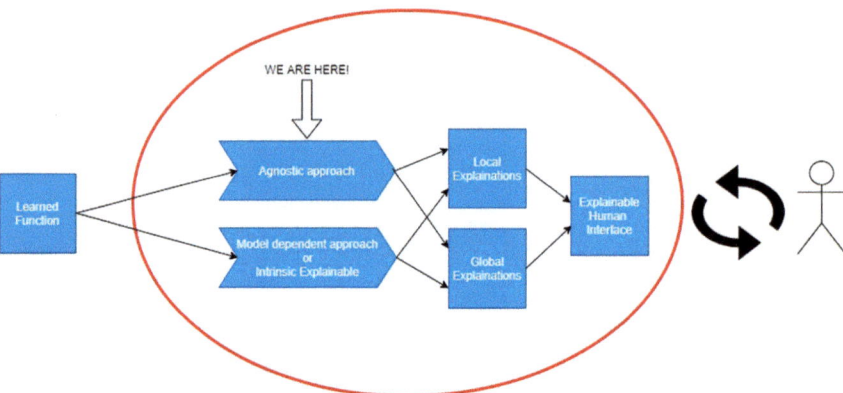

Fig. 5.1 XAI flow: agnostic approach

We will see that XAI methods and training best practices use similar methodologies, and any methodology that speeds up training a Deep Learning model and makes it more accurate overlaps with an XAI method.

For simplicity, we will make examples taken from Computer Vision, and we will face the problem of creating a "good explainable model": we will start from an agnostic point of view, then gradually, we will try to open the black box leveraging the more advanced techniques in this area (Fig. 5.1).

5.1 Agnostic Approach

We begin our discussion with the agnostic approach (Guidotti et al., 2020) which is the approach in which we perturb our model's input without exploiting the internal functioning of Deep Learning models. Only in the following sections we will cover gradient-based methods.

5.1.1 Adversarial Features

It is not possible to tackle the discussion of Deep Models without properly framing the problems and difficulties they entail.

For concreteness, we begin with a series of conceptual experiments.

Let's think about the most classic classics, a model to classify Dogs' and Cats' images.

By treating the model as a Black Box, we do not care at all if the model is a neural network, a decision tree, or even a linear model. We do not care at all if the model has a preprocessing of the image. If you internally transform it to grayscale, do

5.1 Agnostic Approach

some feature extractions, image convolutions, or so on. Think of the model in the abstract, without any look at its implementation.

Surely you will think that it is enough to have images that have been correctly labeled to solve the classification problem. Yet a question that often goes unnoticed is that of the origin of the picture.

Let's take two images, one of a dog and one of a cat (Fig. 5.2).

We do not go into the details of the camera used, but obviously, we tend to photograph dogs in the open air and cats indoors.

Dogs are usually stationary, and cats are always *on the move,* so we will have to use different exposure times. Also, dogs and cats have different sizes, and therefore, magnification will also be different.

An excellent Black Box can exploit this information from the photo through the SNR, color balance, depth field, and noise of texture. The challenge of learning causal rather than spurious features has been extensively discussed in recent theoretical work (Bottou, 2019). Thus a good Black Box can learn to distinguish dogs and cats from features that are absolutely not the features used by a human being.

We have already met the example of the classification of the wolf through the snow. Here we are dealing with the same problem, but we are trying to generalize the concept. Many of the features we have listed are legitimate features naturally occurring in the photos.

In the following, we will call Adversarial Features those features that distort learning, typically reducing the ability to generalize the Black Box.

It is easy to imagine that a classification carried out by the Field of View alone, for example, would give that a pony would be classified as a dog.

The training of a black box for image recognition must therefore include preprocessing so that the model does not learn from adversarial features. The most used approach to limit this risk is to randomize the image sources as much as possible and resort to augmentations.

Fig. 5.2 Adversarial Features in images. Courtesy of L.Bottou for the design

5.1.2 Augmentations

Augmentations are a very common approach in Computer Vision to train models.

It is simply a matter of varying the input images through a series of transformations: rotations, translations, enlargements, color and contrast variations, vertical and horizontal scaling, flipping, and so on. They are so common in computer vision that every Deep Learning framework has its own built-in augmentations.

Clearly presenting every possible rotation of a photo to a model in the training phase will force the model to use features that do not depend on the orientation. So a cat will remain a cat even if viewed upside down.

Traditionally a correspondence has always been made between the amount of data and the complexity of a model.

So, for example, complex models of High Capacity in the sense of Vapnik-Chervonenkis dimension (Cohn and Tesauro, 1992) with many parameters require large amounts of data for their training, and the meaning of augmentation would be to increase this amount of data so as not to make the model overfit.

When we talk about Deep Neural Networks, we will see that this approach, luckily, is not completely correct.

Both for regularizations that are imposed on the weights and for phenomena such as the Double Descent (Nakkiran et al., 2019), for now, let's not talk about architecture. Let's just say that augmentations are useful for increasing the database available in the training phase, but they must not be done in a completely random way because they could actually confuse the model by adding noise. They should be aimed at providing the model with robust information.

And this allows a more generalizable training and often more and often faster.

Let's clarify these concepts with another mental experiment.

5.1.3 Occlusions as Augmentations

Suppose we want to train a Black Box to recognize Oranges and Apples (Fig. 5.3).

Fig. 5.3 How do we distinguish an apple and an orange from the skin's texture, color, or stem? *(photo by vwalatke)*

5.1 Agnostic Approach

The Black Box could learn to distinguish the two fruits from the shape, from the texture, or for example, from the stem.

The fruit petiole is an absolutely legitimate feature that uniquely identifies the fruit, but it is not a robust feature in the sense that it does not give us robust training. In fact, it is enough to turn one of the two fruits to make the petiole disappear and prevent the model from making its prediction.

It should be noted here that the robustness of a feature is a different concept from the feature importance that we talked about in previous chapters.

Feature importance is the confidence that a model that has already been trained has in the importance of a feature and how much the feature is relevant for the calculation of the output.

Now let's take two models one that has learned to distinguish fruit from petiole and one that has learned to distinguish fruit based on peel; both consider respectively petiole or peel as very important features, and both models have the ability to correctly predict the fruit, but the model that uses the petiole is less robust, so in front of new examples it will be less generalizable.

We talk about overfit when a model fails to generalize well on new cases, so the robustness of a model is to build a model that, in some sense, has not overfitted in learning from its training set.

There are several techniques we can use to force a Black Box to learn to distinguish oranges from the skin and not from the stem.

One could be to provide the model with images with different levels of blur in the image in order to make it less easy to use the stem as a feature, but this method is definitely destructive for the image.

Another technique that can be used is that of **occlusion** (Zeiler, 2013) in which some information has been cut off from the images: as an example, a small gray square patch could be randomly applied to the image.

The stem has been cut off to force the model to walk the path of recognition through the peel. This does not reduce the accuracy of the model but increases its robustness and generality (Fig. 5.4).

Fig. 5.4 The occlusion idea as an augmentation technique: random grey rectangles forces the model to rely more on robust features suche as skin's texture

5.1.4 Occlusions as an Agnostic XAI Method

This is a fantastic example of correspondence between good training and XAI because, starting from the occlusion, an XAI method can be built to understand what the Black Box is looking at.

If using the occlusion in the training phase, we forced the Black Box not to learn by looking at the finer details; now, we can take a pretrained Black Box and question it on images content using occlusions as an Xai method.

The model has already been trained and fixed, so we don't care in this phase of training it in a robust way; we want to understand which details of the image are most significant for the class's attribution or to evaluate the importance of some groups of pixels.

With the application of the occlusion, we have varied the input, so a priori we will have a different output.

An analysis with occlusions is equivalent to sliding the patch square in each position along the image and assessing how much the output is distorted from time to time.

This is the most classic implementation of the algorithm, which obviously will take the longer, the smaller the patch. A slightly smarter solution is to approach the problem as an optimization problem and apply a metaheuristic algorithm to quickly evaluate the areas where the square patch will give larger differences for the expected class.

The result of the analysis will be a heatmap or a saliency map that will indicate the most sensitive parts of the model.

And this analysis is independent of the type of model or its internal functioning, so the explanation is agnostic.

But now, let's try to apply it to a real model. We take a pretrained model in Keras and visualize the most important parts using a library.

In Keras, the best-known libraries are tf-explain, DeepExplain, eli5, while for PyTorch users, there is the beautiful Captum.[1]

Here we will use tf-explain before we install the library by using pip

```
pip install tf-explain
```

then we load the libraries

```
import tensorflow as tf
from tf_explain.core.occlusion_sensitivity import
OcclusionSensitivity
```

---

[1] https://github.com/sicara/tf-explain
https://github.com/marcoancona/DeepExplain
https://eli5.readthedocs.io/en/latest/tutorials/keras-image-classifiers.html
https://github.com/pytorch/captum

5.1 Agnostic Approach

and we import a pretrained ResNet50 model on the ImageNet dataset

```
if __name__ == "__main__":
    model = tf.keras.applications.resnet50.ResNet50(
        weights="imagenet", include_top=True
    )
```

For the rest of the code, we load a specific image and launch two instances of the explainer, one to identify the areas of the image that most likely provide the class tabby cat and the other to identify the areas that most indicate the dog class (Fig. 5.5).

```
IMAGE_PATH = "./dog-and-cat-cover.jpg"
    img = tf.keras.preprocessing.image.load_img(IMAGE_PATH,
target_size=(224, 224))
    img = tf.keras.preprocessing.image.img_to_array(img)

    model.summary()
    data = ([img], None)

    tabby_cat_class_index = 281
    dog = 189

    explainer = OcclusionSensitivity()
    # Compute Occlusion Sensitivity for patch_size 10
    grid = explainer.explain(data, model, tabby_cat_class_
index, 10)
    explainer.save(grid, ".", "occlusion_sensitivity_10_cat.png")

    # Compute Occlusion Sensitivity for patch_size 10
    grid = explainer.explain(data, model, dog, 10)
    explainer.save(grid, ".", "occlusion_sensitivity_10_dog.png")
```

Fig. 5.5 Original image

Fig. 5.6 Occlusions used to highlight relevant features respectively for the tabby cat and the dog classes

This is the original image and following the explanations for the Tabby Cat and Common Dog classes, respectively (Fig. 5.6).

The parts in yellow are the most important.

Since the model of explanation is the occlusion, we can interpret these images by saying that the model needs to look at the textures of eyes and head to recognize if there is a dog or a cat in the photo. So we can say that, relative to other pixels in the image, those are the most important pixel for the explanation.

We do not doubt that the techniques for the agnostic explanation will become more and more refined and faster in the near future, for example, allowing us to better weigh the joint presence of different classes in the same image.

In the next paragraphs, we will go deeper in terms of explaining and speeding up the process by abandoning the Black Box paradigm and opening the model. This will allow us to use differential methods and see the information stored in the model.

5.2 Neural Networks

In this section, we will make a brief review of what neural networks are and how they work. The inner working will introduce us to differential methods.

5.2.1 The Neural Network Structure

A neural network is a machine learning model that mimics the functioning of the brain in a simplified model.

Mathematically, this idealization takes place through a graph in which information starts from the input features and arrives at the output features by crossing

5.2 Neural Networks

computational nodes. The most used NN scheme is Feed-Forward or DAG (Direct Acyclic Graph), where information flows from input to output without any loop (Fig. 5.7).

In this respect, a (nonrecurring) Neural Network is undoubtedly different from a brain because in the brain, the connections are also recurrent, and there are different types of memory, short-term and long-term.

Each computational node is typically a linear combination of the inputs then passed through a nonlinear activation function, typically a ReLU function.

This nonlinearity guarantees that the model as a whole is not simply a linear combination of the input features, but, as we will see also in the seventh chapter, the output will remain locally mainly linear in respect of the weights.

Typically the nodes (neurons) are arranged in successive layers. A network with only two layers, the internal (hidden) one and the output one, is called a Shallow Network. A neural network with two or more internal layers is called Deep.

The current substantial development of Deep Learning is based precisely on the enormous capabilities of Deep networks to learn and generalize patterns.

What makes a neural network so efficient?

Layer topology allows us to imagine a neural network as a sequence of several computational blocks. Let's take a Shallow network; the transition from the internal state to the output occurs simply through a linear combination. So we can imagine the shallow network as a linear model that has as inputs, not directly the input

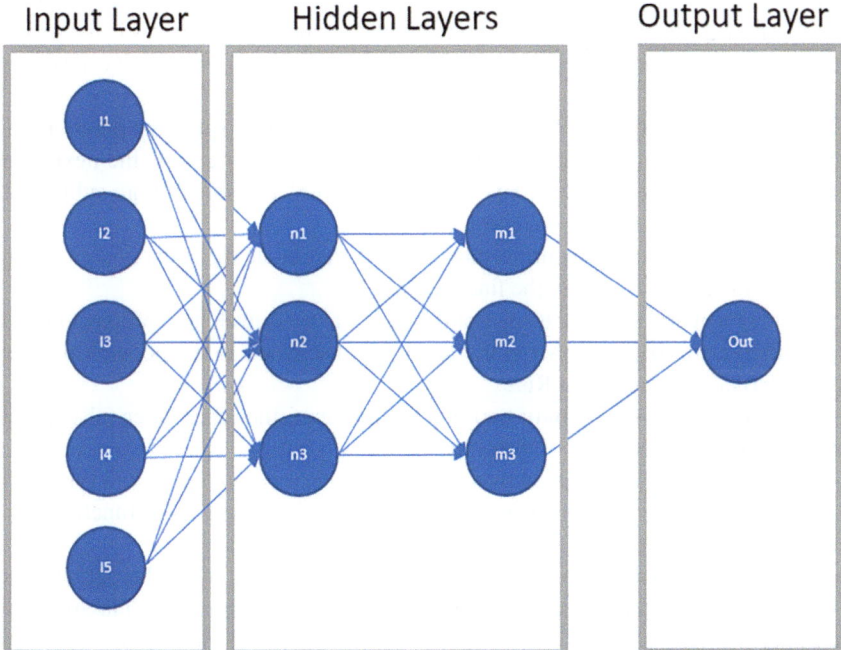

Fig. 5.7 A generic feed forward network/DAG

features but a convenient transformation of the features into a suitable space of basis functions.

If it is possible to train such a model and if the basis functions are flexible enough, the internal representation (latent representation) assumes a feature extractor's role.

That is, It can transform the initial dataset (an arbitrarily complex task) to a simpler one. So the recognition of an image behaves in the last layer like a standard logistic regression: the highly nonlinear problem is now mapped to a linear one.

It is remarkable that a Neural Network can extract its features by itself and train a neural network, for example, through backpropagation, is equivalent to simultaneously training the classifier and the features extractor.

The Universality Theorem of Cybenko (1989) and Hornik (1991) assures us that Shallow Neural Networks are universal approximators of any continuous function. The problem (NP-hard) is to find the exact values of weights to obtain such an approximation.

In practice, for a large class of NN, the search for weights (optimization) through optimizers such as SGD (Stochastic Gradiente Descent) and Adam (Adaptive Moment Estimation) is often surprisingly fast and effective.

The breakthrough for the transition to computer vision was discovering a layer type particularly suited to image processing: **convolutional filters**.

Convolutional filters make learning robust to translations of the objects in the images and reduce noise in images. Also, there are far fewer connections between layers in respect of fully connected layers, so fewer weights to train.

Furthermore, you can think of convolutional filters as an analog of the Fourier transforms: they immediately allow you to find periodicity and patterns in the image.

The layered structure of the neural network allows for successive levels of abstraction.

We can think of the first layer as a sort of classifier that splits the points in the image, so the first layer "sees" straight shapes and uniform colors, the next layer "sees" angles, and so on. Each layer responds to (i.e., classifies) more subtle features in the images giving more complex forms.

For this reason, we find neurons that tend to differentiate their functionality more and more the closer we get to the final layers.

Training a neural network using gradient descent is equivalent to propagating the recognition error that is shown in the last layers backward in the network through partial derivatives and Chain Rule.

Let's take a network of only two layers: the input, the inner layer (also called *hidden layer*), and the layer of the output.

The input will be represented by a vector x. A linear combination will transform the input into the first layer of neurons. A nonlinear sigma activation function (or a ReLU) will be applied to model the response of each neuron.

The output of each neuron will be mapped to the second layer (the output), and a new activation function will be applied (typically softmax in a classification problem.

In the formula, all two layers of Neural Networks can be simply expressed as

$$y = \sigma\left(W_2\ \sigma\left(W_1 x\right)\right) \tag{5.1}$$

For suitable weight matrices W_1 and W_2.

In analogy with the logistic regression, the choice activation function σ was historically a sigmoid. In more recent times, the sigmoid has been replaced by the ReLU for speed and convergence of the backpropagation method.

For fun, the reader can refer to https://playground.tensorflow.org.

5.2.2 Why the Neural Network Is Deep? (vs Shallow)

At this point, it is essential to return to the distinction between Shallow and Deep networks. Historically the difficulty of training deep networks with only the SGD coupled with the Universality theorem of Cybenko and Hornik that states that even Shallow Networks are universal had discouraged the widespread of Deep networks.

Furthermore, from the point of view of Explainability, a Shallow network is almost linear so it can be easily explained with the methodologies we have already dealt with.

So, why move to deeper networks?

Let's take the example of the parity function.

The parity function is a function that has N bits in input and one output. The output is 0 if we have an even number of 1 in input, or it is one if we have an odd number of 1.

It can be shown that using a neural network with logic gates; a shallow network can exactly reproduce the parity function with 2^n neurons in its intermediate layer. Instead, a Deep Network of n layers can reproduce the parity function with just one neuron per layer and, therefore n neurons.

So Deep Neural Networks are more compact, and this compactness is essentially due to the fact that each layer learns the next level of abstraction.

For this reason, Deep Learning people usually say that shallow networks don't learn the function but merely approximate it.

This property is matched by a fundamental property of Deep networks. Computational complexity in training a Deep network decreases exponentially with the depth of the network. We call this property Bengio's conjecture. You can read a proof in Mhaskar et al. (2019).

Of course, we are talking about ideal networks, but, from a theoretical point of view, having the possibility to train enormously deep networks allows us to circumvent or mitigate the curse of dimensionality in model training.

The summary is that we cannot hope that the networks remain Shallow for the sake of interpretability; instead, Deep Neural Networks are here to stay and become even bigger and deeper (like in Gpt-3), so we need to find new ways to explain them.

The good news is that we expect that in the future, the explanation of Deep Neural Networks will cope with the successive levels of abstraction.

For fun, you Can visualize the ResNet50 NN we have explained in the previous paragraph in its 25.5 M parameters' glory.

```
from tensorflow.keras.utils import plot_model
plot_model(model, to_file='model_plot.png', show_shapes=True,
show_layer_names=True
```

5.2.3 Rectified Activations (and Batch Normalization)

To train a very Deep Network using Gradient Descent, a series of chained partial derivatives have to be calculated. The idea is that the error is propagated back from the output back to the input, updating all the weights in the process.

All the theory can be explained and derived in terms of the flux of the gradient going back from the output to the input. The real breakthrough of the theory was to recognize that the main obstacle to the training of a very deep network was the sigmoid function that was used as an activation function.

Specifically, the logistic function has almost zero derivatives away from the origin. When a neuron is in these conditions, the flow no longer propagates backward through it, and the weights of connections entering the neuron are no longer changed: the neuron dies.

Nowadays, Neural Networks are trained with ReLU function

$$\sigma(x) = \text{ReLU}(x) = \max(x, 0) \tag{5.2}$$

The ReLu function is always positive and for positive numbers has a derivative exactly of one preserving the flux.

So using the ReLU activation, we can train very deep networks.

Alas, the nonlinearity, and the discontinuity of the ReLU function also give a very jagged loss function with many local minima.

So to speed up training, we have to resort to regularization techniques or by adding explicit terms that limit the norm of weights or with specific methods of neural networks such as dropout, early stopping, and Batch Normalization.

We dwell briefly on Batch Normalization, and during the rest of the chapter, we will understand why this technique is so effective.

Batch normalization consists of normalizing the batches of samples on which the network must train layer by layer.

To clarify this concept, let us take the case of the loss function of a simple linear model. The best practice dictates that you normalize the samples before training the model because we make the loss function more regular.

In the case of a loss function with nonnormalized features, the gradient descent tends to make many more steps. The single-step has a jagged zig-zag behavior that makes training more difficult.

Thus normalizing the samples before training the model facilitates the convergence of the model.

Batch normalization does precisely this but layer by layer. And It works like magic.

In the original paper, the authors argued that the merit of the method's success was the reduction of the covariate shift, but we will see later that batch normalization has another significant effect, the reduction of gradient shattering.

5.2.4 Saliency Maps

Armed with this technical knowledge, we can go on with the more properly XAI part.

In computer vision, it is common to speak of Saliency Maps, that is, maps that indicate which parts of an image capture the attention of the person who looks at them; in our case, the observer is replaced by a model that is examining them.

We have already dealt with an example of Saliency Map with occlusion analysis, now armed with the ability to use the internal workings of a network, we will talk about gradient-based saliency maps that are more accurate and faster to calculate than in the pure Black Box approach.

The naïve idea on the description of the response of a neural network is to make a sensibility analysis or to evaluate how the response changes as the input changes. Following the article by Simonyan et al. (2013), we can calculate

$$\text{Sensibility} = \left(\frac{\Delta \text{Output}}{\Delta \text{Input}} \right)^2 \tag{5.3}$$

This is equivalent to propagating information backward from the output class to the individual pixels of the image.

But the use of the ReLU as an activation function has the consequence that the DNNs as a whole are difficult to differentiate. They are locally linear in pieces, so their derivative is constant in pieces. The result of a simple analysis through the derivative will be very inaccurate and cannot discriminate between different classes (Fig. 5.8).

Several partial solutions to this problem have been created and each of them uses brilliant and powerful ideas we list a few.

5.3 Opening Deep Networks

5.3.1 Different Layer Explanation

In a computer vision neural network, the image is usually first processed by convolutional filters and then crosses fully connected layers and a softmax to get the probabilities of the classes.

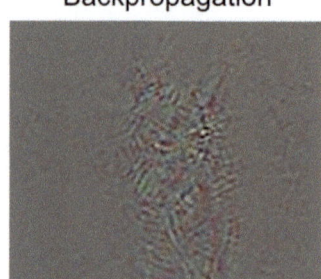

Fig. 5.8 Directly backpropagating the output class probability to the input image pixels is very inaccurate and cannot discriminate between different classes

We have already said that convolutional filters have the remarkable property of being invariant by translation, so it is not important to know where a cat will be in the image; they will still react to its presence.

The fully connected layers, on the other hand, are excellent for decoding the output of the convolutional layers but completely break invariance due to translations.

From this, we can understand the usefulness of having methodologies that explain a Neural Network's working not starting from its last layer's output but starting from the output of any layer or even from disconnected sets of neurons or filters.

The next methods, among the most used, will take the input–output directly from the last convolutional layer.

5.3.2 *CAM (Class Activation Maps) and Grad-CAM*

The CAM methodology was one of the first to be devised and also one of the most widespread.

Class Activation Maps of Zhou et al. (2016) help us understand which regions of an image affect the output of a Convolutional Neural Network.

The technique is based on a heatmap that highlights, as in the agnostic case, the pixels of the image that push a model to associate a certain class, a certain label, to the image.

It is noteworthy that the layers of a CNN behave in this case as unsupervised object detectors.

The implementation of the CAM technology is based on the properties of the global average pooling that were added after the last convolutional layer to decrease the size of the image and reduce the parameters so as to reduce overfitting.

The global average pooling layer works in the following way.

5.3 Opening Deep Networks

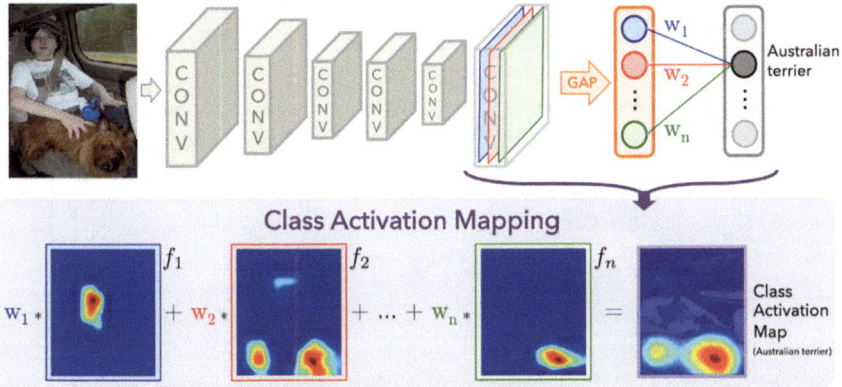

Fig. 5.9 The Global Average Pooling gives the Class Activation Map as a linear combination of features

Each image class in the dataset is associated with an activation map, and the GAP layer calculates the average of each feature map.

We can see the representation in the image (Fig. 5.9).

The assumption of the CAM model is that the final score can always be expressed as a linear combination of the global pooled average of feature maps.

The CAM procedure is thus removing the last fully connected layers, applying a GAP to the last convolutional layer and training the weights from the reduced layer to the classes.

The linear combination allows us to have the final visualization.

This procedure, although efficient, has the drawback of changing the structure of the network and retraining it; furthermore, it can only be used by applying it only starting from a convolutional layer and therefore is not applicable in all architectures.

An evolution of the CAM is the Grad-CAM of Selvaraju et al. (2017). Grad-CAM does not retrain the network. Evaluate the weights of the linear combination, starting from the value of the gradient at the exit from the convolutional layer, and then apply a ReLU function to regularize it (Fig. 5.10).

To obtain the heatmap, Grad-CAM calculates the gradient of y^c (the probability of class c) with respect to the feature map A of the convolutional layer.

These backpropagation gradients are averaged in the sense of the GAP to obtain the weights α_k^c —In formula (5.4) the summation represents the global average pooling and the partial derivates are the gradients via backprop

$$\alpha_k^c = \frac{1}{Z} \sum_i \sum_j \frac{\partial y^c}{\partial A_{ij}^k} \tag{5.4}$$

$$L_{\text{Grad-CAM}}^c = \text{ReLU}\left(\sum_k \alpha_k^c A^k\right) \tag{5.5}$$

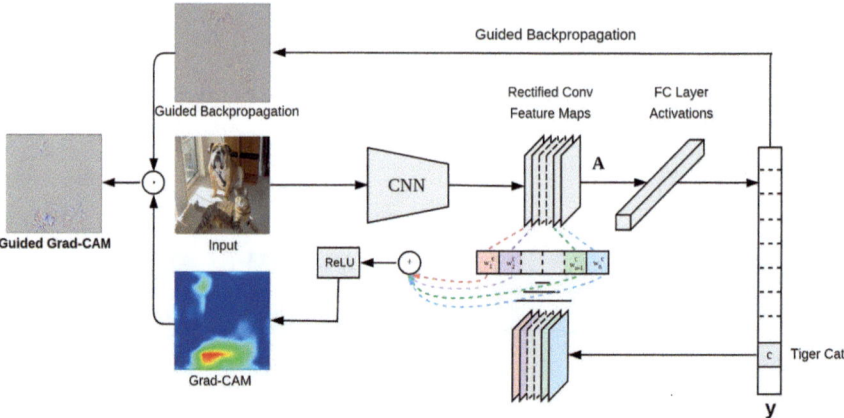

Fig. 5.10 Grad-Cam schema as: Grad-CAM: Visual Explanations from Deep Networks via Gradient-based Localization

The final contribution comes from an adjusted linear combination of the individual feature maps.

The main drawback of the Grad-CAM is the numerical instability that derives from the aforementioned gradient problems of neural networks.

5.3.3 DeepShap/DeepLift

Where do the gradient problems come from?

We have already mentioned that using the ReLU as an activation function, the neural network is almost always locally flat, and the gradient itself is discontinuous.

To this phenomenon is added another known as Shattered Gradients Problem (Balduzzi et al. (2017)).

The correlation between gradients in normal neural networks declines exponentially with depth until a white noise pattern is obtained.

A partial solution to the Shattered Gradients Problem is to "cure" the activation functions by ensuring that the gradient flow through them does not distort.

Currently, the reference method is that of Deep Learning Important Features or DeepLIFT Shrikumar et al. (2017).

DeepLift is a method that decomposes the prediction of a single-pixel neural network. This is done by carrying out the backpropagation of the contribution of all neurons in the network for each input feature. DeepLift compares the activation of each neuron with its reference activation and evaluates the importance of each contribution, starting from this difference. DeepLIFT can reveal dependencies that may be hidden in other approaches, in fact, unlike other gradient-based criteria, it can

5.3 Opening Deep Networks

also flow information through neurons with a zero gradient, a partial solution to the Shattered Gradient problem.

Unfortunately, to achieve all this, it is necessary to replace each activation function with that of the DeepLift, so the solution via DeepLift, unless you retrain the network, is a solution through a surrogate model.

Let's use the DeepLIFT implementation in the SHAP library for a practical example taking a cue from the keras tutorial we train a model on the MNINST dataset for the classification of the figures and then we launch the deep explainer

```python
# DeepShap using DeepExplainer
# ...include code from https://github.com/keras-team/keras/blob/
master/examples/mnist_cnn.py

from __future__ import print_function
import keras
from keras.datasets import mnist
from keras.models import Sequential
from keras.layers import Dense, Dropout, Flatten
from keras.layers import Conv2D, MaxPooling2D
from keras import backend as K

batch_size = 128
num_classes = 10
epochs = 1

# input image dimensions
img_rows, img_cols = 28, 28

# the data, split between train and test sets
(x_train, y_train), (x_test, y_test) = mnist.load_data()

if K.image_data_format() == 'channels_first':
    x_train = x_train.reshape(x_train.shape[0], 1, img_rows,
img_cols)
    x_test = x_test.reshape(x_test.shape[0], 1, img_rows,
img_cols)
    input_shape = (1, img_rows, img_cols)
else:
    x_train = x_train.reshape(x_train.shape[0], img_rows, img_
cols, 1)
    x_test = x_test.reshape(x_test.shape[0], img_rows, img_
cols, 1)
    input_shape = (img_rows, img_cols, 1)

x_train = x_train.astype('float32')
```

```
x_test = x_test.astype('float32')
x_train /= 255
x_test /= 255
print('x_train shape:', x_train.shape)
print(x_train.shape[0], 'train samples')
print(x_test.shape[0], 'test samples')

# convert class vectors to binary class matrices
y_train = keras.utils.to_categorical(y_train, num_classes)
y_test = keras.utils.to_categorical(y_test, num_classes)

model = Sequential()
model.add(Conv2D(32, kernel_size=(3, 3),
                 activation='relu',
                 input_shape=input_shape))
model.add(Conv2D(64, (3, 3), activation='relu'))
model.add(MaxPooling2D(pool_size=(2, 2)))
#model.add(Dropout(0.25))
model.add(Flatten())
model.add(Dense(128, activation='relu'))
model.add(Dropout(0.5))
model.add(Dense(num_classes, activation='softmax'))

model.compile(loss=keras.losses.categorical_crossentropy,
              optimizer=keras.optimizers.Adadelta(),
              metrics=['accuracy'])

model.fit(x_train, y_train,
          batch_size=batch_size,
          epochs=epochs,
          verbose=1,
          validation_data=(x_test, y_test))
score = model.evaluate(x_test, y_test, verbose=0)
print('Test loss:', score[0])
print('Test accuracy:', score[1])
```

after training we launch the deep explainer.

```
#DeepShap using DeepExplainer


import shap
import numpy as np

# select a set of background examples to take an expectation over
```

5.3 Opening Deep Networks

```
background = x_train[np.random.choice(x_train.shape[0], 100,
replace=False)]

# explain predictions of the model on four images
e = shap.DeepExplainer(model, background)
# ...or pass tensors directly
# e = shap.DeepExplainer((model.layers[0].input, model.
layers[-1].output), background)
shap_values = e.shap_values(x_test[1:5])

# plot the feature attributions
shap.image_plot(shap_values, -x_test[1:5])
```

In Fig. 5.11 we can see all the pixels for or against the number's attribution to some class. So in the first image, we see a circle formed by blue pixels that are the missing pixel for the zero class's proper attribution. In the third image, where the two is correctly classified in the two-class, the picture is full of red pixels.

5.3.4 Integrated Gradients

While the methods we've discussed so far help identify important regions in the input space, they can suffer from saturation problems where the gradients become very small in certain regions. To address this issue, Sundararajan et al. (2017) proposed Integrated Gradients, a powerful attribution method with strong theoretical foundations.

Integrated Gradients (IG) works by accumulating gradients along a path from a baseline input (typically a zero or black image in computer vision) to the actual input. The method satisfies important axioms that other methods may not, including:

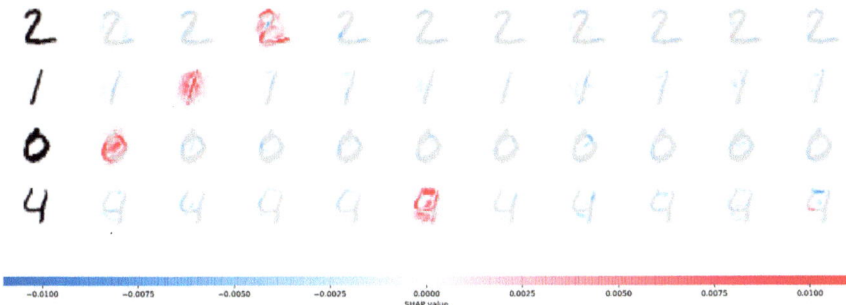

Fig. 5.11 DeepShap gives an attribution score for the class coloring the pixels in the images

1. **Sensitivity**: If an input and baseline differ in one feature and the output changes, the difference should be attributed to that feature.
2. **Implementation Invariance**: Attributions are identical for functionally equivalent networks.
3. **Completeness**: The attributions sum to the difference between the output at the input and the baseline.

Mathematically, the Integrated Gradients attribution for a particular pixel is defined as:

$$IG(x)_i = (x_i - x_i') \times \int_{\alpha=0}^{1} \left[\partial F / \partial x_i \left(x' + \alpha \times (x - x') \right) \right] d\alpha$$

Where x is the input image, x' is the baseline image, F is the model's prediction function, and the integral captures how the gradient changes as we move from baseline to input. In practice, this integral is approximated using a finite sum:

```
def integrated_gradients(model, image, baseline, target_class,
steps=50):
    """
    Compute Integrated Gradients for an image.

    Args:
        model: The neural network model
        image: The input image
        baseline: The baseline image (typically black)
        target_class: The class for which to compute attributions
        steps: Number of steps for the integral approximation

    Returns:
        Attributions for each pixel
    """
    # Generate interpolated images
    alphas = np.linspace(0, 1, steps)
    interpolated_images = [baseline + alpha * (image - baseline)
for alpha in alphas]

    # Compute gradients for each interpolated image
    gradients = []
    for interp_img in interpolated_images:
        # Prepare input for gradient computation

input_tensor = torch.tensor(interp_img, requires_grad=True)

        # Forward pass
```

5.3 Opening Deep Networks

```
        output = model(input_tensor.unsqueeze(0))
        score = output[0, target_class]

        # Backward pass
        score.backward()
        gradient = input_tensor.grad.numpy()
        gradients.append(gradient)

    # Average the gradients
    avg_gradients = np.average(gradients, axis=0)

    # Scale by the input difference
    integrated_grads = (image - baseline) * avg_gradients

    return integrated_grads
```

A compelling demonstration of Integrated Gradients' value comes from dermatology, where researchers were developing a system to identify melanomas from skin lesion images. The model showed impressive 92% accuracy on internal validation but dropped to 75% when deployed in a new hospital. This performance gap prompted investigation using attribution methods (Metta et al., 2023).

Integrated Gradients revealed a startling insight: the model was not primarily focusing on the lesion characteristics but on the presence of surgical rulers that appeared in many melanoma images in the training set. Since malignant lesions are routinely measured for documentation, rulers frequently appeared alongside melanomas but rarely with benign lesions. The model had learned this spurious correlation rather than focusing on intrinsic characteristics of the lesions themselves.

This discovery enabled targeted interventions: the training data was augmented with examples breaking this correlation, and data augmentation techniques were applied to synthetically add and remove rulers from images. After retraining, the model's performance in the new hospital improved to 88%, and more importantly, Integrated Gradients confirmed that the model now focused primarily on the lesion itself rather than contextual artifacts.

Compared to GradCAM, Integrated Gradients offers several significant advantages:

Theoretical foundation: IG satisfies important axioms including completeness (attributions sum to the difference between output and baseline) and sensitivity (features with zero impact receive zero attribution).

Architecture independence: Unlike GradCAM, which requires convolutional layers, IG works with any differentiable model, including Vision Transformers and other nonconvolutional architectures.

Pixel-level precision: IG provides attributions at the same resolution as the input image, rather than being limited by feature map resolution.

Signed attributions: IG indicates not just which pixels were important but whether they contributed positively or negatively to the classification.

Extensions of the basic Integrated Gradients approach have further enhanced its utility. SmoothGrad-IG combines IG with noise-based averaging to produce more visually coherent attributions. XRAI (eXplanation with Ranked Area Integrals) builds upon IG by segmenting the image and aggregating attributions within semantically meaningful regions.

The development of Integrated Gradients exemplifies how theoretical advances in attribution methods can address practical challenges in model interpretation. By providing more accurate and reliable explanations of model behavior, these methods enable more effective model debugging, enhance trust in deployed systems, and facilitate scientific discovery by revealing patterns the model has identified.

As computer vision models continue to expand into high-stakes domains like medical imaging, autonomous vehicles, and security applications, the ability to accurately understand their decision-making becomes increasingly crucial. Integrated Gradients and its extensions represent a significant step toward making these complex models more transparent and trustworthy.

5.3.5 TracIn: Tracing Training Data Influence

While the methods we've explored so far focus on explaining a model's predictions by examining the input and the model itself, an alternative approach is to understand which training examples were most influential for a particular prediction. This is where TracIn (Tracing Influence) by Pruthi et al. (2020) comes in.

TracIn addresses a key question: "Which training examples were most responsible for a model's prediction on a specific test instance?" This approach offers a different perspective on explainability, connecting a model's behavior to its training data rather than just analyzing its internal mechanics.

The core idea of TracIn is to approximate the effect of removing a training example by using the training trajectory information. It computes the influence of a training example on a test example by measuring the similarity of their gradients, weighted by the learning rate.

Mathematically, the TracIn score for a training example z_train with respect to a test example z_test is calculated as:

$$\text{TracIn}_{\text{slim}}(z\_train, z\_test) = \sum^{c} \eta_k \nabla \theta L(z\_train, \theta_k)^T \nabla \theta L(z\_test, \theta_k)$$

Where θ_k represents model parameters at checkpoint k, η_k is the learning rate at that checkpoint, L is the loss function, and the summation is taken across multiple checkpoints during training. The score essentially measures the similarity between the gradients produced by the training and test examples, weighted by the learning rate.

This approach has profound implications for understanding model behavior, particularly in identifying sources of bias or error. A significant application emerged in medical imaging, where researchers were developing a diagnostic system for

5.3 Opening Deep Networks

identifying skin conditions from photographs. The model performed well on the development test set but showed unexpected degradation when deployed at certain clinical sites.

Using TracIn, the team identified that a subset of images from a particular hospital had disproportionate influence on specific misclassifications. Further investigation revealed that this hospital used a slightly different imaging protocol, creating a distinctive artifact in their images that the model had learned to associate with certain diagnoses. This bias would have remained hidden without the ability to trace prediction influences back to specific training examples.

After correcting this bias by retraining with an augmented dataset that decorrelated the imaging artifacts from the diagnoses, the model's accuracy improved by 4.2% in external validation, with the most significant improvements occurring precisely at the clinical sites where performance had previously lagged.

The implementation of TracIn requires access to model checkpoints and gradients throughout training:

```python
def tracin_score(model_checkpoints, learning_rates, train_example, test_example, loss_fn):
    """Calculate the TracIn score between a training and test example"""
    score = 0.0

    for checkpoint, lr in zip(model_checkpoints, learning_rates):
        # Load model at this checkpoint
        model.load_state_dict(checkpoint)

        # Calculate gradient for training example
train_loss = loss_fn(model(train_example[0]), train_example[1])
        train_grad = torch.autograd.grad(train_loss, model.parameters())

        # Calculate gradient for test example
test_loss = loss_fn(model(test_example[0]), test_example[1])
        test_grad = torch.autograd.grad(test_loss, model.parameters())

        # Calculate inner product of gradients
        similarity = sum(torch.sum(g1 * g2) for g1, g2 in zip(train_grad, test_grad))

        # Accumulate weighted similarity
        score += lr * similarity

    return score
```

Beyond identifying biases, TracIn has proven valuable for several interpretability applications. In natural language processing, it has been used to identify which training sentences most strongly influenced specific translations, helping to diagnose and correct systematic errors in machine translation systems.

In content moderation systems, TracIn has provided crucial insights into why certain benign content is incorrectly flagged as problematic. By identifying the training examples responsible for false positives, system developers can refine the training data to improve specificity without sacrificing sensitivity.

TracIn also facilitates data quality assessment by identifying training examples with outsized influence, which might represent outliers, mislabeled instances, or particularly informative cases. This capability enables more focused data curation, where resources can be directed toward verifying and potentially correcting the most influential examples rather than reviewing the entire dataset.

While extremely powerful, TracIn does present certain limitations and challenges. The approach is computationally intensive, requiring access to and processing of model checkpoints throughout training. It also assumes that the model is trained using gradient-based optimization, which may not always be the case. Additionally, the interpretation of influence scores requires careful consideration of the model architecture and training procedure.

Despite these challenges, TracIn represents a significant advancement in our ability to understand the relationship between training data and model behavior. By making explicit the connections between specific training examples and predictions, it provides a complementary perspective to feature-based interpretability methods, offering a more complete picture of how machine learning models function and why they sometimes fail.

5.4 A Critic of Saliency Methods

5.4.1 What the Network Sees

So far, through salience methods, we have seen how to highlight the features of an image that most influence the output of the model.

However, this way of proceeding has an intrinsic limit, which is clearly illustrated by a sentence by Cynthia Rudin (2019):

> Saliency maps are often considered to be explanatory. Saliency maps can be useful to determine what part of the image is being omitted by the classifier, but this leaves out all information about how relevant information is being used. Knowing where the network is looking within the image does not tell the user what it is doing with that part of the image

We can say Salience tells us what the network sees, not what the network thinks.

That is, it gives us feature importance but in no way illustrates the process by which that feature is analyzed.

We will try to answer this limit in this section by citing a beautiful work in terms of gradient flow analysis and in the next section by unsupervised methods.

5.4.2 *Explainability Batch Normalizing Layer by Layer*

A fascinating article by Chen et al. (2020) allows us to return to the often underestimated importance of batch normalization. As we mentioned in the past paragraphs, batch normalization was introduced to prevent the input of neurons from being too large and facilitate the convergence of the hanging and regularizing the loss function making it more spherical near the minimum.

Rudin's work allows us, once again, to show how a methodology designed to accelerate the learning process has its roots in Explainable AI.

That is, how each request to have an Explainable AI method at a deeper level leads to more robust and effective learning.

Which Explainable AI request does Batch normalization respond to?

Let's think of the classic example of CIFAR10: classifying images in ten different classes.

Rudin wonders if it is possible to follow the flow of the "concept" of class between layer and layer and if it is possible to build a new type of neural network that layer by layer manages to keep concepts (classes) as separate as possible.

To do this, all he does is use a batch normalization procedure and reorganize the information in each layer (appropriately applying linear transformations).

Let's not go deep into the math involved. We are satisfied with the idea (Fig. 5.12).

Thus it trains a special type of network that is able to follow the flow of class information from input to output.

This work not only shows us how neural networks can be trained with the requirement of interpretability but also tells us that batch normalization is an indispensable requirement for this interpretability. We can even think of Explainability as a regularizing procedure.

Take this paragraph as a glimpse of the future: we don't know that the neural networks of tomorrow will be like, but we bet they will be interpretable!

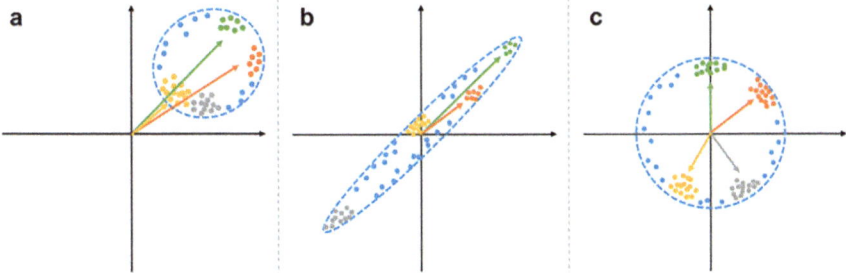

Fig. 5.12 The data distributions in the latent space. (**a**) The data are not mean centered (**b**) the batch normalized but not decorrelated (**c**) the data are fully decorrelated (whitened). In (**c**) the unit vectors can represent concepts

5.5 Unsupervised Methods

In this section, we will see how unsupervised methods can help us analyze what the network has learned.

5.5.1 Unsupervised Dimensional Reduction

Nonsupervised methods are a field in active development.

In this section, we will limit ourselves to a classic use of them: dimensional reduction. Dimensional reduction is a very powerful method to control the curse of dimensionality by reducing the number of features of the samples, and at the same time, it allows us to draw representations of our datasets in low dimensional spaces.

Let's take a quick example using two of the best-known techniques:

Principal Component Analysis (PCA) and t-SNE. PCA is a linear transformation that maps our dataset in one that produces new features along the dimensions of the largest variance.

t-SNE instead is a nonlinear method; it leverages Kullback–Leibler divergence so that the statistical properties of the samples in the high dimensional space and the low dimensional one are the same.

Using sklearn we will map the 64-dimensional dataset of small images of digits (8×8 pixels each) to a two-dimensional space.

```
import numpy as np
import matplotlib.pyplot as plt
from sklearn import manifold, decomposition, datasets

digits = datasets.load_digits(n_class=4)
X = digits.data
y = digits.target
n_samples, n_features = X.shape

def plot_lowdim(X, title=None):
    x_min, x_max = np.min(X, 0), np.max(X, 0)
    X = (X - x_min) / (x_max - x_min)

    plt.figure(figsize=(8,6))
    ax = plt.subplot(111)
    for i in range(X.shape[0]):
        plt.text(X[i, 0], X[i, 1], str(y[i]),
                 color=plt.cm.Set1(y[i] / 10.),
                 fontdict={'weight': 'bold', 'size': 9})
    plt.xticks([]), plt.yticks([])
    if title is not None:
```

5.5 Unsupervised Methods

```
        plt.title(title)

#-----------------------------------------------------------------
# PCA

X_pca = decomposition.TruncatedSVD(n_components=2).fit_
transform(X)
plot_lowdim(X_pca,"PCA")

# t-SNE
tsne = manifold.TSNE(n_components=2, init='pca', random_state=0)
X_tsne = tsne.fit_transform(X)

plot_lowdim(X_tsne,"t-SNE")


plt.show()
```

We can see from the images that similar figures are grouped, and intuitively we can get an idea of how much t-SNE does a better job by leaving a larger space between the clusters (Fig. 5.13).

5.5.2 Dimensional Reduction of Convolutional Filters

To help understand what activation atlases are, we illustrate a work by Karpathy (2014).

Karpathy uses t-SNE as an unsupervised method of investigating neural networks.

He extracts the 4096-dimensional output from the seventh layer of the famous AlexNet A. Krizhevsky et al. (2017) and feeds it to t-SNE.

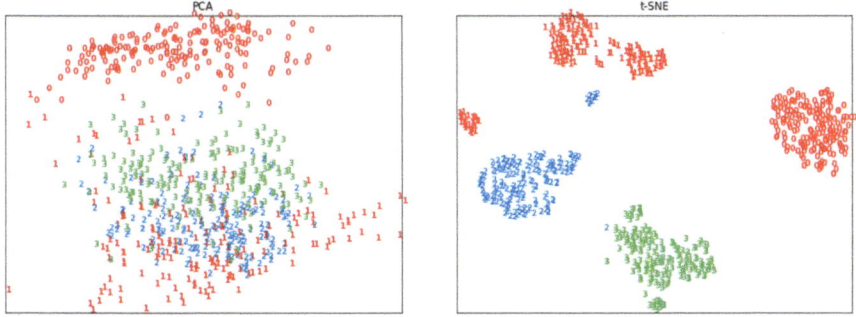

Fig. 5.13 PCA and t-SNE dimensional reduction reducing from 8x8 = 64 dimensions to two. t-SNE clustered MINST digits more effectively

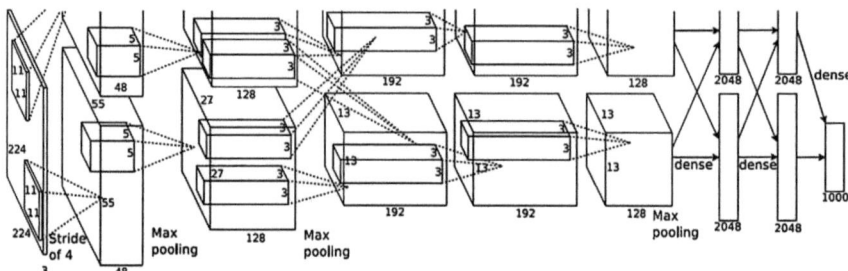

Fig. 5.14 AlexNet Architecture. A succession of convolutional filters and Max pooling reductions to other convolutional filters and at the end of the network two ful connected dense layers outputting to the output 1000 class vector with a softmax function

Now the output of t-SNE can be seen as a possibly unique hash code of the image seen by the neural network. In his work, Karpathy uses this code to classify we have a two-dimensional representation of the connected la layer with relative clustering (Fig. 5.14).

And by showing images at NN, we see how they are mapped in 2d space.

We will say that, by using the dimensional reduction algorithms, we can create an atlas of the images shown on a CNN network (Fig. 5.15).

This representation allows us to understand how the network thinks without detailing it. The network shows nearby images that it considers similar, and this allows us both to isolate errors in the dataset (for example, a number 1 in the same cluster of numbers 5) and to empirically evaluate the effectiveness of learning. It is heartening to see in the previous figure that all the animals are combined, but this could be the effect of some adversarial feature rather than an actual semantic division.

To know what the network really thinks, we need a more refined tool.

5.5.3 Activation Atlases: How to Tell a Wok from a Pan

With the CAM method, we have seen how it is possible to extract information from individual convolutional filters by aggregating them locally, and therefore reducing their dimensionality through a simple arithmetic average with GAP. Then we have seen how to make class predictions with the base of reduced filters.

With the t-Sne method, we have seen how to dimensionally reduce an entire image by transforming it through the network, taking the filter values of all the convolution filters and using them as a hashed encoding of the image and then organizing the images of the train set into a two-dimensional grid. The two-dimensional grid showed us the similarity between images with respect to some metric learned from the neural network in class recognition.

Now let's go to the next step.

In recent joint work, DeepMind and OpenAi Carter et al. (2019) have used size reduction to classify not the training set images but directly the convolutional filters.

5.5 Unsupervised Methods

Fig. 5.15 An Atlas of images using the dimensional reduction technique (Karpathy, 2014)

That is, they used size reduction not so much to see how CNN classifies things rather the density of the filters used to recognize things: an atlas of activation maps (Fig. 5.16).

Activation atlases not only allow us to see what the network thinks but also to more carefully sampling the activation functions that will be used most frequently.

After calculating which activation functions are close to each other, the functions in the same cell are averaged. Of the mediated activation functions, it searches which images came from with a feature visualization technique.

The result from a theoretical point of view is amazing but let's also see a practical application.

Suppose that as a Data Scientist, you may be asked to find some erroneous training in a model seeking counterfactual answers.

The Visualization Atlas Technique can be a very powerful one because it can revert the search to the way the model looks at the images.

Say we want to search what the network thinks seeing a wok (or a frying pan) or, more precisely, what maximally excites CAM to output the answer.

In the meantime, we can do a search of which are the classes of activations that respond the most when we show images of wok and frying pan to the network.

Now we can see that the Activation Atlas shows us many examples of a wok, and in some of them, there are noodles (Fig. 5.17).

Fig. 5.16 One million images are given to the network, one activation function per image. Activations are dimensionally reduced to two dimensions. Similar activations close together

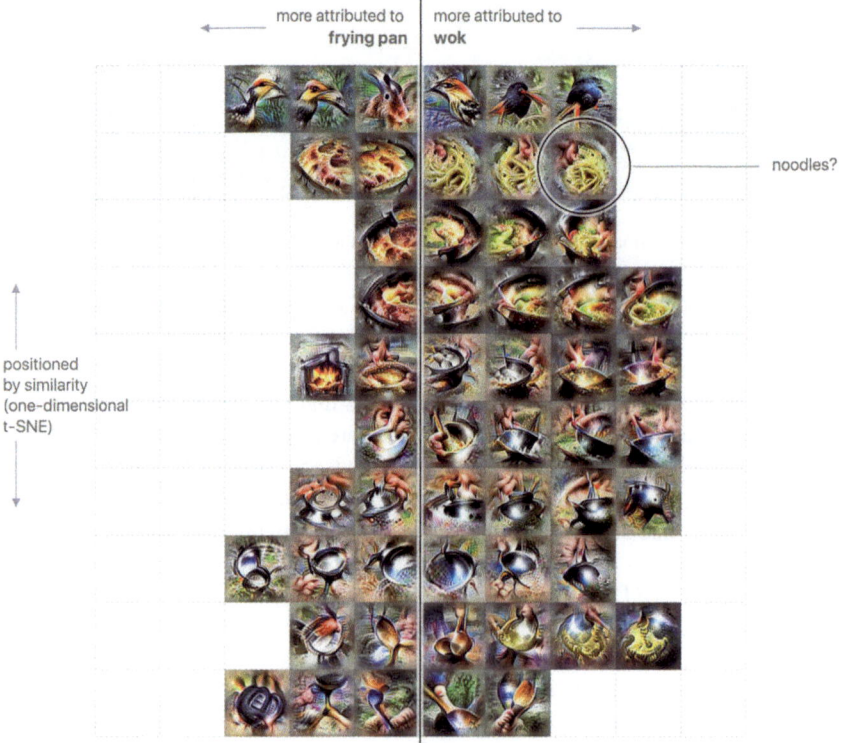

Fig. 5.17 Is it a frying pan or a wok? These are the filters most strongly activated by the presence of frying pans and woks

5.6 Summary

1.	frying pan	76.5%
2.	wok	15.8%
3.	stove	5.4%
4.	spatula	1.0%
5.	Dutch oven	0.5%
6.	mixing bowl	0.2%

1.	wok	63.2%
2.	frying pan	35.1%
3.	spatula	0.6%
4.	hot pot	0.5%
5.	mixing bowl	0.1%
6.	stove	0.1%

Fig. 5.18 We have a perfect noodle discriminator

If we think of our model as a discriminator, say of wok and frying pans, we can see that the noodles are absent in frying pans' visualization atlas.

Perhaps we have found a counterfactual feature, and in fact, by overlaying a sticker with noodles on the photos of a frying pan, we can make the neural network believe that it sees woks (Fig. 5.18).

Alas, our neural network is actually a wok discriminator.

5.6 Summary

- In the context of the agnostic approach we have introduced a parallelism between robust learning and Explainability.
- In the context of the agnostic approach we have introduced a parallelism between robust learning and Explainability.
- We introduced adversarial features and augmentations.
- We have made the parallel between occlusions used as augmentations and an agnostic XAI method.
- We introduced neural networks by presenting their structure, equations, and convergence problems.
- We answered the question about the need for deep neural networks and introduced batch normalization.

- We opened the neural networks by introducing the concept of salience the explanation starting from different layers and the CAM (Class Activations Maps) and grad cam methods.
- We have introduced DeepLift as a solution to the gradient shattering problem.
- We carried out a critique of salience methods by presenting a work on interpretability via batch normalization.
- We introduced unsupervised methods that use size reduction for the semantic representation of what the network has learned.
- We have introduced the Activation Atlases capable of showing us how the web sees the world.

References

Balduzzi, D., Frean, M., Leary, L., Lewis, J. P., Ma, K. W. D., & McWilliams, B. (2017). The shattered gradients problem: If Resnets are the answer, then what is the question? *arXiv preprint arXiv:1702.08591*.

Bottou, L. (2019). *Learning representations using causal invariance*. Institute for Advanced Studies talk at https://www.youtube.com/watch?v=yFXPU2lMNdk&t=862s

Carter, S., Armstrong, Z., Schubert, L., Johnson, I., & Olah, C. (2019). Activation atlas. *Distill, 4*(3), e15.

Chen, Z., Bei, Y., & Rudin, C. (2020). Concept whitening for interpretable image recognition. *arXiv preprint arXiv:2002.01650*.

Cohn, D., & Tesauro, G. (1992). How tight are the Vapnik-Chervonenkis bounds?. *Neural Computation, 4*(2), 249–269.

Cybenko, G. (1989). Approximation by superpositions of a sigmoidal function. *Mathematics of Control, Signals, and Systems, 2*(4), 303–314.

Guidotti, R., Monreale, A., Matwin, S., & Pedreschi, D. (2020). Black box explanation by learning image exemplars in the latent feature space. In *Machine learning and knowledge discovery in databases* (pp. 189–205). Springer.

Hornik, K. (1991). Approximation capabilities of multilayer feedforward networks. *Neural Networks, 4*(2), 251–257.

Karpathy, A. (2014). *t-SNE visualization of CNN codes*. Retrieved from https://cs.stanford.edu/people/karpathy/cnnembed/

Krizhevsky, A., Sutskever, I., & Hinton, G. E. (2017). Imagenet classification with deep convolutional neural networks. *Communications of the ACM, 60*(6), 84–90.

Metta, C., Beretta, A., Guidotti, R., Yin, Y., Gallinari, P., Rinzivillo, S., & Giannotti, F. (2023). Improving trust and confidence in medical skin lesion diagnosis through explainable deep learning. *International Journal of Data Science and Analytics, 17*(2), 189–205.

Mhaskar, Liao, & Poggio. (2019). *Learning functions: When is deep better than shallow*. Retrieved from https://arxiv.org/abs/1603.00988

Nakkiran, P., Kaplun, G., Bansal, Y., Yang, T., Barak, B., & Sutskever, I. (2019). Deep double descent: Where bigger models and more data hurt. *arXiv preprint arXiv:1912.02292*.

Pruthi, G., Liu, F., Kale, S., & Sundararajan, M. (2020). Estimating training data influence by tracing gradient descent. *Advances in Neural Information Processing Systems, 33*, 19920–19930.

Rudin, C. (2019). Stop explaining black box machine learning models for high stakes decisions and use interpretable models instead. *Nature Machine Intelligence, 1*(5), 206–215.

Selvaraju, R. R., Cogswell, M., Das, A., Vedantam, R., Parikh, D., & Batra, D. (2017). Grad-cam: Visual explanations from deep networks via gradient-based localization. In *Proceedings of the IEEE International Conference on computer vision* (pp. 618–626). IEEE.

References

Shrikumar, A., Greenside, P., & Kundaje, A. (2017). Learning important features through propagating activation differences. *arXiv preprint arXiv:1704.02685*.

Simonyan, K., Vedaldi, A., & Zisserman, A. (2013). Deep inside convolutional networks: Visualising image classification models and saliency maps. *arXiv preprint arXiv:1312.6034*.

Sundararajan, M., Taly, A., & Yan, Q. (2017). Axiomatic attribution for deep networks. In *International Conference on machine learning* (pp. 3319–3328). PMLR.

Vapnik, V. (2000). *The nature of statistical learning theory*. Springer.

Zeiler, M. (2013). *Visualizing and understanding convolutional networks*. Retrieved from https://arxiv.org/abs/1311.2901

Zhou, B., Khosla, A., Lapedriza, A., Oliva, A., & Torralba, A. (2016). Learning deep features for discriminative localization. In *Proceedings of the IEEE conference on computer vision and pattern recognition* (pp. 2921–2929). IEEE.

Chapter 6
Additive Models for Interpretability

6.1 Additive Models: When Interpretability Meets Predictive Power

Machine learning models have increasingly demonstrated extraordinary capabilities in prediction tasks across numerous domains. However, as they grow in complexity and accuracy, they often become less interpretable, functioning essentially as "black boxes" where the internal logic remains opaque to human understanding. This tension between predictive power and interpretability represents one of the fundamental challenges in modern artificial intelligence, particularly in high-stakes domains like healthcare, finance, and criminal justice.

Consider a healthcare scenario: A physician needs to predict a patient's risk of developing heart disease. A machine learning model provides a 95% accurate prediction, classifying the patient as high-risk. When the physician asks why this classification was made—which specific factors contributed to this high-risk assessment—the model offers no explanation. Despite its accuracy, this lack of transparency severely limits the model's practical utility in clinical settings where justification for decisions is ethically and legally required.

This interpretability gap not only affects practitioners' trust but also impacts their ability to integrate model predictions into their decision-making processes. As Rudin (2019) argues, "In high-stakes decisions, interpretability is a prerequisite for trust." Without understanding why a model has made a particular prediction, domain experts cannot effectively reconcile algorithmic suggestions with their professional judgment, and individuals affected by these decisions cannot contest or understand their basis.

Additive models emerge as a compelling solution to this fundamental challenge, offering an effective balance between predictive performance and interpretability. Rather than processing inputs through opaque layers of computation, additive models represent the relationship between predictors and outcomes as a sum of functions, each capturing the effect of a single feature or a carefully selected interaction.

This architectural choice provides clear benefits for interpretability: Each feature's contribution to the prediction can be visualized, quantified, and communicated (Fig. 6.1). For the physician in our example, an additive model would explicitly show how each risk factor—age, blood pressure, cholesterol levels, family history—contributes to the overall heart disease risk score. These contributions can be presented as intuitive graphs showing how risk varies across the range of each feature.

The practical impact of this transparency extends beyond mere explanation. When stakeholders understand not just what the prediction is but why it was generated, they can implement targeted interventions addressing specific factors. The physician might focus treatment on the most significant contributors to risk; the patient might better understand which lifestyle modifications would most effectively reduce their risk profile.

6.2 Generalized Additive Models (GAM): Clarity in Complex Decisions

Generalized Additive Models (GAMs) represent the foundational framework for interpretable nonlinear modeling. First introduced by Hastie and Tibshirani (1986), GAMs extend traditional regression by replacing linear terms with smooth functions applied to individual predictors, while maintaining the additive structure that ensures interpretability.

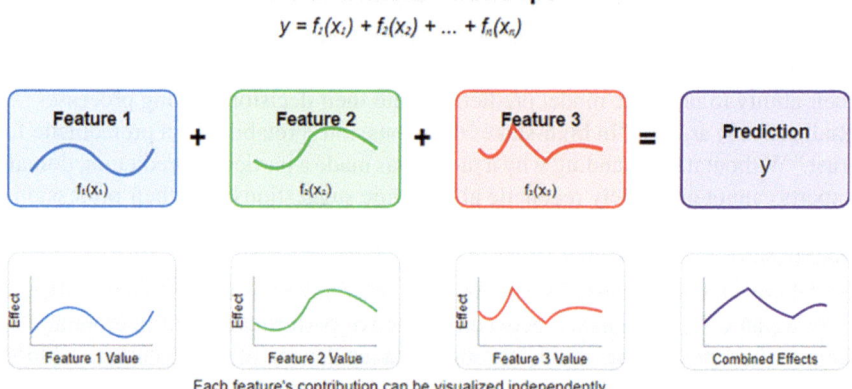

Fig. 6.1 Additive Model as composition of prediction terms by feature

6.2 Generalized Additive Models (GAM): Clarity in Complex Decisions

The mathematical formulation of GAMs is elegant in its simplicity:

$$y = \beta_0 + f_1(x_1) + f_2(x_2) + \cdots + f_n(x_n)$$

Where β_0 is a constant intercept term, and each function f_i captures the potentially nonlinear relationship between the predictor x_i and the target variable. This additive structure ensures that each feature's contribution to the prediction can be isolated and examined independently, a critical property for interpretability.

While linear models constrain these relationships to straight lines through coefficients ($y = \beta_0 + \beta_1 x_1 + \beta_2 x_2 + ...$), GAMs offer significant flexibility in how each function f_i is represented. Typical implementations use splines, local regression methods, or other nonparametric techniques that can adapt to complex patterns in the data without imposing rigid structural assumptions.

The link between GAMs and generalized linear models appears in how they handle different response distributions. For continuous outcomes, an identity link function is typically used. For binary classification, a logistic link function transforms the additive components into probabilities. This flexibility allows GAMs to address various prediction tasks while maintaining their interpretable structure.

A seminal application of GAMs in healthcare was documented by Caruana et al. (2015) at Seattle Children's Hospital, where these models were implemented to predict pneumonia risk and hospital readmission. The study demonstrated how GAMs could capture complex nonlinear relationships that were clinically significant yet had been missed by traditional linear approaches.

One of the most striking findings was the relationship between age and pneumonia risk, which followed a U-shaped pattern: Both very young patients and elderly individuals showed significantly elevated risk compared to middle-aged adults. While this general pattern aligned with clinical intuition, the GAM quantified the relationship precisely, showing exactly how risk varied across different age groups. Similar nonlinear relationships were identified for vital signs including blood pressure, oxygen saturation, and respiratory rate.

The fitted GAM not only provided accurate risk predictions but also generated interpretable visualizations showing each feature's contribution to the prediction. These partial dependence plots became valuable tools for communication among clinical teams, providing an intuitive representation of how each factor influenced the model's assessment.

For healthcare providers, these visualizations transformed how risk was understood and communicated. Rather than receiving an opaque risk score, clinicians could examine precisely how each clinical measurement contributed to the patient's overall risk profile. This transparency facilitated more targeted interventions focused on the most significant risk factors.

The practical benefits of this interpretability extended beyond clinical decision-making to model validation and improvement. When predictions diverged from clinician expectations, the model's transparency allowed immediate investigation into which feature contributions were driving the unexpected result. In several cases, this process identified previously overlooked patterns in the data that represented genuine clinical insights.

While GAMs demonstrated clear advantages for interpretability, they also revealed limitations. The strictly additive structure precludes modeling interactions between features—situations where the combined effect of two factors differs from the sum of their individual effects. In clinical contexts, such interactions are common: The significance of a particular symptom might vary dramatically depending on a patient's age or comorbidities.

6.3 GAM with Interactions: Caruana's Innovation (GA²M)

While Generalized Additive Models offer exceptional interpretability through their strictly additive structure, this same property imposes a significant limitation: the inability to model interactions between features. In many domains, particularly healthcare, the effect of one variable frequently depends on the value of another—a phenomenon that cannot be captured by summing independent feature contributions.

To address this limitation while preserving interpretability, Caruana et al. (2015) introduced Generalized Additive Models with Interactions (GA²M), extending the GAM framework to selectively incorporate pairwise interactions between features. This advancement is formalized mathematically as:

$$y = \beta_0 + f_1(x_1) + f_2(x_2) + \cdots + f_n(x_n) + f_{12}(x_1, x_2) + f_{13}(x_1, x_3) + \cdots + f_{ij}(x_i, x_j)$$

The model maintains the original additive terms but supplements them with interaction terms $f_{ij}(x_i, x_j)$ that capture how pairs of features jointly influence the outcome. This structure strikes a balance between expressive power and interpretability—more flexible than pure GAMs but still far more transparent than typical black-box models.

The clinical relevance of this approach was demonstrated in the pneumonia risk prediction study at Seattle Children's Hospital. The GA²M model identified critical interactions that had remained invisible to traditional GAMs. For instance, the relationship between respiratory rate and risk was found to vary substantially with patient age—an elevated breathing rate carried different implications in pediatric patients compared to adults. Similarly, the combination of moderate fever and oxygen desaturation indicated significantly higher risk than what would be predicted by simply adding their individual effects.

These interaction effects were not merely statistical artifacts but represented clinically meaningful patterns that aligned with medical knowledge. By capturing these relationships explicitly, GA²M models provided more accurate risk assessments while maintaining complete transparency about the factors driving each prediction.

The implementation of GA²M presents an algorithmic challenge: with n features, there are $n(n-1)/2$ possible pairwise interactions. For even moderately sized datasets with 20 features, this yields 190 potential interaction terms—far too many to include without compromising the model's interpretability and risking overfitting.

Lou et al. (2013) addressed this challenge by developing an efficient ranking procedure to identify the most significant interactions. First, a standard GAM is fitted to capture main effects for each feature. Residuals from this model are calculated, representing variation not explained by main effects. All possible pairwise interactions are then evaluated based on their ability to predict these residuals. Interactions are ranked by their predictive power, and only the top k are incorporated into the final model.

This selective approach preserves interpretability by including only the most important interactions—typically between 5 and 20 even for datasets with dozens of features. Each selected interaction can be visualized as a heatmap showing how different combinations of values for the two features affect the prediction, providing an intuitive representation of complex relationships.

The performance improvements from this approach are substantial yet balanced. In the pneumonia prediction task, GA²M achieved approximately 3–5% higher accuracy compared to standard GAMs, significantly narrowing the gap with black-box methods. More importantly, these gains came without sacrificing the transparency that makes additive models valuable in clinical settings.

Tan et al. (2018) conducted a comprehensive evaluation comparing GA²M with other machine learning approaches across various datasets. Their analysis showed that GA²M models consistently achieved 96–98% of the accuracy of state-of-the-art ensemble methods while remaining fully interpretable. This finding challenges the conventional assumption that interpretability necessarily comes at a substantial cost to predictive performance.

6.4 GAM for Fairness: Explicit Bias Mitigation

The application of machine learning in domains with significant social impact has raised critical concerns about fairness and bias. Traditional approaches to mitigating algorithmic bias often involve simply excluding sensitive attributes (such as race, gender, or age) from the training data. However, this approach suffers from two fundamental limitations: it fails to address proxy variables that correlate with protected attributes, and it ignores the legitimate signal these attributes might contain.

Generalized Additive Models offer a more nuanced and effective approach to algorithmic fairness through their inherent feature-level transparency. Their additive structure allows for explicit identification and targeted mitigation of bias without sacrificing predictive performance or removing potentially useful information. This capability represents a significant advancement in developing models that are both accurate and equitable.

The key insight enabling this approach is that bias can be modeled as a specific component within the additive structure:

$$y = \beta_0 + f_1(x_1) + f_2(x_2) + \cdots + f_{bias}(x_{bias}) + \cdots + f_n(x_n)$$

Once the bias component $f_{bias}(x_{bias})$ is identified and estimated, it can be neutralized while preserving the legitimate predictive signal from other features. Mathematically, this correction can be implemented as:

$$y_{corrected} = y_{original} - f_{bias}(x_{bias}) + \text{mean}(f_{bias}(x_{bias}))$$

This formulation effectively "zeroes out" the contribution of the biased component by replacing its varying effect with its average effect across all observations. The approach preserves the overall calibration of the model while removing the differential impact of the sensitive attribute.

Wadsworth et al. (2018) demonstrated this technique in the context of recidivism prediction, where racial bias has been a significant concern. Rather than excluding race-related variables entirely, they identified the direct and proxy contributions of race within the GAM framework, then neutralized these specific components. The resulting model maintained strong predictive performance while substantially reducing disparate impact across racial groups.

The advantages of this approach over simple variable exclusion are significant. It explicitly acknowledges and quantifies bias rather than pretending it doesn't exist. It retains legitimate signal that may correlate with sensitive attributes. It provides transparency about how much bias existed and how it has been mitigated. And it allows for targeted intervention at exactly the point where bias enters the model.

A particularly illuminating application of this technique comes from the domain of lending. Kamiran and Calders (2012) examined how postal codes—frequently used in credit scoring—can serve as proxies for racial and socioeconomic information. Using a GAM, they isolated the component of postal code influence that correlated with protected attributes, neutralizing this specific contribution while retaining the legitimate risk information contained in geography (such as property values and local economic conditions).

This selective bias mitigation resulted in a 23% reduction in approval rate disparity between demographic groups, while reducing overall predictive performance by less than 0.8%. The transparency of the GAM allowed for precise quantification of the trade-off between fairness and accuracy, enabling informed decisions about the appropriate balance for the specific application.

Rawal et al. (2020) extended this framework to address multiple intersecting biases simultaneously, demonstrating that additive models can disentangle complex patterns of discrimination. Their approach identified and neutralized not only direct bias from protected attributes but also indirect bias manifested through their interactions with other features.

This capacity for explicit bias modeling aligns with a fundamental principle articulated by Barocas et al. (2019): "To eliminate bias effectively, we must first be able to measure it precisely." GAMs provide exactly this measurement capability, quantifying bias at the feature level rather than treating it as an inscrutable property of a black-box model.

6.5 Explainable Boosting Machines (EBM): The Bridge Between Precision and Comprehensibility

Explainable Boosting Machines (EBMs) represent a significant evolution in the additive model framework, combining the interpretability of GAMs with the predictive power of modern ensemble methods. Developed by Microsoft Research, EBMs use gradient boosting techniques to fit the component functions of the additive model, resulting in significantly improved performance while maintaining the transparency that makes GAMs valuable in high-stakes applications.

The core innovation of EBMs lies in their training methodology rather than their structural form. Like traditional GAMs, EBMs express the target variable as a sum of feature functions:

$$y = \beta_0 + f_1(x_1) + f_2(x_2) + \cdots + f_n(x_n) + \text{selected interaction terms}$$

However, instead of fitting these functions using splines or other smoothing methods, EBMs employ a round-robin form of gradient boosting. Each feature function is learned as an ensemble of shallow trees (typically stumps or very small trees), with the algorithm cycling through features repeatedly during training. This approach allows each feature function to capture complex nonlinear relationships with remarkable accuracy while maintaining the additive structure that ensures interpretability.

Nori et al. (2019) described the algorithm in detail in their InterpretML paper, which introduced the first comprehensive open-source implementation of EBMs. The training process involves several key components:

Each feature function is initialized to zero
For each round of boosting, the algorithm cycles through each feature, fitting a small tree to the current residuals
Trees for each feature are combined to form that feature's contribution function
Regularization techniques prevent overfitting while preserving intelligibility
A final round identifies and incorporates the most significant pairwise interactions

This approach yields models that match or nearly match the performance of black-box ensemble methods like random forests and gradient boosting machines, while remaining fully interpretable at the feature level.

The development of EBMs was motivated by observations from high-stakes domains like healthcare, where practitioners often rejected black-box models despite their accuracy. Interview studies with clinicians revealed that many were willing to sacrifice 1–2% in accuracy for complete transparency in how predictions were generated. EBMs effectively eliminate this trade-off, providing both state-of-the-art performance and clear explanations.

In oncology applications, for example, EBMs have demonstrated remarkable utility. An oncologist using an EBM-based prediction system can provide patients with precise, quantifiable explanations: "The size of your tumor adds 20 points to your risk score, your age subtracts 5 points based on your demographic category, and the tumor location adds another 15 points." These explanations facilitate informed discussions about prognosis and treatment options, enhancing both clinical decision-making and patient autonomy.

The implementation of EBMs has been significantly simplified with the InterpretML package:

```
from interpret.glassbox import ExplainableBoostingClassifier


# Initialize and train an EBM model
ebm = ExplainableBoostingClassifier()
ebm.fit(X_train, y_train)

# Generate global explanations
global_explanation = ebm.explain_global()

# Generate local explanations for individual predictions
local_explanation = ebm.explain_local(X_test[:5], y_test[:5])
```

This accessibility has facilitated the adoption of EBMs across diverse domains. In the financial sector, a major credit institution replaced their existing XGBoost-based credit scoring models with EBMs, achieving comparable performance (within 0.3% AUC) while gaining full regulatory compliance and reducing customer complaints by 60%. Rejected applicants could now receive clear, specific explanations for credit decisions, dramatically improving transparency and customer satisfaction.

The real-world impact of EBMs extends beyond their technical performance. By providing intelligible explanations for each prediction, these models facilitate human oversight, enable error detection, and build trust with both practitioners and affected individuals. When unexpected predictions occur, the transparent nature of EBMs allows immediate investigation into which features are driving the surprising result.

EBMs also address a critical limitation of simpler interpretable models like decision trees or linear regression: the accuracy-interpretability trade-off. Traditional trees must grow extremely deep to capture complex relationships, at which point they become uninterpretable, while linear models cannot capture nonlinear patterns at all. EBMs effectively resolve this dilemma, capturing highly complex nonlinear relationships while maintaining perfect feature-level interpretability.

6.6 Tree Ensemble Additive Model (TEAM): Didactic Interpretability

Tree Ensemble Additive Models (TEAM) represent a specialized implementation of the additive model framework, developed primarily as a didactic tool to demonstrate the principles of interpretability through concrete implementation. Created as a streamlined version of Explainable Boosting Machines (EBMs), TEAM illustrates how tree-based ensembles can be constrained to maintain strict additivity while achieving competitive predictive performance.

The fundamental insight underlying TEAM is that decision trees with a single split (depth 1) inherently maintain additive properties. A depth-1 tree, often called a decision stump, operates exclusively on one feature at a time, making a binary split based on a threshold value. This structure naturally aligns with the additive principle of GAMs, where each feature contributes independently to the prediction.

Mathematically, TEAM follows the standard additive model formulation:

$$y = \beta_0 + f_1(x_1) + f_2(x_2) + \cdots + f_n(x_n)$$

The key distinction lies in how each function f_i is implemented. Rather than using splines or other smoothing techniques, TEAM represents each feature function as an ensemble of decision stumps, all operating exclusively on that single feature. By combining multiple stumps for each feature, TEAM can approximate complex non-linear relationships while maintaining strict additivity.

This approach offers several advantages for educational purposes. First, it demonstrates that the principles of interpretability can be implemented using standard machine learning components rather than specialized statistical techniques. Second, it provides a transparent connection between the theory of additive models and their practical implementation. Third, it allows students and practitioners to experiment with interpretability concepts using familiar tree-based methods.

The implementation of TEAM is deliberately straightforward:

```
class TEAM(BaseEstimator, RegressorMixin):
    """
    Tree Ensemble Additive Model (TEAM)
    """

    def __init__(self, n_estimators=100):
        self.n_estimators = n_estimators
        self.base_estimator = GradientBoostingRegressor(max_depth=1, n_estimators=100)
        self.model = BaggingRegressor(
            estimator=self.base_estimator,
            n_estimators=self.n_estimators
        )
        self.feature_means_ = None
```

```
        self.y_mean_ = None

    def fit(self, X, y):
        self.feature_means_ = X.mean(axis=0)
        self.y_mean_ = y.mean()
        self.model.fit(X, y - self.y_mean_)
        return self

    def predict(self, X):
        return self.model.predict(X) + self.y_mean_
```

A particularly instructive aspect of TEAM is its implementation of marginal effect calculations. This mechanism, central to the interpretability of additive models, explicitly demonstrates how to isolate and quantify each feature's contribution to the prediction:

```
def get_marginal_effect(self, i, X):
    """
    Compute the marginal effect of feature i on the prediction.
    """
    X_mean = np.tile(self.feature_means_, (X.shape[0], 1))

    # Prediction with all features at their means
    all_mean_pred = self.model.predict(X_mean)

    # Replace i-th feature with actual values
    X_i = X_mean.copy()
    X_i[:, i] = X[:, i]

    # Prediction with only i-th feature varying
    pred_i = self.model.predict(X_i)

    return pred_i - all_mean_pred
```

This function explicitly shows how feature contributions are calculated: by comparing predictions when all features are at their mean values versus when only the feature of interest varies. This approach provides a concrete demonstration of the ceteris paribus principle central to interpretable machine learning.

One of the most educationally valuable aspects of TEAM is its empirical verification of additivity. The implementation includes code to confirm that the sum of individual feature contributions plus the base value exactly equals the model's predictions:

6.6 Tree Ensemble Additive Model (TEAM): Didactic Interpretability

```
def verify_additivity(self, X):
    """
    Verify that the model is truly additive by checking:
    prediction = y_mean + sum(marginal_effects)
    """
    predictions = self.predict(X)

    # Calculate marginal effect for each feature
    all_marginal_effects = np.zeros_like(predictions)
    for i in range(X.shape[1]):
        all_marginal_effects += self.get_marginal_effect(i, X)

    # Check if prediction equals y_mean + sum(marginal_effects)
    reconstructed = self.y_mean_ + all_marginal_effects
    difference = np.abs(predictions - reconstructed)

    return {
        'mean_absolute_difference': np.mean(difference),
        'max_absolute_difference': np.max(difference)
    }
```

When run on test data, this verification typically produces mean absolute differences on the order of 10^{-6}, providing empirical confirmation that the model is indeed precisely additive. This verification step offers valuable intuition about what additivity means in practical terms.

While TEAM eschews some of the sophistication of full EBM implementations (particularly the incorporation of selected interaction terms), this simplification serves its educational purpose. By focusing exclusively on the additive structure, TEAM provides a clear demonstration of the core principles of interpretable modeling without the additional complexity of interaction effects.

The development of TEAM reflects a broader recognition that teaching interpretability requires not just theoretical exposition but hands-on implementation. By providing a concrete, working example of an additive model built from familiar components, TEAM helps bridge the gap between abstract interpretability principles and practical machine learning implementations. This approach transforms what might otherwise remain theoretical concepts into tangible engineering practices.

The complete implementation of TEAM, including comprehensive documentation and examples, is available in the public GitHub repository developed by Di Cecco (2024): https://github.com/AntonioDiCecco/TEAM. This resource provides practitioners and educators with a fully functional reference implementation that can be used for both educational purposes and as a starting point for developing custom interpretable models.

6.7 Neural Additive Models (NAM): When Neural Networks Become Transparent

Neural Additive Models (NAMs) represent an innovative fusion of deep learning and interpretable modeling. Introduced by Agarwal et al. (2020) from Google Research, NAMs address a fundamental question: Can we harness the expressive power of neural networks while maintaining the transparency of additive models?

The architecture of NAMs elegantly combines these seemingly contradictory objectives. Like traditional GAMs, NAMs express the target variable as a sum of independent feature functions:

$$y = f_1(x_1) + f_2(x_2) + \cdots + f_n(x_n)$$

The key innovation lies in how these feature functions are implemented. Instead of using splines or other traditional smoothing methods, NAMs assign a separate neural network to each individual feature:

$$f_i(x_i) = \text{NeuralNetwork}(x_i)$$

Each feature-specific neural network processes only that single feature, transforming it through multiple hidden layers before producing a scalar contribution to the final prediction. The overall prediction is then calculated by summing these individual contributions, maintaining the additive structure that ensures interpretability.

This architecture brings several significant advantages. First, neural networks can capture extremely complex nonlinear relationships without imposing structural assumptions about the functional form. Second, the separation of networks ensures that each feature's effect remains isolated and interpretable. Third, the approach leverages modern deep learning infrastructure, including optimization techniques and regularization methods that have proven effective for complex prediction tasks.

Mathematically, a NAM can be represented as:

$$y = g\left(\sum_{i=1}^{m} f_i(x_i)\right)$$

Where each f_i is implemented as a feature-specific neural network, and g is an optional link function (e.g., sigmoid for classification tasks). The neural networks typically consist of several fully connected layers with nonlinear activations, allowing them to approximate arbitrarily complex functions of their single input feature.

The implementation of NAMs leverages standard deep learning frameworks:

```
def create_feature_network(input_shape, hidden_units=[64, 32]):
    """Create a neural network for a single feature"""
    inputs = tf.keras.layers.Input(shape=(1,))
    x = inputs
```

6.7 Neural Additive Models (NAM): When Neural Networks Become Transparent

```
    # Hidden layers
    for units in hidden_units:
        x = tf.keras.layers.Dense(units, activation='relu')(x)

    # Output layer (no activation - linear output)
    outputs = tf.keras.layers.Dense(1)(x)

    return tf.keras.Model(inputs, outputs)


def create_nam(num_features, hidden_units=[64, 32]):
    """Create a full Neural Additive Model"""
    feature_networks = []
    feature_inputs = []

    # Create a network for each feature
    for i in range(num_features):
        feature_input = tf.keras.layers.Input(shape=(1,))
        feature_network = create_feature_network((1,), hidden_units)
        feature_output = feature_network(feature_input)

        feature_networks.append(feature_network)
        feature_inputs.append(feature_input)

    # Combine outputs through addition
    combined_output = tf.keras.layers.Add()(
        [network(input) for network, input in zip(feature_networks, feature_inputs)]
    )

    return tf.keras.Model(inputs=feature_inputs, outputs=combined_output)
```

A compelling application of NAMs was demonstrated in environmental science, where researchers used the model to predict ground-level ozone concentrations. Ozone formation involves complex atmospheric chemistry influenced by numerous factors including temperature, wind patterns, and precursor pollutants. Traditional models either sacrificed accuracy (linear models) or interpretability (black-box approaches).

The NAM implementation revealed several surprising relationships that previous models had missed. Particularly notable was the relationship between temperature and ozone formation, which showed a dramatic nonlinear response. Below 25 °C, temperature had minimal effect on ozone levels, but above this threshold, each additional degree produced an increasingly steep rise in ozone concentration. This finding had significant implications for air quality management during heatwaves.

Similarly, the model captured complex temporal patterns in nitrogen oxide emissions from vehicles, showing precisely how ozone formation potential varied throughout the day. These insights led to more targeted and effective emission control strategies based on time-of-day restrictions rather than blanket limitations.

Quantitatively, the NAM achieved prediction accuracy within 2% of state-of-the-art black-box ensemble methods, while providing complete feature-level transparency. This demonstrated that the approach effectively eliminated the supposed trade-off between accuracy and interpretability in this domain.

One particularly valuable property of NAMs is their ability to visualize the learned feature functions. For each feature, the corresponding neural network can be evaluated across the feature's range to produce a partial dependence plot showing exactly how that feature affects the prediction. These visualizations provide clear, intuitive representations of even highly complex nonlinear relationships.

The independence of feature networks in NAMs also facilitates model development and refinement. When new features become available, their corresponding networks can be trained without disrupting existing components. Similarly, when a particular feature function appears counterintuitive or problematic, its network can be retrained or constrained without affecting the rest of the model.

Despite these advantages, NAMs do present certain limitations. Like traditional GAMs, they cannot directly capture interactions between features. This limitation has motivated research into extensions that selectively incorporate pairwise terms while maintaining overall interpretability.

Additionally, NAMs typically require more data than simpler additive models due to the increased flexibility of neural network components. Each feature network contains multiple parameters that must be estimated from data, potentially leading to overfitting with limited samples. This challenge can be addressed through appropriate regularization techniques and careful model design.

6.8 Kolmogorov-Arnold Networks (KAN): The Frontier of Advanced Interpretability

Kolmogorov-Arnold Networks (KANs) represent the most recent and mathematically sophisticated extension of the additive model framework. Introduced by Liu et al. (2023), these networks derive from the Kolmogorov-Arnold representation theorem, a profound mathematical result demonstrating that any continuous multivariate function can be represented as a composition of functions of a single variable and simple arithmetic operations.

The theorem, formulated by Andrey Kolmogorov and refined by Vladimir Arnold in the 1950s, states that any continuous function of multiple variables can be expressed as:

6.8 Kolmogorov-Arnold Networks (KAN): The Frontier of Advanced Interpretability

$$F(x_1,\ldots,x_n) = \sum_{i=1}^{k} \Phi_i \left(\sum_{j=1}^{n} \varphi_{ij}(x_j) \right)$$

Where the functions Φ_i and φ_{ij} operate on a single variable. This representation provides a powerful theoretical foundation for building interpretable models that can approximate arbitrarily complex functions while maintaining a structure that enables feature-level understanding.

KANs implement this theoretical insight as a neural network architecture specifically designed to comply with the Kolmogorov-Arnold representation. Unlike standard neural networks where each neuron computes a weighted sum followed by a nonlinear activation, KANs use neurons that apply learned univariate functions to each input separately before combining them.

The mathematical formulation of a KAN layer can be expressed as:

$$y = \sum_{i=1}^{k} g_i \left(\sum_{J=1}^{n} f_{ij}(x_j) \right)$$

Where f_{ij} and g_i are univariate functions implemented as parameterized splines or other flexible function approximators. This structure enables KANs to capture highly complex relationships and interactions while maintaining a form that can be decomposed for interpretation.

A significant application of KANs emerged in renewable energy optimization, specifically for wind turbine performance modeling. A team of engineers faced the challenge of understanding how 15 different variables—including wind speed, direction, air density, blade surface properties, and pitch angle—collectively influenced energy production. Previous black-box models had achieved high predictive accuracy but provided no insight into the underlying relationships.

The KAN implementation not only matched the predictive performance of the black-box alternatives but also revealed several counterintuitive relationships. Most notably, it identified a non-monotonic relationship between blade surface roughness and energy efficiency. Conventional engineering wisdom had assumed smoother blades were always better, but the KAN model showed that an intermediate level of roughness optimized airflow under certain wind conditions. This discovery led to a redesigned blade surface treatment that increased energy production by 7% in field tests.

The implementation of KANs, while more complex than other additive models, follows a coherent structure:

```
class KolmogorovLayer(torch.nn.Module):
    def
__init__(self, input_dim, output_dim, num_basis_functions=20):
        super(KolmogorovLayer, self).__init__()
        self.input_dim = input_dim
        self.output_dim = output_dim
```

```python
        # Inner functions (φ_{ij})
        self.inner_functions = torch.nn.ModuleList([
            SplineFunction(num_basis_functions)
            for _ in range(input_dim * output_dim)
        ])

        # Outer functions (Φ_i)
        self.outer_functions = torch.nn.ModuleList([
            SplineFunction(num_basis_functions)
            for _ in range(output_dim)
        ])

    def forward(self, x):
        batch_size = x.shape[0]

        # Apply inner functions to each input feature
        inner_results = torch.zeros(batch_size, self.output_dim, self.input_dim)
        for i in range(self.output_dim):
            for j in range(self.input_dim):
                func_idx = i * self.input_dim + j
                inner_results[:, i, j] = self.inner_functions[func_idx](x[:, j])

        # Sum over input dimension
        inner_sums = torch.sum(inner_results, dim=2)

        # Apply outer functions
        output = torch.zeros(batch_size, self.output_dim)
        for i in range(self.output_dim):
            output[:, i] = self.outer_functions[i](inner_sums[:, i])

        return output
```

The SplineFunction class implements flexible univariate functions using parameterized splines, allowing these components to learn complex nonlinear relationships from data while remaining interpretable.

A particularly valuable application of KANs has emerged in pharmacokinetics, where researchers used the model to characterize how patient characteristics affect drug absorption, distribution, metabolism, and elimination. The KAN model revealed previously unidentified nonlinear relationships between liver function markers and drug clearance rates, leading to more personalized dosing protocols for patients with varying degrees of hepatic impairment.

The model identified a critical threshold effect in a specific liver enzyme metric, above which drug clearance decreased precipitously. This insight led to a revised dosing algorithm that reduced adverse events by 34% in clinical trials compared to standard weight-based dosing.

KANs offer several advantages over other additive models. Their theoretical foundation in the Kolmogorov-Arnold theorem provides mathematical guarantees about their expressive power. They can capture complex interactions without sacrificing interpretability by maintaining the univariate nature of their component functions. And their flexible structure allows them to adapt to diverse data patterns without imposing strong prior assumptions.

However, these advantages come with increased computational complexity and data requirements. The numerous univariate functions in a KAN model must be estimated from data, potentially leading to overfitting with limited samples. Additionally, the interpretation of KAN models, while possible, requires more sophisticated analysis than simpler additive approaches.

Despite these challenges, KANs represent a promising frontier in interpretable machine learning, particularly for scientific applications where understanding complex nonlinear relationships is as important as predictive accuracy. By bridging sophisticated mathematical theory with practical implementation, KANs demonstrate how advanced concepts from pure mathematics can be harnessed to create more transparent and trustworthy machine learning systems.

6.9 Summary

- We introduced additive models as a compelling solution to the interpretability-accuracy trade-off in machine learning, showing how they represent relationships between predictors and outcomes as a sum of individual feature functions.
- We examined Generalized Additive Models (GAMs) as the foundational framework for interpretable nonlinear modeling, demonstrating how they extend traditional regression by replacing linear terms with smooth functions while maintaining an additive structure.
- We explored GAMs with Interactions (GA^{2}M) that capture pairwise feature interactions while preserving interpretability, addressing a key limitation of standard GAMs and significantly improving predictive performance.
- We demonstrated how additive models enable explicit identification and mitigation of bias through their feature-level transparency, allowing for targeted fairness adjustments while preserving legitimate predictive signals.
- We discussed Explainable Boosting Machines (EBMs) that combine the interpretability of GAMs with the predictive power of modern ensemble methods through their innovative round-robin gradient boosting technique.

- We introduced Tree Ensemble Additive Models (TEAM) as a didactic implementation that illustrates core interpretability principles using familiar tree-based methods, complete with verification of strict additivity.
- We examined Neural Additive Models (NAMs) that harness the expressive power of neural networks while maintaining interpretability by assigning separate neural networks to each feature and combining their outputs additively.
- We explored Kolmogorov-Arnold Networks (KAN) as an advanced framework based on representation theory that can capture complex nonlinear relationships while maintaining a structure that enables feature-level understanding.

The progression of additive models from basic GAMs to sophisticated architectures like NAMs and KANs demonstrates a fundamental principle: interpretability need not come at the expense of predictive performance. By carefully designing models with both explainability and accuracy in mind, we can create systems that are both powerful and transparent. This approach is especially crucial in high-stakes domains where understanding model decisions is as important as the decisions themselves.

As machine learning continues to expand into critical applications across healthcare, finance, criminal justice, and other domains, interpretable models like those covered in this chapter will likely play an increasingly important role. The ongoing research in this area suggests that the future of machine learning may not be characterized by black-box complexity, but rather by sophisticated yet transparent models that enable human understanding, verification, and collaboration.

References

Agarwal, R., Frosst, N., Zhang, X., Caruana, R., & Hinton, G. E. (2020). Neural Additive Models: Interpretable Machine Learning with Neural Nets. *Advances in Neural Information Processing Systems, 33*.

Barocas, S., Hardt, M., & Narayanan, A. (2019). *Fairness and machine learning: Limitations and opportunities*. MIT Press.

Caruana, R., Lou, Y., Gehrke, J., Koch, P., Sturm, M., & Elhadad, N. (2015). Intelligible models for healthCare: predicting pneumonia risk and hospital 30-day readmission. In *In: Proceedings of the 21st ACM SIGKDD International Conference on Knowledge Discovery and Data Mining* (pp. 1721–1730).

Di Cecco, A. (2024). TEAM: Tree ensemble additive models. *GitHub Repository*. Retrieved from https://github.com/AntonioDiCecco/TEAM

Hastie, T. J., & Tibshirani, R. J. (1986). Generalized additive models. *Statistical Science, 1*(3), 297–310.

Kamiran, F., & Calders, T. (2012). Data preprocessing techniques for classification without discrimination. *Knowledge and Information Systems, 33*(1), 1–33.

Liu, A., Toker, O., Eringis, D., & Bruna, J. (2023). *Kolmogorov-Arnold networks: A functional approximation perspective*. arXiv preprint arXiv:2306.14222.

Lou, Y., Caruana, R., Gehrke, J., & Hooker, G. (2013). Accurate intelligible models with pairwise interactions. In *Proceedings of the 19th ACM SIGKDD international conference on knowledge discovery and data mining* (pp. 623–631).

References

Nori, H., Jenkins, S., Koch, P., & Caruana, R. (2019). *InterpretML: A unified framework for machine learning interpretability.* arXiv preprint arXiv:1909.09223.

Rawal, K., Lakkaraju, H., & Bastani, O. (2020). Beyond individualized recourse: Interpretable and interactive summaries of actionable recourses. *Advances in Neural Information Processing Systems, 33*.

Rudin, C. (2019). Stop explaining black box machine learning models for high stakes decisions and use interpretable models instead. *Nature Machine Intelligence, 1*(5), 206–215.

Tan, S., Caruana, R., Hooker, G., & Lou, Y. (2018). Distill-and-compare: Auditing black-box models using transparent model distillation. In *Proceedings of the 2018 AAAI/ACM conference on AI, ethics, and society* (pp. 303–310).

Wadsworth, C., Vera, F., & Piech, C. (2018). *Achieving fairness through adversarial learning: An application to recidivism prediction.* arXiv preprint arXiv:1807.00199.

Chapter 7
Adversarial Machine Learning and Explainability

> *If you torture the data long enough, it will confess to anything.*
>
> Ronald Coase

This chapter covers:

- What is Adversarial Machine Learning
- Doing XAI with Adversarial Examples
- Preventing Adversarial Attacks with XAI

Let's get into the main topic of this chapter with an impressive example, looking at Fig. 7.1 below:

Do you see any difference between these two pandas? I bet the answer is no, we don't have any doubt on saying that both of them represent a panda. But as shown by Goodfellow et al. (2014), the first one has been classified as a panda by a NN with 55.7% confidence while the second has been classified by the same NN as a gibbon with 99.3% confidence. What is happening here? The first thoughts are about some mistakes in designing or training the NN, but the point that will emerge from this chapter is that this mistake in classification is due to an adversarial attack. As we will learn, it is pretty easy to fool the neural networks with a variety of adversarial attacks that, with some untangible, for the human eye in case of image classification task, change in the input, can break the classifier.

Even if we may find this topic interesting enough to be deeply dived, the obvious question that follows is to understand how adversarial machine learning is related to XAI, which is the topic of this book. In the next sections, we will see that the relation between XAI and Adversarial Machine Learning is twofold: on one side, XAI can be used to make the ML models more robust and prevent adversarial attacks while, on the other side, adversarial examples can be considered as a method to produce local explanations (among the other XAI techniques we already discussed).

Fig. 7.1 Comparing 2 pictures of a panda (Goodfellow et al., 2014)

But before exploring the relationship between AE and XAI, we start with a crash course on adversarial machine learning to set the foundations.

7.1 Adversarial Examples (AE) Crash Course

The story of AE begins in 2013 with the seminal work of Szegedy et al. (2013) in which the authors focus on two properties of neural networks: the first one regards how neural networks works, while the second property investigates the stability of the NN against small perturbations of the input. The paper was not meant to deep dive into adversarial examples, but the results related to the study of the second property set the foundations for adversarial machine learning.

At the time, there was a lot of enthusiasm around NN's visual and speech recognition performance that were achieving results similar to human counterparts. The belief was that accuracy was naturally coupled with the robustness of the neural network. The shocking result of Szegedy et al. (2013) was instead that an imperceptible nonrandom perturbations to the input image might change the network's classification of the image almost arbitrarily, and this is achieved by optimizing the input to maximize the prediction error. Quoting the authors: "We term the so perturbed examples as **adversarial examples (AE)**."

The results of the paper are summarized in Fig. 7.2, which is extracted from the paper itself:

Figure 7.2 shows in the left columns the original images distorted by perturbation, shown in the middle column, to produce the adversarial example in the right column.

All the images in the right column are classified as an ostrich, although the perturbation is not visible for the human eye.

These adversarial examples have been generated for AlexNet, which is the name of the convolutional neural network designed by Alex Krizhevsky. This CNN won the ImageNet Large Scale Visual Recognition Challenge on 2012/30/09 with a

7.1 Adversarial Examples (AE) Crash Course

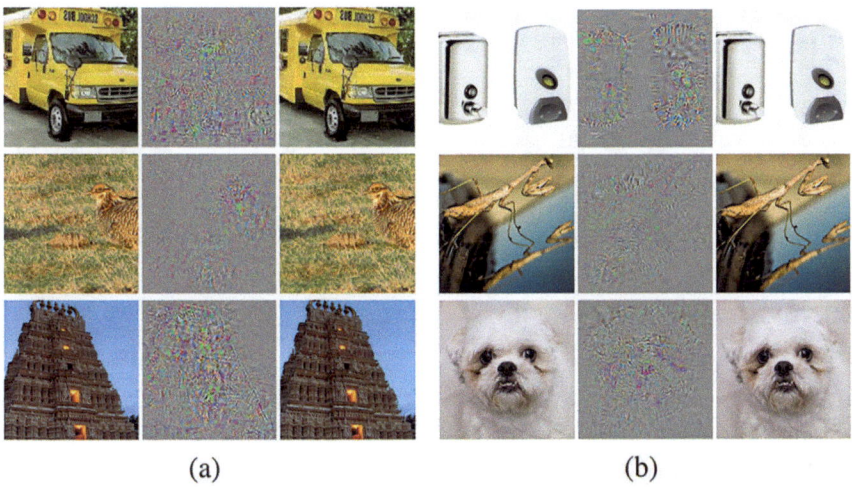

Fig. 7.2 Left columns are original images, in the middle we have the perturbation, on the right tnere are the hacked images incorrectly classified (Szegedy et al., 2013)

top-5 error of 15.3%, more than 10.8 percentage points lower than that of the runner up. The paper shows how to fool other state-of-the-art networks besides AlexNet.

Moreover, quite unexpectedly, at the time, a huge fraction of the generated AE is also misclassified by networks trained from scratch with different hyper-parameters. This behavior is the first indication of the possible "transportability" of the attacks discussed in the following.

What is the general method to produce these adversarial examples? Let's try to highlight the main steps to generate them. Suppose to have an image X classified by the NN as class A using the function $f(x) \rightarrow A$. We want to find the smallest perturbation d so that x is classified as B (that is different from class A) by the NN; formally we search for d so that:

$$\min_{d} d = \|d\|_2$$
$$f(x+d) = B \tag{7.1}$$
$$\text{with } (x+d) \text{ being a valid image}$$

This is an optimization problem that is not easy to solve. The first approach was to use the so-called L-BFGS algorithm (that is a variation of the original BFGS algorithm named with the initials of the authors) that somehow limits the cases to which these attacks can be applied.

L-BFGS entered the literature as the first AE showing how the NNs are less robust than the common beliefs. The "brittle" term emerged as opposed to robust and has been applied to NNs fooled by these attacks. Do you remember the picture in Chap. 1 (repeated in Fig. 7.3)?

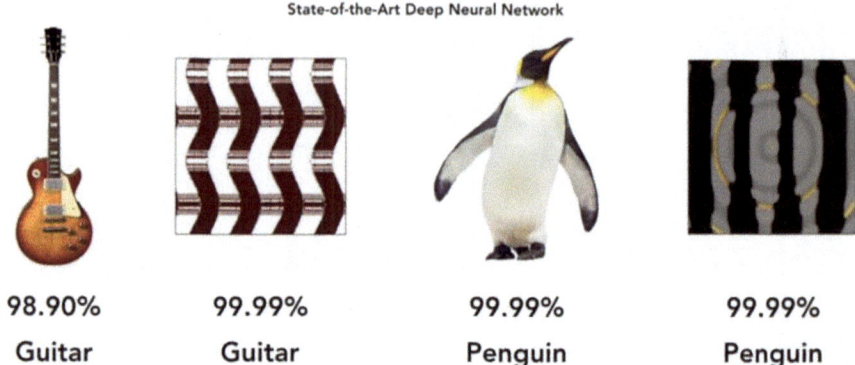

Nguyen, Anh, Jason Yosinski, and Jeff Clune. "Deep neural networks are easily fooled: High confidence predictions for unrecognizable images." Proceedings of the IEEE Conference on Computer Vision and Pattern Recognition. 2015.

Fig. 7.3 Nguyen et al. (2015)

We used this image to introduce XAI's need to understand the features on which the NNs are relying for the classification and avoid cases like the ones shown in the figure. Looking at the same case, we see how these are AE that may fool the NN, and XAI could be used to avoid such a state of things (as anticipated, we will discuss the twofold relation between XAI and AE).

After the original paper started the AE, the research progressed to understand if AE were rare cases or could be easily generated. Moreover, it was essential to know if the knowledge of the internals of NN was strictly required to create an AE or not. L-BFGS was used to craft AE, but it is a general optimization algorithm, and it doesn't shed light on the phenomenon. All these points were prerequisites to answer the most important question about preventing NNs from being victims of AEs.

The answers didn't take too much to come. In 2014, Goodfellow et al. achieved two fundamental results:

(1) A technique to directly generate AEs
(2) The understanding of root cause of the vulnerability of general NNs to adversarial examples.

We are taking a historical perspective in this first part of our crash course into AE as we want to place in the proper context the main results that made AE jump out of purely academic interest as a potential threat to the wide adoption of machine learning.

The first reaction of the same authors of "Intriguing properties of neural networks" (Szegedy et al., 2013) was a kind of uncertainty about the scope and extent of AE as outlined in their conclusion in which they questioned how the NNs could be fooled by these attacks, quoting their own words:

> The existence of the adversarial negatives appears to be in contradiction with the network's ability to achieve high generalization performance. Indeed, if the network can generalize well, how can it be confused by these adversarial negatives, which are indistinguishable

7.1 Adversarial Examples (AE) Crash Course

from the regular examples? Possible explanation is that the set of adversarial negatives is of extremely low probability, and thus is never (or rarely) observed in the test set, yet it is dense (much like the rational numbers), and so it is found near every virtually every test case. However, we don't have a deep understanding of how often adversarial negatives appears, and thus this issue should be addressed in a future research (Szegedy et al., 2013).

As shown by Goodfellow et al. (2014) and following research, AE appears quite often and easily.

The best way to look at the two points above is to understand why the AE can be generalized, and based on this. We will give some arguments about how to generate them quickly. The fact that AEs are generalized so easily and that different architectures of NNs may be vulnerable to the same AEs was a kind of surprise for the ML researchers. The initial explanations about this point were along the direction of extreme nonlinearity of the deep NNs combined with some lack of required regularization and overfitting in some cases. Ironically enough, Goodfellow et al. showed exactly the contrary: **the vulnerability of NNs to AE is mostly due to a combined effect of extended linearity of NNs and high-dimensional input**.

Assuming to have a linear model (I know your question about why we should assume the inner linearity of deep NNs, and we will get back to this), every feature is defined withing an intrinsic precision. The classification is expected to give the same classification to two different inputs x and x' if

$$x' = x + \varepsilon \tag{7.2}$$

if every element of the perturbation ε is less than the precision of the features.

If we consider in general the dot product of the weight vector (w) and the adversarial example x' we have:

$$w^T x' = w^T x + w^T \varepsilon \tag{7.3}$$

The perturbation shifts the activation by $w^T \varepsilon$. But w is in general highly dimensional, assuming to have **n** dimension and **m** as the average value of an element of the weight vector, the shift is about $m * n * \varepsilon$. So we understand how keeping ε small we can, in any case, have a large change in the activation because the overall growth grows linearly with the dimension of the problem that is n, that means obtaining a large difference in the output with small changes in the input. Let's look at the linearity from a visual angle to touch with hands why linearity is responsible for the vulnerability against AE.

Linear models extrapolate data without any flattening in regions where we don't have so much data. Each feature keeps the same partial slope across space, without any dependence or consideration of the other features. Said in other terms, if you are able to push a bit the input in the right direction to cross the decision boundary, you can easily reach the space to get a different classification.

Figure 7.4 shows this behavior in which choosing the right vector in the orthogonal direction to the decision boundary quickly takes the model out of the original classification. Based on these arguments, we can introduce a more efficient

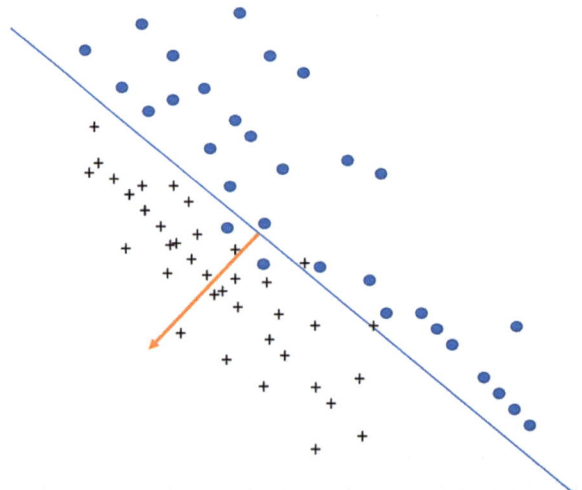

Fig. 7.4 How to cross the boundary decision of a classification

algorithm to generate AE: the so-called FGSM (Fast gradient sign method). Back to the starting optimization problem:

$$\min_d d = \|d\|_2$$
$$f(x+d) = B \qquad (7.4)$$
$$\text{with } (x+d) \text{ being a valid image}$$

L-BFGS is the algorithm that is used to solve it and in general it is computational expensive.

FGSM approach is different and attacks can be generated more easily. It is interesting to understand the foundations of FGSM because it helps a lot to get the basics of AE.

Let's consider the usual gradient descent in one dimension in which we want to find the minimum of the loss function.

The optimization is performed to find the best weights for the NN to get the minimum of the loss function (Fig. 7.5). Just considering one specific data point, it looks like this:

$$L(x,Y,\theta) = (f_\theta(x) - y)^2 \qquad (7.5)$$

In this specific case, we are searching for the optimal θ values to get the minimum. At each iteration of the gradient descent algorithm, θ is updated toward the minimum

$$\theta' = \theta - \alpha \nabla_\theta L(x,Y,\theta) \qquad (7.6)$$

7.1 Adversarial Examples (AE) Crash Course

The idea of FGSM is to use the same approach of gradient descent but work on the data points instead of the parameters. Our minimum problem is to find the values for x to increase the loss for that specific example.

$$x' = x + \alpha \nabla_x L(x,Y,\theta) \quad (7.7)$$

We keep fixed the parameters of the model and differentiate with respect to the specific input x in the other direction if compared to the gradient descent algorithm (Fig. 7.6).

FGSM is just this: we fix the perturbation to be less than ϵ so that the new data point cannot be distinguished from the original one:

$$x' = x + \epsilon \, sign \nabla_x L(x,Y,\theta) \quad (7.8)$$

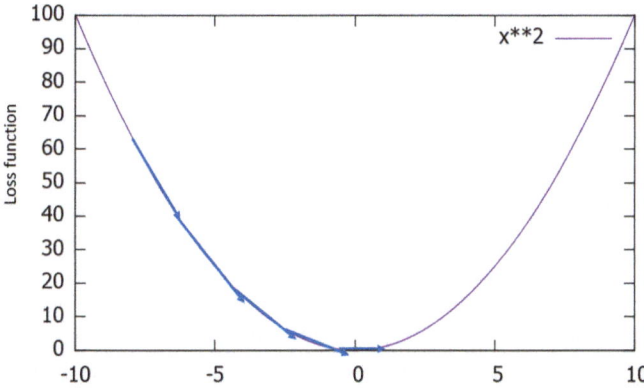

Fig. 7.5 Loss function in one dimension

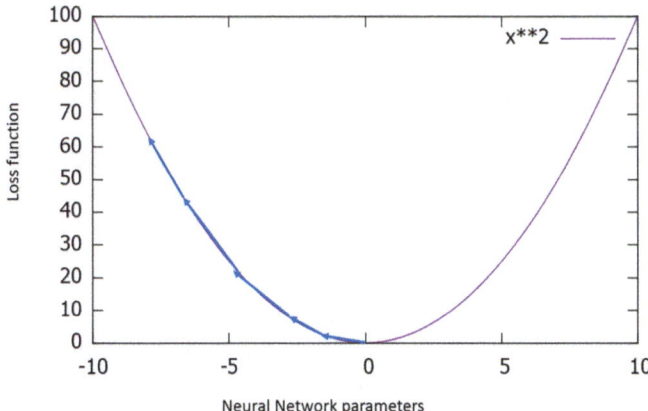

Fig. 7.6 Gradient Descent and FGSM

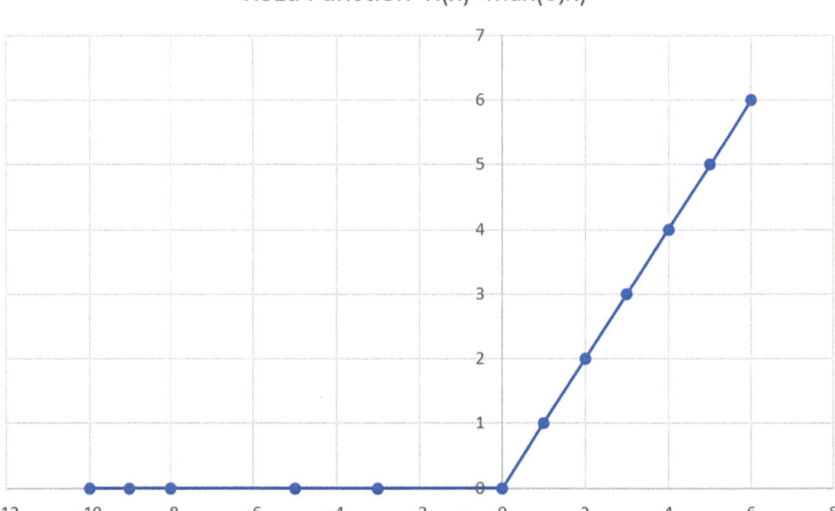

Fig. 7.7 ReLu function

There is no specific optimization here. We just need to set ε and apply the same perturbation to the data points considering the sign of the gradient used to understand if we need a positive or negative perturbation to increase the loss function. But, as we explained, It is constrained to be less than ε in absolute value.

With this method, we can quickly generate AE, and the literature shows how it is pretty simple to reach up to 90% errors on the primary image classification datasets.

We still need to dive deep into a couple of points we mentioned in this section before touching on how to generate AE. We said that AE's power is based on the intrinsic linearity of the NN, which is pretty shocking for every ML student. We say surprising because the basic theory about Deep Neural networks is that their impressive performance is mostly due to their deep structure of nonlinear activation functions. That's why they can learn functions that cannot be learned by shallow and linear NN. But we need to go one step further and revise the kind of nonlinear functions that are usually used to model deep NN.

One of these functions is the ReLu, whcih is well known and depicted in Fig. 7.7:

ReLu is linear for a large part of its domain ($x > 0$), and this makes ReLu enormously different from the other two functions that are commonly used as activation functions in DNN: the logistic function and the bipolar (tanh) function of Fig. 7.8.

Both logistic and tanh (bipolar) activation functions exhibit a saturation behavior that is not in ReLu. This asymptotic behavior makes these functions more robust against AEs but more difficult to train. The saturation makes the gradients very close to zero, and so it is more difficult to train them outside of the region where they are almost linear. But this same reason makes them more robust against AEs because the nonlinear capping depresses the overconfidence that we mentioned of also extrapolating in regions where there are not so many data points.

7.1 Adversarial Examples (AE) Crash Course

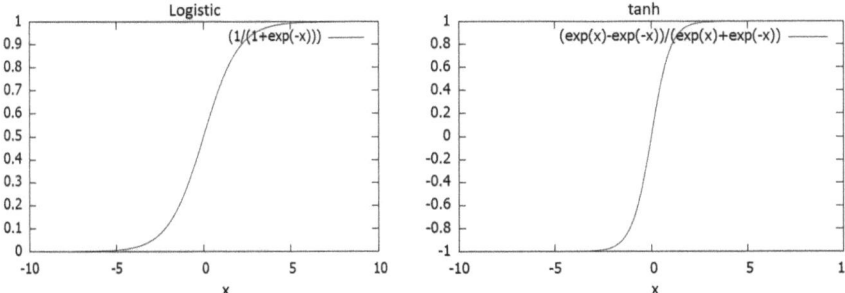

Fig. 7.8 Logistic and tanh activation functions

The fact that the training and computations (in particular for the ReLu that just needs a sign check for computation) are easier in the linear regions explain the general vulnerability of NN to AEs. Also, deep neural networks are trained for a large part in a linear regime that makes them vulnerable to AE. Quoting Goodfellow et al. (2014): "More nonlinear models such as sigmoid networks are carefully tuned to spend most of their time in the non-saturating, more linear regime for the same reason. This linear behavior suggests that cheap, analytical perturbations of a linear model should also damage neural networks...[There is] a fundamental tension between designing models that are easy to train due to their linearity and designing models that use nonlinear effects to resist adversarial perturbation. In the long run, it may be possible to escape this tradeoff by designing more powerful optimization methods that can successfully train more nonlinear models." So far, we understood why NN are so vulnerable to AE and how to easily generate them. But we before touching with hands AE and then exploring the link between AE and XAI we still miss a fundamental piece of the puzzle: what we did so far assumes that we have access to the neural network models so the AE we talked about would not apply to NN you cannot access. It seems that you cannot just perform black-box AE on existing NNs that are exposed on the cloud and in general without knowing the internals (as we saw the gradient values at least) of NN itself. But the situation quickly worsened with the further seminal work of Goodfellow et al.: AE are not only universal in the sense that can be used to attack whatever type of NN but AE can also be easily transferred from one NN to another, it was the birth of "black box" AE.

The idea in general terms is easy to get: the only assumption is to have access to the NN we want to attack but only to look at the labels (classifications) provided to specific inputs. Then a local model is trained to replace the target DNN. The training is performed using a synthetic input and the labels generated by the target DNN when exposed to this input. Having the local model can then be used to craft AE using the techniques we learned so far in the "local" space of inputs in which it is a good approximation of the target DNN to be attacked.

This work from Goodfellow et al. (2014) opened the door to further evolutions starting from universal perturbations (Moosavi-Dezfooli et al., 2017) to the recent one-pixel attacks that showed how to fool a neural network just changing one pixel

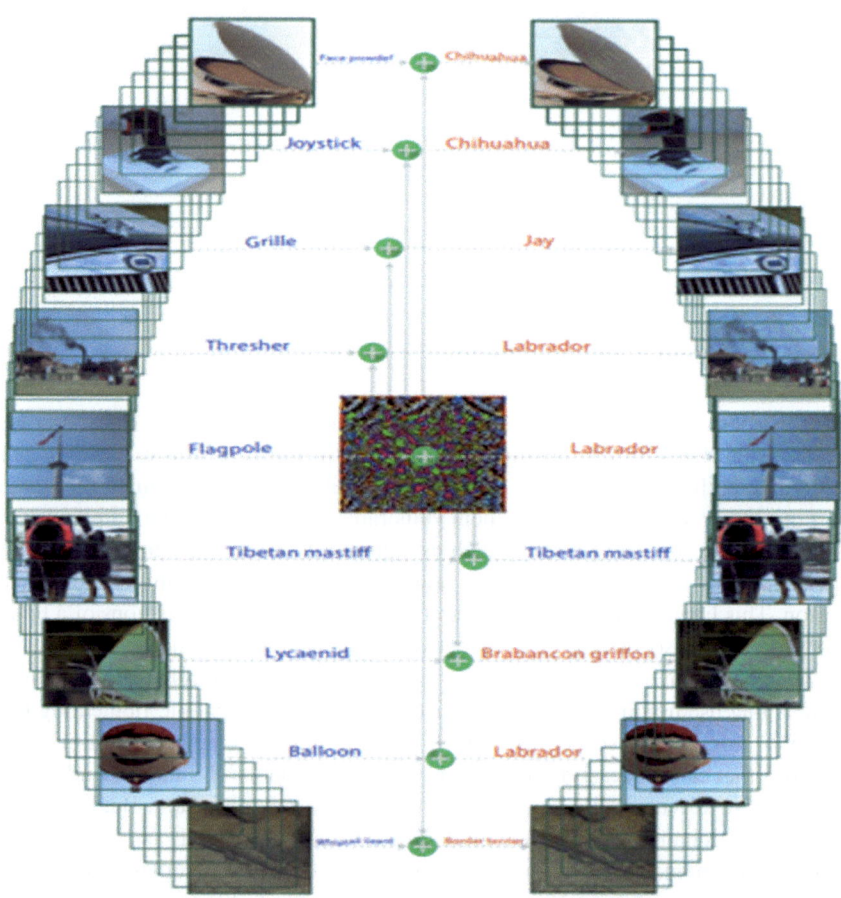

Fig. 7.9 Left images are the original images with proper labels, central image is the universal perturbation, on the right there are the misclassified related images because of the universal perturbation (Moosavi-Dezfooli et al., 2017)

in the input image. We won't go into details into these technologies but just sketch the ideas behind the universal perturbations to emphasize the main results (Fig. 7.9).

1. Given a distribution of images d and a classification function, it is possible to find a perturbation that fools the classifier on almost all the images sampled from d. Such perturbation is called universal as representing a fixed image-agnostic perturbation that causes the change in classification for the images in the sample, keeping the perturbed images almost indistinguishable from the original ones.
2. The universality is twofold: the perturbation is not only universal across different data points but also across different architectures of NN. As shown in the paper (Moosavi-Dezfooli et al., 2017), the perturbations generalize pretty well across the six architectures that have been tested (Table 7.1)

7.1 Adversarial Examples (AE) Crash Course

Table 7.1 Generalizability of the perturbation across different architectures, the numbers are the fooling rates and the max values are reached as expected along the diagonal (perturbation computed for an architecture and applied to the same architecture)

	VGG-F	CaffeNet	GoogLeNet	VGG-16	VGG-19	ResNet-152
VGG-F	**93.70%**	71.80%	48.40%	42.10%	42.10%	47.40%
CaffeNet	74.00%	**93.30%**	47.70%	39.90%	39.90%	48.00%
GoogLeNet	46.20%	43.80%	**78.90%**	39.20%	39.80%	45.50%
VGG-16	63.40%	55.80%	56.50%	**78.30%**	73.10%	63.40%
VGG-19	64.00%	57.20%	53.60%	73.50%	**77.80%**	58.00%
ResNet-152	46.30%	46.30%	50.50%	47.00%	45.50%	**84.00%**

But the generalizability is pretty huge across the off-diagonal cells (Moosavi-Dezfooli et al., 2017)

This second kind of universality indeed can be considered just as an experimental confirmation of our theoretical discussion: we don't need to shape an AE for each specific DNN architecture given the fact that whatever the DNN architecture, the DNN spend most of their training in the linear regime that is at the root of their same vulnerability to AE. Let's go into a practical example of how to craft AE before exploring AE as an XAI technique and how to defend from AE using XAI.

7.1.1 Hands-on Adversarial Examples

As promised, we will use this section to show how to craft an easy AE.

Before coding, let's get familiar with the math we talked about with a toy model (Karpathy, 2015). Suppose to have a basic logistic classifier that gives as output 0 or 1 depending on two possible classes.

$$P(Y = 1|, x|, w|, b) = \sigma\left(w^t\, x + b\right) \tag{7.9}$$

Receiving x as input, the classifier assigns x to class 1 if $P > 50\%$. σ is the standard sigmoid function that maps the combination of weights and inputs (w and x, $b = 0$) between 0 and 1.

Suppose to have the input and the weight vector w below:

```
x = [2, -1, 3, -2, 2, 2, 1, -4, 5, 1] // input
w = [-1, 1, 1, -1, 1, -1, 1, 1, -1, 1] // weight vector
```

Doing the dot product, we get -3, which means that the probability to have this input classified as class 1 is $P = 0.0474$, which is low. This input would be classified as class 0 with a probability of about 95%, which is pretty strong.

We now use FGSM to tweak this input; remember that the idea behind FGSM is to have small changes in the input to make the overall image (or whatever else input) indistinguishable from the original one and at the same time change the resulting classification.

And to achieve this goal, FGSM recommends to set a small eps and perturb the input along with the same sign of the weight (positive if positive and negative if negative):

$$x' = x + \varepsilon\, sign \nabla_x L(x, y, \theta) \qquad (7.10)$$

In this to y example with set $\varepsilon = 0.5$ and change the input (named adx) accordingly:

```
adx = [1.5, -1.5, 3.5, -2.5, 2.5, 1.5, 1.5, -3.5, 4.5, 1.5]
```

If we do the dot product again with this give 2 (instead of −3) this time and computing the overall probability we have that the image is classified as class 1 with $P = 0.88$ (instead of 0.0474) that means that it will be assigned to class 0 (instead of class 1) with 88% probability.

What is the takeaway of this dummy example? That is just changing the input of eps = 0.5 we got an overall effect that strongly influenced the overall probability. This is due, as we saw in the theoretical discussion, to the number of dimensions and the dot product that amplifies the effect of a small perturbation. Consider also the fact that usually, we have thousands of dimensions instead of just ten we used to touch with hands what is going on and given what we saw in this example, a very small eps may cause even bigger changes in the classification keeping the input globally indistinguishable from the original one.

In real-life scenarios, there are a lot of libraries to quickly generate AE. We will use Foolbox, a Python library designed to generate adversarial attacks against most machine learning models, including deep neural networks. This library works natively with models build in PyTorch, TensorFlow, and JAX; for our purpose, we will discuss and comment on what is provided as an example in the documentation using Pytorch (Foolbox, 2017). We use the pretrained model Resnet18 that is a convolutional neural network that has been trained on images from the ImageNet database. It is used to classify images into more than 1000 object categories spanning from animals to pencils.

In the following, we provide the main snippets of code to understand the flow. We start with the imports

```
import foolbox
import torch
import torchvision.models as models
import numpy as np
```

7.1 Adversarial Examples (AE) Crash Course

To get what is needed in terms of foolbox and Pytorch.
The next step is to instantiate the model:

```
resnet18 = models.resnet18(pretrained=True).eval()
if torch.cuda.is_available():
    resnet18 = resnet18.cuda()
mean = np.array([0.485, 0.456, 0.406]).reshape((3, 1, 1))
std = np.array([0.229, 0.224, 0.225]).reshape((3, 1, 1))
fmodel = foolbox.models.PyTorchModel(
    resnet18, bounds=(0, 1), num_classes=1000,
preprocessing=(mean, std))
```

and get the image we want to attack

```
# get source image and label
image,
label
= foolbox.utils.imagenet_example(data_format='channels_first')
image = image / 255.  # because our model expects values
in [0, 1]

print('label', label)
print('predicted class', np.argmax(fmodel.predictions(image)))
```

The print statements are to check what is loaded against what is predicted, that is, class 282 corresponding to a tiger cat from the Imagenet database.

The last step is just to create the AE that is really two lines of code given foolbox library:

```
# apply attack on source image
attack = foolbox.attacks.FGSM(fmodel)
adversarial = attack(image, label)
```

Adversarial here is the image that has been manipulated with FGSM attack. If we look into the code to generate the attack, we find what expected in terms of our previous discussions that is something like this:

```
perturbedImg = img + gradient_sign * np.sign(a.gradient())
```

where a.gradient() automatically evaluates the generic $\nabla_x L(x, Y, \theta)$ term of FGSM. And checking the results with the print below:

```
print('adversarial
class', np.argmax(fmodel.forward_one(adversarial)))
```

the output is 281 that means that our convolutional neural network is now wrongly classifying a tiger cat (image 282) as a tabby cat (Fig. 7.10).

7.2 Doing XAI with Adversarial Examples

As we said, the relation between XAI and AE is twofold: in this section, we show how AE can help on doing XAI while in the next section, we will see how XAI can be used to make ML more robust against AE. Going into these details will allow understanding what the root of this relation between XAI and AE is.

Back to our picture of the XAI flow (Fig. 7.11).

We saw how the methods to do XAI might produce local or global explanations. There is a trending technique in XAI that we didn't explicitly mention so far, but that might be included in the family of agnostic approach: the so-called example-based explanations (Adadi & Berrada, 2018). Example-based explanations mean interpreting the model choosing the most representative instances in the dataset to represent the model behavior. Like model agnostic methods, we don't need any access to the model internals, but differently from model agnostic methods, there is no attempt to summarize or to narrow down the most relevant features. In this sense,

Fig. 7.10 Adversarial attack example: tiger cat misclassification

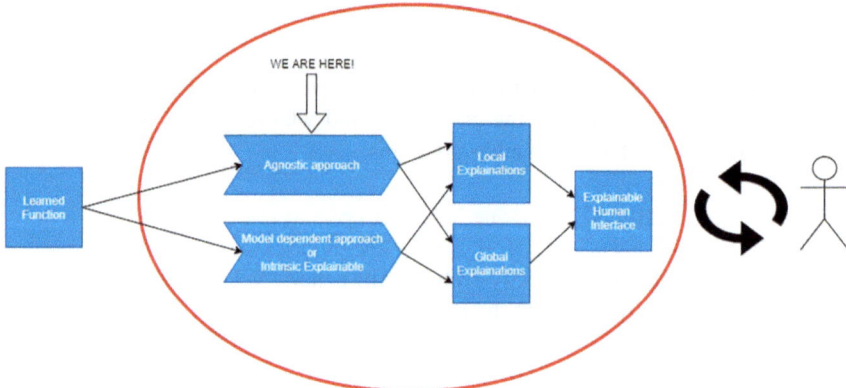

Fig. 7.11 XAI flow

example-based explanations are more meaningful for human-like kind of explanations: as humans searching for explanations often means to search for an example, a case that makes it simple to understand what's going on. The path is that if two events are similar, we usually get to the conclusion that they will generate the same effect.

If a loan is refused to one of our friends, we try to compare our situation to his one to understand what the criteria are for a loan to be granted or not. The example is more tangible than trying to understand and get explanations for the whole ML algorithm. There two main types of example-based explanations are the prototypes and counterfactual explanations. We will see how AE can be considered a specific case of counterfactual explanations.

Prototypes are what you might expect precisely from the meaning of the word: we search for instances looking from the most representative ones. The explanations are then generated by looking at the similarity of other data points to those chosen as prototypes. To avoid generalizations, we don't want; a prototype is often coupled with a "criticism" that, on the contrary, is a data point that is not well represented by any prototype. Prototypes are usually identified through clustering algorithms like k-means.

We mentioned prototypes for completeness, even if we are more interested in counterfactual explanations and their relation with XAI. Counterfactual explanations mean to search for the minimal conditions for the minimal changes to apply to a specific input that would have caused a different decision for that input. In our example, assuming to have a loan that has been accepted, we would search for the smallest change in one or more features to have the loan rejected.

This is again an example-based explanation, but it is different from prototypes. Prototypes are data points that exist in the data set, while a counterfactual example is a data point that is not present in the data set, an event that didn't happen, something on which the ML model has not been trained or tested.

Following this path, we can now understand how this is related to adversarial examples that can be considered a specific type of counterfactual example-based explanation. Do you remember what an AE is? We searched for the minimal changes for a data point to fool a ML model. That is precisely the same kind of reasoning we followed to explain counterfactual explanations. Learning to fool a NN can be considered an XAI method in the sense that is learning what to touch, how to change the features to change the prediction is to learn how the ML model works. As for agnostic methods, considering the ML as a black box, the creation of AE helps us to understand "why" and "how" a specific input received a specific classification and how this input would be sensitive to a small change in the most important features that produced that classification.

We said that differently from prototypes, counterfactual instances do not exist in the dataset. This should be a kind of Deja-vu for the reader because we deep dived counterfactual reasoning from a different angle in Chap. 6 already. It is worth to recall some important topics that we already covered there to have the full picture.

We talked about the ladder of causation that is part of Pearl's seminal work on causation (Pearl & Makenzie, 2019).

We know that counterfactuals are at the top of the ladder as to get there, we need to deal with imaging and retrospective climbing from interpretability to full explainability (remember that we saw in Chap. 6 that the law of physics can be considered a kind of counterfactual assertions).

Doing XAI with counterfactual example-based explanations is a human-friendly method of producing explanations because humans are naturally inclined to make sense of things answering questions like "What if I had acted differently?". The alert about this is that the world in which something went differently does not exist, and so we need to get a full causal model to deal with such a state of things.

Having this background in mind and having identified AE as a specific type of counterfactual example-based explanation, we can look at how to generate counterfactual explanations with the methods we learned in XAI.

We saw in Chap. 4 the power of SHAP method to generate explanations for a single instance. We can adopt the same approach in this context and use SHAP to generate counterfactual explanations.

Remember that the Shapley value Ψ_{ij} for feature "j" and instance "i" is how much the specific feature contributed to the classification of instance "i" if compared to the average prediction of the dataset; so that Shapley values can be used to understand which factors contribute more or against a particular classification.

Following the work of Rathi (2019) we can use this algorithm to generate counterfactuals with SHAP. The idea is to use SHAP to answer P-contrast questions that are of the form 'Why [predicted-class] not [desired class]?'. We want to deep dive for a specific data point; we calculate the Shapley values for every possible target class. The negative Shapley values contribute negatively to the target classification, while positive values do the vice-versa. We can break the P-contrast question into two parts: Why P? and why not Q? We get Shapley values for P and Q classes and use those that work against the classification of the category selected to obtain counterfactual datapoints.

Given the datapoint, we estimate its Shapley values for each of the possible target classes. The negative Shapley values indicate the features that have negatively contributed to the specific class classification and vice-versa.

The algorithm implements this flow: the starting point is to identify the desired class (Q), the predicted class (P), and the data point. Shapley values are calculated for each target class to produce counterfactual explanations. This approach has been tested on IRIS dataset and Wine Quality dataset. As reported in the paper, in the Iris dataset, the answer to the basic question "Why 0, not 1" produced as an explanation that the petal width has strongly influenced the result 0 while the petal length has driven the counterfactual classification with label 1.

In this case, the explanation points out that to change the classification from 0 to 1, the target feature is the petal length. Generally speaking, we narrow down the features that work against the classification of the desired category; we can also get the counterfactual data that are related to contrastive explanations and give real examples of what needs to be changed to achieve a specific output. These data points represent the counterfactual answer to the contrastive query.

It is pretty different from the standard approach to generate AE as a counterfactual because the method we saw relied on a fixed $\in$ to provide a small but widespread perturbation of the features for the single data point. But the root of the problem is the same; we are challenging our methods to understand what makes a flower that specific kind of flower and discover how to change it.

In the following section, we will explore the other direction of the link and defend against AE using XAI (instead of doing XAI using AE) to close the loop.

7.3 Defending Against Adversarial Attacks with XAI

Keeping in mind what we learned about AE, the obvious question that emerges is how to defend against AE making the ML models more robust.

Our scope is to investigate the relation between AE and XAI, and we already saw how AE could be considered a specific type of example-based explanations like counterfactuals. There are several approaches for defenses against AE, and we will focus on the one that uses XAI itself to defend against AE. Just to mention the general approaches, there are four main types:

(1) Data augmentation: this idea is to add AE as part of the training to make the model more robust. In this way, the model is trained against that specific type of AE, but the apparent limitation is that this method would require the knowledge of all the possible attacks to be exhaustive.
(2) Defensive Distillation: in ML literature, Distillation is used as a general method of reducing the size of DNN architectures to decrease the request on computational resources. The high-level idea is to extract information from the original DNN and transfer it to a second DNN of reduced dimensionality. Defensive Distillation is a variation of this method to increase the robustness of a DNN against AE. This is obtained with a 2-phase procedure similar to Distillation but aimed to enhance resilience against perturbations instead of compression of the DNN.
(3) "Detector" subnetwork: in this approach (on detecting adversarial perturbations) the original DNN is not changed, but a small additional detector subnetwork is trained on a binary classification to distinguish original input from the ones containing AE.
(4) Adversarial training: one of the most promising approaches. In this approach, all the AE found are used to augment the trainset. This procedure can be applied recursively, obtaining increasingly robust models to AE attacks.

Using XAI methods to defend from AE is an additional emerging approach that cannot be classified in these four main families of defenses.

The background to understand how XAI can be used is outlined by the work of Ilyas et al. (2019) in which AE is considered as an intrinsic property of the dataset itself. The authors introduce the concept of robust and non-robust features. Non-robust features are features that are highly predictive but are very sensitive to every change in the input. We can consider them as details that would not be generically used by a human being to perform a classification task.

On the contrary, robust features are again highly predictive but are not impacted by small input changes. In the case of a car, we may consider wheels' presence as a robust feature just to give an example. For our purposes, the behavior of robust/non-robust features against small changes of input is fundamental for AE. AEs are crafted searching for non-robust features so that a small change in the input produces a considerable change in the values of these highly predictive features. A direct attack to a robust feature would not be feasible because it would require a more significant change to the input easily discovered by an observer.

Given the above, Fidel et al. (2020) showed how to use SHAP to leverage the difference between robust and non-robust features and defend against AE. While in the previous section, we used SHAP to generate example-based explanations (kind of AE), in this case, use SHAP from the opposite direction that is to make the ML model more resistant against AE. However, it is important to note that XAI methods themselves can be vulnerable to adversarial manipulation. Recent work has demonstrated that post-hoc explanation methods like LIME and SHAP can be systematically fooled by adversarial attacks designed specifically to target the explanation generation process (Slack et al., 2020). This highlights the need for robust explanation methods that are themselves resistant to adversarial perturbations.

Considering that XAI's main objective is to interpret ML models providing the relative importance of the features in determining the output, the hypothesis is that we may apply XAI to discriminate AE from original inputs. The idea is that the classification of a normal input should rely more on robust features if compared to the classification of AE that is likely to rely on non-robust features attacked to change the output classification.

We use SHAP to have a ranking of the relative importance of features to identify AE leveraging the different SHAP signatures.

Figure 7.12 adapted from the work of Fidel et al. (2020), clearly shows the proposed solution. On the left and right parts of the figures are normal examples of cats

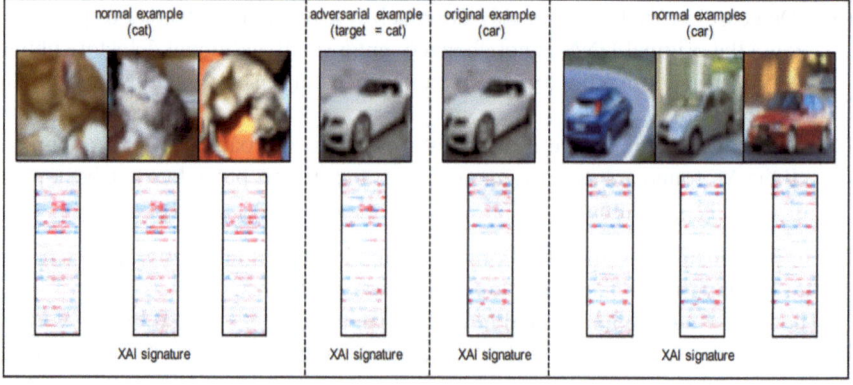

Fig. 7.12 Different images with different SHAP signatures. Normal and hacked examples are compared based on their SHAP signatures (Fidel et al., 2020)

7.3 Defending Against Adversarial Attacks with XAI

and cars. In the middle there is an original input for a cat that is hacked with a PGD L2 attack to create an AE that has as target class a cat.

Up to now, there is nothing new, in the sense that, as expected, the original example of a car and the adversarial example are indistinguishable for a human observer.

But looking below the figures, we can see the SHAP signature is added to each figure so that the SHAP value of each neuron **i** for target class **j** is provided in different colors. The red pixels are the positive contributions toward the target class; the blue ones are the negative contributions with an intensity that depends on the absolute value of the contribution itself.

In particular, the transparent pixels don't contribute to the classification that is what we see in the dedicated section of Chap. 4 for SHAP. This time we can look at these signatures from a different angle. A visual review of the figure is enough to get the main flow of the method.

Each image has a similar SHAP Signature if compared with images of the same type. All the cats on the left and the cars on the right have a matching red pixel pattern (the positive ones).

The standard cars have five evident and strong rows, three rows on the top, one in the middle, and two on the bottom of the SHAP diagram. The cats on the left are in the same situation; the SHAP signatures are different from the cars, but they share the same red pixel pattern in the middle. The exciting part comes if we look at the SHAP signatures of the adversarial example. Out of the original five-strong rows of the original car image, the car AE kept only two rows while we don't see any correspondence with the other 3 SHAP rows.

The cited paper shows how the three rows that disappeared in the AE are the non-robust features: AE attacked the non-robust features highly predictive but very sensible to small

input changes. And the AE also shows the middle pattern of red pixels similar to the cat image. As for the car's original image, the AE transferred only the cat's non-robust features to avoid the robust ones.

The classifier produces the output for the AE based on a mixture of features coming car and cat images, and the method proposed by the paper to recognize the AE is to rely on the SHAP signatures. We just presented the idea of Fidel et al. and the basic concepts. The paper further shows on real data how the method works in a real-case scenario.

For our purpose, it is important to get a deep connection between XAI and feature robustness. We used SHAP to generate AE and the same SHAP method to detect AE narrowing down the non-robust features as SHAP signatures of the AE themselves.

Before closing this chapter, we would further emphasize the theoretical connection between AE and XAI. As we saw, most agnostic methods rely on input gradients to select the most important features for a model and produce explanations. And at the same time, we saw how AE exploit these gradients to understand where the NNs are more vulnerable to small changes in the input. Adversarial gradients are directions where small perturbations generate big output variations, as we understood from our linear regimes discussion. To defend against such perturbation, the idea is to reduce the variation of the outputs around adversarial gradients to smooth the function learned by the NN to generalize better outside the domain of training. But smoothing these gradients means also making them more interpretable in explaining model predictions closing the loop between XAI and AE: robustness of the model against AE helps XAI and XAI helps defending against AE.

This chapter is very rich in content and essential ideas that are not easy to digest. There is a lot of theory. Despite our attempts to make this theory backed up by real-case scenarios and straightforward examples, we are aware that we are moving away toward a more "research" like material. The most important takeaway is the deep relation between XAI and adversarial examples, and the hope is that this relation is clear in terms of the main ideas that can be deep dived if needed in the literature we pointed out.

7.4 Summary

- What are Adversarial Examples
- Generate Adversarial Examples
- Transport AE from a specific ML model to a generic one
- Create universal Adversarial Examples
- Use AE to do XAI
- Use XAI to defend against AE

References

Adadi, A., & Berrada, M. (2018). Peeking inside the black-box: A survey on Explainable Artificial Intelligence (XAI). *IEEE Access, 6*, 52138–52160.

Fidel, G., Bitton, R., & Shabtai, A. (2020, July). When explainability meets adversarial learning: Detecting adversarial examples using SHAP signatures. In *2020 International Joint Conference on Neural Networks (IJCNN)* (pp. 1–8). IEEE.

Foolbox (2017). *Foolbox 2.3.0*. Retrieved from https://foolbox.readthedocs.io/en/v2.3.0/user/examples.html

Goodfellow, I.J., Shlens, J., & Szegedy, C. (2014). *Explaining and harnessing adversarial examples*. arXiv preprint arXiv:1412.6572.

Ilyas, A., Santurkar, S., Tsipras, D., Engstrom, L., Tran, B., & Madry, A. (2019). Adversarial examples are not bugs, they are features. In *Advances in neural information processing systems* (pp. 125–136).

Karpathy, A. (2015). *Breaking linear classifiers on ImageNet*. Retrieved from http://karpathy.github.io/2015/03/30/breaking-convnets/

Moosavi-Dezfooli, S. M., Fawzi, A., Fawzi, O., & Frossard, P. (2017). Universal adversarial perturbations. In *Proceedings of the IEEE conference on computer vision and pattern recognition* (pp. 1765–1773).

Nguyen, A., Yosinski, J., & Clune, J. (2015). Deep neural networks are easily fooled: High confidence predictions for unrecognizable images. In *Proceedings of the IEEE conference on computer vision and pattern recognition* (pp. 427–436).

References

Pearl, J., & Makenzie, D. (2019). *The book of why*. Penguin. eBook Edition.

Rathi, S. (2019). *Generating counterfactual and contrastive explanations using SHAP*. arXiv preprint arXiv:1906.09293.

Slack, D., Hilgard, S., Jia, E., Singh, S., & Lakkaraju, H. (2020, February). Fooling lime and shap: Adversarial attacks on post hoc explanation methods. In *Proceedings of the AAAI/ACM conference on AI, ethics, and society* (pp. 180–186).

Szegedy, C., Zaremba, W., Sutskever, I., Bruna, J., Erhan, D., Goodfellow, I., & Fergus, R. (2013). *Intriguing properties of neural networks*. arXiv preprint arXiv:1312.6199.

Chapter 8
Explainability of Language Models (XAI and LLM)

> *"I meant," said Ipslore bitterly, "what is there in this world that truly makes living worthwhile?"*
> *Death thought about it. "CATS," he said eventually. "CATS ARE NICE."*
>
> —Terry Pratchett

8.1 Introduction

In the increasingly complex landscape of artificial intelligence, Large Language Models (LLMs) are revolutionizing the way we interact with machines. These models, such as BERT, GPT-4, LLaMA, Claude, and PaLM, demonstrate a remarkably natural ability to understand and generate text. However, their intrinsic complexity makes it difficult to comprehend exactly how they make decisions - a crucial aspect for critical applications in medicine, law, and finance.

As these models grow in scale and capability, exceeding billions of parameters, the challenge of explaining their inner workings becomes increasingly pressing. This opacity raises significant concerns: How can we trust systems we don't fully understand? What biases might they encode? How can we ensure their reliability in high-stakes contexts?

The field of Explainable AI (XAI) offers a path forward, providing frameworks and techniques to peek inside these "black boxes." This chapter explores how XAI approaches can be applied to large language models, examining both their theoretical foundations and practical implementations. We will progressively work through the evolution of language models, from recurrent architectures to modern transformers, and explore specialized techniques for making their decisions more transparent and interpretable (Doshi-Velez & Kim, 2017).

The chapter is structured in eight sections:

1. Evolution of Sequential Models: From RNNs to GRUs
2. The Attention Mechanism: A Conceptual Revolution
3. Transformer Architecture: Beyond Sequentiality
4. BERT and Encoder-Based Models: Understanding Context
5. Vision Transformers: From NLP to Computer Vision
6. Multimodal Models: Integrating Text, Images, and Audio
7. Explainability Techniques for Language Models
8. Case Studies and Practical Applications

Through this journey, we aim to bridge the gap between the impressive capabilities of modern language models and our ability to understand, verify, and trust their outputs.

8.2 Evolution of Sequential Models: From RNNs to GRUs

8.2.1 The Challenge of Sequentiality

Imagine reading a book. By the time you reach the middle of a sentence, your brain has already processed all previous words, creating a context that helps you interpret what you're reading. But how can we replicate this capability in a machine learning model?

Traditional feed-forward neural networks have a fundamental limitation: they process each input independently, as if isolated from the rest. It's like trying to understand the meaning of a word without considering the sentence it's in. This limitation becomes particularly problematic for natural language processing, where meaning emerges from the relationships between words across time.

For this reason, Recurrent Neural Networks (RNNs) were introduced in the 1980s. These architectures represented the first serious attempt to endow neural networks with a form of memory appropriate for sequential data.

8.2.2 Recurrent Neural Networks: An Artificial Memory

RNNs represent the first attempt to provide neural networks with a form of "memory." The idea is as simple as it is elegant: each time the network processes a new input, it considers what it has seen previously through a "hidden state" that is continuously updated.

8.2 Evolution of Sequential Models: From RNNs to GRUs

Mathematically, we can describe this process as:

$$h_t = \tanh\left(W_{hh} h_{t-1} + W_{xh} x_t + b_h\right)$$

where h_t is the hidden state at time t, x_t is the current input, and the W matrices are the weights that the network must learn.

This seemingly simple formula conceals a profound insight: the network is effectively creating a compressed representation of everything it has seen up to that point. It's as if it were taking notes while reading, continuously updating them with new information. This approach proved particularly effective for tasks requiring temporal context, such as speech recognition and language modeling.

8.2.3 The Vanishing Gradient Problem

However, traditional RNNs have a fundamental problem that becomes evident when analyzing their behavior during training. When we try to make the network learn long-term dependencies (for example, connecting information distant in the text), the error signal tends to "vanish" during backpropagation.

To understand this phenomenon, imagine repeatedly multiplying a number <1 by itself: the result quickly becomes very small. The same happens with gradients in RNNs: as they propagate backward through many time steps, they become so small that they don't allow effective weight updates. This "vanishing gradient problem" severely limited the practical utility of basic RNNs for capturing long-range dependencies.

Mathematically, the gradient at time t depends on the product of the partial derivatives of all previous hidden states:

$$\frac{\partial L}{\partial h_t} = \prod_{i=t+1}^{T} \frac{\partial h_i}{\partial h_{i-1}}$$

This product tends to zero exponentially with T, making it impossible to learn long-term dependencies.

8.2.4 LSTM: Selective Memory

To solve this problem, in 1997 Hochreiter and Schmidhuber proposed Long Short-Term Memory (LSTM) networks. The key insight is that, instead of passing all information through the same "path," we can create "gates" that decide which information to pass, which to forget, and which to update.

An LSTM unit contains three gates:

1. **The forget gate**: decides what information to discard from the cell state

$$f_t = \sigma\left(W_f\left[h_{t-1}, x_t\right] + b_f\right)$$

2. **The input gate**: decides what new information to add

$$i_t = \sigma\left(W_i\left[h_{t-1}, x_t\right] + b_i\right)$$

$$\tilde{c}_t = \tanh\left(W_c\left[h_{t-1}, x_t\right] + b_c\right)$$

3. **The output gate**: decides which parts of the cell state to output

$$o_t = \sigma\left(W_o\left[h_{t-1}, x_t\right] + b_o\right)$$

The cell state is updated according to:

$$c_t = f_t \odot c_{t-1} + i_t \odot \tilde{c}_t$$
$$h_t = o_t \odot \tanh(c_t)$$

It's as if each LSTM unit were a small processor with its own memory, capable of autonomously deciding what is important to remember and what can be forgotten. This architecture proved remarkably effective, becoming the standard approach for sequence modeling tasks for many years.

8.2.5 GRU: Simplifying Without Losing Power

Gated Recurrent Units (GRUs), introduced by Cho et al. represent an attempt to simplify the LSTM architecture while maintaining its main advantages. The GRU combines the forget gate and the input gate into a single "update gate," and merges the cell state with the hidden state.

$$z_t = \sigma\left(W_z\left[h_{t-1}, x_t\right] + b_z\right)$$

- update gate

$$r_t = \sigma\left(W_r\left[h_{t-1}, x_t\right] + b_r\right)$$

8.2 Evolution of Sequential Models: From RNNs to GRUs

- reset gate

$$\tilde{h}_t = \tanh\left(W\left[r_t \odot h_{t-1}, x_t\right] + b\right)$$

- candidate hidden state

$$h_t = (1 - z_t) \odot h_{t-1} + z_t \odot \tilde{h}_t$$

- new hidden state

Empirical studies have shown that GRUs can match LSTM performance on many tasks while being computationally more efficient. This combination of simplicity and effectiveness made GRUs particularly popular for applications with computational constraints.

8.2.6 Explainability and Interpretation

From an explainability perspective, RNNs and their variants offer several interesting insights:

Hidden State Analysis: We can visualize how the hidden state evolves over time, identifying which patterns the network has learned to recognize.

Gate Analysis: In LSTMs and GRUs (Fig. 8.1), we can analyze gate values to understand which information the network considers important at each moment.

```
def analyze_lstm_gates(model, input_sequence):
    """
    Analyzes the gate values of an LSTM
    for a given input sequence
    """
```

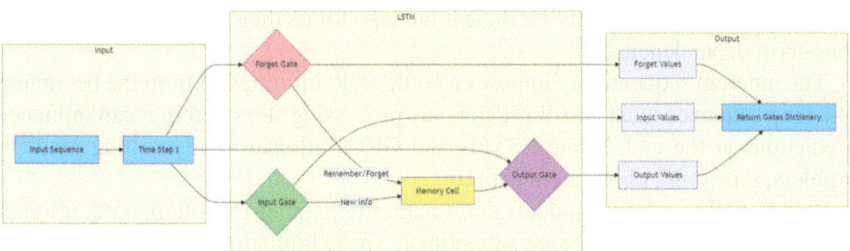

Fig. 8.1 LSTM Gate Analysis Flow Diagram: The process illustrates how an input sequence is processed through the forget gate (pink), input gate (green), and output gate (purple), interacting with the memory cell (yellow) to produce gate activation values. The diagram shows the data flow path through LSTM gates and their interaction with the memory cell state, capturing the temporal dependencies in the sequence

```
    forget_gates = []
    input_gates = []
    output_gates = []
    for t in range(len(input_sequence)):
        # Forward pass up to time t
        output, (h, c) = model(input_sequence[:t+1])
        # Extract gate values
        f = model.lstm.forget_gate(h)
        i = model.lstm.input_gate(h)
        o = model.lstm.output_gate(h)
        forget_gates.append(f.detach())
        input_gates.append(i.detach())
        output_gates.append(o.detach())
    return {
        'forget': forget_gates,
        'input': input_gates,
        'output': output_gates
    }
```

Gradient Analysis: By studying how gradients propagate through the network, we can identify which parts of the input sequence have a greater influence on the final prediction.

Research by Linzen et al. demonstrated that LSTMs can learn hierarchical syntactic dependencies when trained on appropriate tasks, providing insights into how these models capture linguistic structure.

8.2.7 Limitations and Transition to Transformers

Despite the innovations of LSTMs and GRUs, these models maintain a fundamental limitation: they process the input sequentially, one element at a time. This not only makes them computationally inefficient but also limits their ability to capture very long-term dependencies.

The inherent sequentiality imposes a bottleneck: information from the beginning of a long sequence must pass through many processing steps before it can influence predictions at the end. While LSTMs and GRUs mitigate the vanishing gradient problem, they don't eliminate it entirely.

Additionally, recurrent models don't naturally parallelize well, making training on modern GPU/TPU hardware suboptimal. These limitations prompted researchers to seek alternative architectures that could overcome these constraints while preserving or enhancing the ability to model sequential data.

As we will see in the next section, the attention mechanism would represent a conceptual revolution, allowing models to directly consider the relationships between distant elements in the sequence, without having to pass through all intermediate states. This breakthrough, culminating in the Transformer architecture, would set the stage for the dramatic advances in language modeling that followed.

8.3 The Attention Mechanism: A Conceptual Revolution

8.3.1 From Sequential Processing to Attention

In the previous section, we examined how RNNs and their sophisticated variants (LSTM and GRU) process sequences element by element, maintaining a hidden state that acts as memory. This approach, while intuitive, doesn't fully reflect how humans process information. When reading a sentence, we don't strictly process it from left to right maintaining a single memory state: our attention jumps back and forth, focusing on the most relevant elements to understand the meaning (Cheng et al., 2023).

The attention mechanism emerged from this insight: instead of forcing the model to compress all information into a single hidden state, we allow it to "look" directly at all parts of the input when making decisions. This fundamental shift in approach was initially proposed by Bahdanau et al. in the context of neural machine translation and was subsequently refined through multiple research efforts (Xu et al., 2015).

8.3.2 The Intuition Behind Attention

Consider the task of translating the Italian sentence "Il gatto che dorme sul divano è nero" into English. A sequential model would need to memorize the entire sentence before starting to produce the translation. With attention, the model can focus on different parts of the input sentence for each word it generates:

To translate "Il" → "The", the model doesn't need particular attention
To translate "gatto" → "cat", the model focuses mainly on "gatto"
To translate "nero" → "black", the model can look directly at "nero", even though it's the last word in the Italian sentence

This approach mirrors human translation strategies, where we often look at specific words or phrases rather than attempting to memorize the entire source text. Empirical studies have shown that attention mechanisms align well with human gaze patterns during translation tasks (Rikters et al., 2017).

8.3.3 The Mathematics of Attention

The attention mechanism can be described through the metaphor of Query, Key, and Value, terminology borrowed from information retrieval systems:

- **Query (Q):** represents "what we're looking for"
- **Key (K):** represents "where we might find it"
- **Value (V):** represents "what we get"

Mathematically, attention is calculated as:

$$\text{Attention}(Q, K, V) = \text{softmax}\left(\frac{QK^T}{\sqrt{d_k}}\right) V$$

Where:—$Q \in R^{n \times d}$: query matrix—$K \in R^{m \times d}$: key matrix—$V \in R^{m \times d}$: value matrix—d_k: dimension of the keys (the scaling factor $\sqrt{d_k}$ prevents vanishing gradients)

This formulation, known as "scaled dot-product attention," represents the most widely used variant of attention, though alternatives such as additive attention and multiplicative attention exist with their own mathematical formulations.

8.3.3.1 Step-by-Step Analysis

To understand the mechanism more deeply, let's break down the calculation:

1. **Compatibility Score Calculation**:
   - QK^T produces an $n \times m$ matrix of scores
   - Each element (i,j) indicates how compatible query i is with key j
   - This matrix effectively measures the relevance of each input element to each output element

2. **Normalization**:
   - Division by $\sqrt{d_k}$ scales the scores to avoid excessively small gradients during backpropagation
   - The softmax converts scores into weights that sum to 1, creating a probability distribution over input elements

3. **Value Weighting**:
   - The normalized weights are used to calculate a weighted average of the values
   - This produces a context vector that emphasizes the most relevant parts of the input

8.3 The Attention Mechanism: A Conceptual Revolution

```
def scaled_dot_product_attention(query, key, value, mask=None):
    """
    Basic implementation of attention
    Args:
        query: tensor of shape (..., seq_len_q, depth)
        key: tensor of shape (..., seq_len_k, depth)
        value: tensor of shape (..., seq_len_v, depth_v)
        mask: optional boolean tensor
    Returns:
        output: tensor of shape (..., seq_len_q, depth_v)
        attention_weights: tensor of shape (..., seq_len_q, seq_len_k)
    """
    matmul_qk = torch.matmul(query, key.transpose(-2, -1))
    # Scaling
    dk = torch.tensor(key.shape[-1], dtype=torch.float32)
    scaled_attention_logits = matmul_qk / torch.sqrt(dk)
    # Masking (optional)
    if mask is not None:
        scaled_attention_logits += (mask * -1e9)
    # Normalization
    attention_weights = F.softmax(scaled_attention_logits, dim=-1)
    # Output
    output = torch.matmul(attention_weights, value)
    return output, attention_weights
```

Research by Vig (2019) demonstrated that different attention patterns emerge when models process different linguistic phenomena, providing insights into how neural networks capture syntactic and semantic relationships (Fig. 8.2).

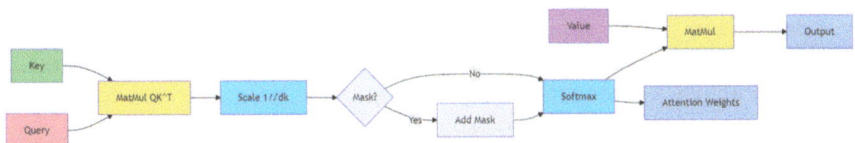

Fig. 8.2 Scaled Dot-Product Attention Flow: Computational pipeline showing matrix operations between Query (Q), Key (K), and Value (V) tensors. The process includes Q-K matrix multiplication, scaling by $1/\sqrt{dk}$, optional masking, softmax normalization, and final value aggregation to produce attention outputs and weights

8.3.4 Multi-Head Attention: Parallel Attention

A single attention head can capture only one type of relationship between elements. Multi-Head Attention, introduced by Vaswani et al., resolves this limitation by allowing the model to look at the sequence from different "perspectives" simultaneously.

The idea is:

1. Create multiple projections of Q, K, and V
2. Calculate attention separately for each head
3. Concatenate the results and project them into the desired space

Mathematically:

$$\text{MultiHead}(Q, K, V) = \text{Concat}(\text{head}_1, \ldots, \text{head}_h) W^O$$

$$\text{where head}_i = \text{Attention}(QW_i^Q, KW_i^K, VW_i^V)$$

This approach allows different heads to specialize in different types of relationships. Empirical studies have shown that attention heads can learn to track syntactic dependencies, coreference relationships, and other linguistic patterns.

```python
class MultiHeadAttention(nn.Module):
    def __init__(self, d_model, num_heads):
        super().__init__()
        assert d_model % num_heads == 0
        self.d_model = d_model
        self.num_heads = num_heads
        self.depth = d_model // num_heads
        # Linear projections for Q, K, V
        self.wq = nn.Linear(d_model, d_model)
        self.wk = nn.Linear(d_model, d_model)
        self.wv = nn.Linear(d_model, d_model)
        self.dense = nn.Linear(d_model, d_model)
    def split_heads(self, x, batch_size):
        x = x.view(batch_size, -1, self.num_heads, self.depth)
        return x.transpose(1, 2)
    def forward(self, query, key, value, mask=None):
        batch_size = query.shape[0]
        # Linear projections and split into heads
        q = self.split_heads(self.wq(query), batch_size)
        k = self.split_heads(self.wk(key), batch_size)
        v = self.split_heads(self.wv(value), batch_size)
        # Scaled attention for each head
```

8.3 The Attention Mechanism: A Conceptual Revolution

```
scaled_attention, attention_weights = scaled_dot_product_
attention(
        q, k, v, mask)
    # Reshape and final projection
    scaled_attention = scaled_attention.transpose(1, 2).
contiguous()
    concat_attention = scaled_attention.view(batch_size, -1,
self.d_model)
    output = self.dense(concat_attention)
    return output, attention_weights
```

Voita et al. (2019) conducted a detailed analysis of multi-head attention, revealing that different heads specialize in different linguistic functions, with some focusing on syntactic relationships and others on semantic connections or positional information (Fig. 8.3).

8.3.5 Self-Attention: Introspective Attention

A particular case of attention is self-attention, where Q, K, and V all derive from the same sequence. This allows the model to discover internal relationships within the sequence itself.

For example, in the sentence "The dog chased the cat because it was hungry," self-attention can help determine that "it" refers to "dog" and not "cat" by analyzing relationships between all words in the sentence. This capability is particularly valuable for tasks requiring contextual understanding, such as coreference resolution and disambiguation (Tenney et al., 2019a, b).

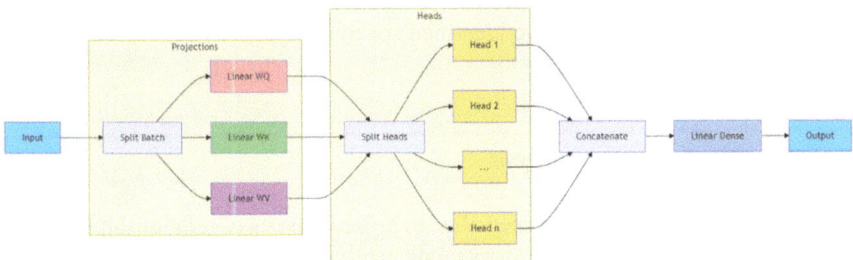

Fig. 8.3 Multi-Head Attention Architecture: Input tensors undergo parallel processing through multiple attention heads. The process includes linear projections (WQ, WK, WV), parallel scaled dot-product attention computation across n heads, concatenation of head outputs, and final linear transformation. Each head operates on a different projection of the input space, enabling the model to attend to different representation subspaces simultaneously

Self-attention operates on the principle that each element in a sequence should be able to attend to all other elements, creating a fully connected graph of relationships. This contrasts with convolutional approaches, which focus on local relationships, and recurrent approaches, which process information sequentially.

8.3.6 Explainability of Attention

Attention offers a unique advantage from an explainability perspective: attention weights can be directly interpreted as measures of relevance. This provides a natural way to understand which input elements influenced each output element.

8.3.6.1 Attention Visualization

```
def visualize_attention(attention_weights, tokens_src,
tokens_tgt):
    """
    Creates a heatmap of attention weights
    """
    plt.figure(figsize=(10, 8))
    sns.heatmap(attention_weights,
                xticklabels=tokens_src,
                yticklabels=tokens_tgt,
                cmap='Blues')
    plt.xlabel('Input Tokens')
    plt.ylabel('Output Tokens')
    plt.title('Attention Weights Visualization')
```

This visualization technique has been widely adopted in the research community to provide insights into model behavior (Vig, 2019). By examining attention patterns (Fig. 8.4), researchers have identified specific linguistic phenomena captured by different models.

8.3.6.2 Attention Head Analysis

Different attention heads often specialize in different types of relationships: - Some heads might focus on syntax—Others might capture semantic relationships (Tenney et al., 2019a, b)—Others still might specialize in specific phenomena such as coreference

8.3 The Attention Mechanism: A Conceptual Revolution

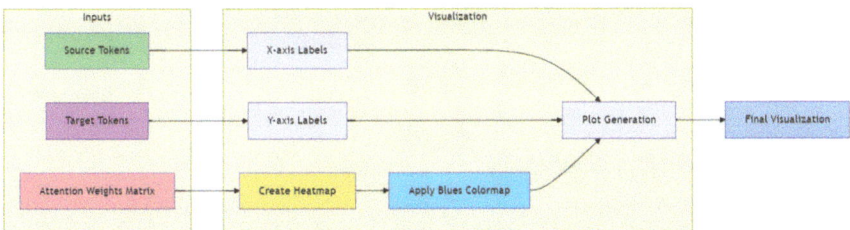

Fig. 8.4 Attention Weight Visualization Pipeline: Transforms attention weight matrices into interpretable heatmaps. The process combines attention weights with source and target token labels, applying a 'Blues' colormap to generate a 2D visualization of attention patterns between input and output sequences

```
def analyze_attention_heads(model, input_text):
    """
    Analyzes the behavior of different attention heads
    """
    # Forward pass
    outputs = model(input_text)
    attention_maps = outputs.attentions
    patterns = []
    for layer_idx, layer_attention in enumerate(attention_maps):
        for head_idx in range(layer_attention.shape[1]):
            head_pattern = classify_attention_pattern(
                layer_attention[:, head_idx, :, :])
            patterns.append({
                'layer': layer_idx,
                'head': head_idx,
                'pattern': head_pattern
            })
    return patterns
```

Research by Belinkov has categorized typical attention patterns into several classes (Fig. 8.5), including diagonal attention (focusing on the token itself), adjacent attention (focusing on neighboring tokens), and global attention (attending broadly across the sequence).

However, caution is warranted when interpreting attention weights as explanations. Jain and Wallace demonstrated that attention weights do not always correlate with feature importance, and alternative attention weights can sometimes produce similar predictions. This has sparked an ongoing debate about the faithfulness of attention as an explanation method.

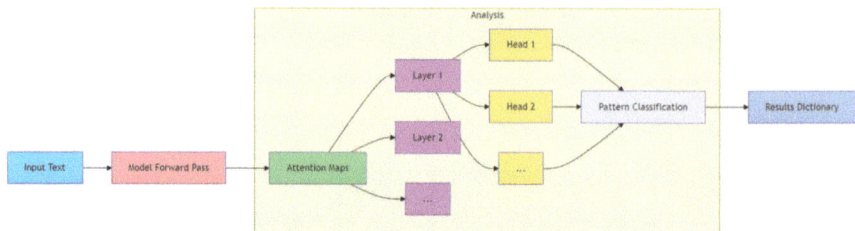

Fig. 8.5 Attention Head Analysis Pipeline: Process flow for examining individual attention heads across model layers. The system processes input text through the model, extracts attention maps, and analyzes each head's attention pattern per layer, producing a structured collection of attention behavior classifications

8.3.7 Limitations and Considerations

Despite its advantages, attention has several limitations:

1. **Quadratic Complexity**: The computational cost grows quadratically with sequence length, limiting applicability to very long sequences (Tay et al., 2020)
2. **Interpretability Not Guaranteed**: Even though attention weights are interpretable, they don't always reflect linguistically significant relationships (Serrano & Smith, 2019)
3. **Pattern Overlap**: With many attention heads, it can be difficult to identify the specific role of each (Voita et al., 2019)
4. **Faithful Attributions**: Attention weights don't necessarily provide faithful explanations of model decisions

Research by Brunner et al. indicates that interpretations based on attention should be treated cautiously, particularly in deep models where the relationship between attention and model outputs becomes increasingly complex.

8.3.8 The Future of Attention

Attention continues to evolve, with several research directions addressing its limitations:

1. **Sparse Attention**: Reduces computational complexity by considering only the most relevant connections
2. **Structured Attention**: Incorporates prior knowledge about data structure (Kim et al., 2017)
3. **Linear Attention**: Reformulations that reduce complexity from quadratic to linear
4. **Faithful Attention**: Mechanisms designed specifically to provide more reliable explanations

Recent work by Tay et al. (2022) surveys the landscape of efficient attention mechanisms, highlighting the trade-offs between computational efficiency and model performance.

In the next section, we'll see how attention becomes the fundamental building block of the Transformer architecture, paving the way for modern Large Language Models.

8.4 Transformer Architecture: Beyond Sequentiality

8.4.1 The Transformer Revolution

In 2017, Vaswani et al. published the seminal paper "Attention is All You Need," introducing an architecture that revolutionized the field of NLP: the Transformer. The fundamental insight was as radical as it was elegant: completely eliminate recurrence and convolution, basing the entire model on the attention mechanism. This architectural innovation addressed the inherent limitations of recurrent models—namely their sequential computation constraints and difficulties in capturing long-range dependencies.

The impact of this work cannot be overstated. Within a remarkably short period, Transformer-based models surpassed the state-of-the-art across virtually all NLP benchmarks (Liu et al., 2021). The architecture's effectiveness stems from its parallel computation capabilities and its ability to model dependencies regardless of their distance in the input sequence.

8.4.2 General Architecture

A complete Transformer consists of two primary components: 1. An encoder that processes the input 2. A decoder that generates the output

Each component comprises multiple identical layers stacked sequentially. The original implementation used 6 layers for both encoder and decoder, though subsequent models have scaled this number significantly, with GPT-3 using 96 layers and PaLM using 118 layers (Chowdhery et al., 2022).

Each layer contains two main sub-components: Multi-Head Attention—Position-wise Feed-Forward Network

These are complemented by residual connections and layer normalization, which facilitate training of deep networks by stabilizing the optimization process.

8.4.3 The Encoder

The encoder transforms the input sequence into a continuous representation that captures contextual relationships. Each encoder layer contains:

1. **Multi-Head Self-Attention**
   - Allows each position to attend to all other positions
   - Uses multiple parallel attention heads
   - Enables the model to jointly attend to information from different representation subspaces

2. **Position-wise Feed-Forward Network**
   - Processes each position independently
   - Consists of two linear transformations with a nonlinearity (originally ReLU, now commonly GELU) in between
   - Can be viewed as a position-wise fully connected feed-forward network

```python
class EncoderLayer(nn.Module):
    def __init__(self, d_model, num_heads, d_ff, dropout=0.1):
        super().__init__()
        self.self_attn = MultiHeadAttention(d_model, num_heads)
        self.feed_forward = nn.Sequential(
            nn.Linear(d_model, d_ff),
            nn.GELU(),
            nn.Linear(d_ff, d_model)
        )
        self.norm1 = nn.LayerNorm(d_model)
        self.norm2 = nn.LayerNorm(d_model)
        self.dropout = nn.Dropout(dropout)
    def forward(self, x, mask=None):
        # Self-Attention with residual connection
        attn_output, _ = self.self_attn(x, x, x, mask)
        x = self.norm1(x + self.dropout(attn_output))
        # Feed-forward with residual connection
        ff_output = self.feed_forward(x)
        x = self.norm2(x + self.dropout(ff_output))
        return x
```

Research by Geva et al. (2021) suggests that the feed-forward layers in Transformers function as key-value memories, storing and retrieving information needed for specific predictions. This provides insights into how these models capture and utilize knowledge (Fig. 8.6).

8.4 Transformer Architecture: Beyond Sequentiality

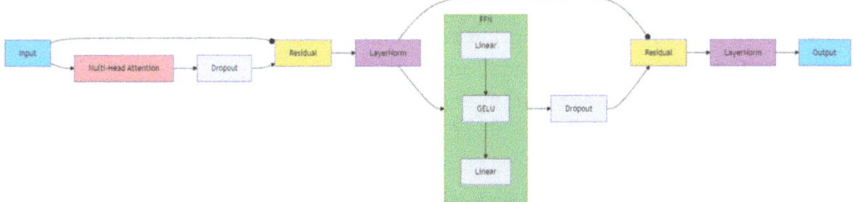

Fig. 8.6 Encoder Layer Architecture: Sequential processing pipeline featuring Multi-Head Attention followed by Feed-Forward Network, with residual connections and layer normalization. Each sublayer (attention and feed-forward) is wrapped with a dropout-residual-normalization sequence, maintaining consistent dimensionality throughout the layer

8.4.4 The Decoder

The decoder generates the output token by token through an autoregressive process. In addition to the encoder components, it adds:

1. **Masked Multi-Head Attention**
   - Prevents attention from looking into the "future"
   - Essential for autoregressive generation
   - Implemented by masking out (setting to $-\infty$) all values in the input of the softmax that correspond to illegal connections

2. **Cross-Attention**
   - Allows the decoder to attend to the encoder's output
   - Fuses input and output information
   - Enables the model to focus on relevant parts of the input sequence when generating each output token

```
class DecoderLayer(nn.Module):
    def __init__(self, d_model, num_heads, d_ff, dropout=0.1):
        super().__init__()
        self.self_attn = MultiHeadAttention(d_model, num_heads)
        self.cross_attn = MultiHeadAttention(d_model, num_heads)
        self.feed_forward = nn.Sequential(
            nn.Linear(d_model, d_ff),
            nn.GELU(),
            nn.Linear(d_ff, d_model)
        )
        self.norm1 = nn.LayerNorm(d_model)
        self.norm2 = nn.LayerNorm(d_model)
        self.norm3 = nn.LayerNorm(d_model)
        self.dropout = nn.Dropout(dropout)
```

```
def
forward(self, x, enc_output, src_mask=None, tgt_mask=None):
    # Masked Self-Attention
    attn1, _ = self.self_attn(x, x, x, tgt_mask)
    x = self.norm1(x + self.dropout(attn1))
    # Cross-Attention with the encoder
    attn2, _ = self.cross_attn(x, enc_output, enc_output, src_mask)
    x = self.norm2(x + self.dropout(attn2))
    # Feed-forward
    ff_output = self.feed_forward(x)
    x = self.norm3(x + self.dropout(ff_output))
    return x
```

Studies by Voita et al. (2019) have shown that different attention heads in the decoder serve distinct purposes: some focus on the current token, others on previous tokens, while others attend to relevant parts of the source sequence (Fig. 8.7).

8.4.5 Key Components for Explainability

8.4.5.1 Positional Encoding

Unlike RNNs, the Transformer has no intrinsic notion of sequential order. This is because self-attention operates on all positions simultaneously. To incorporate positional information, Vaswani et al. introduced positional encodings that are added to the input embeddings:

$$PE(pos, 2i) = \sin\left(pos / 10000^{2i/d_{model}}\right)$$

$$PE(pos, 2i+1) = \cos\left(pos / 10000^{2i/d_{model}}\right)$$

Where *pos* is the position and *i* is the dimension. This formulation has several desirable properties: It allows the model to attend to relative positions, as PE_{pos+k} can be represented as a linear function of PE_{pos}—It has a fixed pattern that extends to sequences longer than those seen during training—It provides a unique encoding for each position

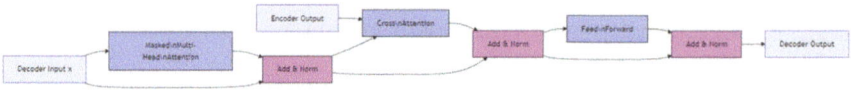

Fig. 8.7 Transformer decoder layer information flow

8.4 Transformer Architecture: Beyond Sequentiality

```
def create_positional_encoding(max_length, d_model):
    pe = torch.zeros(max_length, d_model)
    position = torch.arange(0, max_length, dtype=torch.float).unsqueeze(1)
    div_term = torch.exp(torch.arange(0, d_model, 2).float() *
                         (-math.log(10000.0) / d_model))
    pe[:, 0::2] = torch.sin(position * div_term)
    pe[:, 1::2] = torch.cos(position * div_term)
    return pe
```

Subsequent research has explored learned positional embeddings, relative positional embeddings (Shaw et al., 2018), and rotary position embeddings (Su et al., 2021), each offering different trade-offs between model performance and generalization capabilities.

8.4.6 Layer Normalization

Each sub-layer in the Transformer is followed by a normalization layer:

$$\text{LayerNorm}(x) = \gamma \odot \frac{x - \mu}{\sigma} + \beta$$

Where: μ is the mean of the input features—σ is the standard deviation—γ and β are learnable parameters

Layer normalization serves several important functions: Stabilizes training by reducing internal covariate shift - Makes activations more interpretable by normalizing their distributions—Facilitates analysis of learned patterns by standardizing activation scales

Research by Kovaleva et al. has shown that layer normalization parameters can provide insights into which features the model considers most important for different tasks.

8.4.6.1 Residual Connections

Each sub-layer is wrapped in a residual connection:

$$\text{output} = \text{LayerNorm}(x + \text{Sublayer}(x))$$

These skip connections serve multiple purposes: They facilitate gradient flow during backpropagation, enabling training of very deep networks—They allow for analysis of each layer's contribution to the final representation—They support identification of hierarchical patterns in the model's learned representations

Studies by Veit et al. (2016) and Xiong et al. (2020) demonstrate that residual connections effectively create ensembles of shallower networks, improving both performance and robustness.

8.4.7 Explainability of the Transformer

The Transformer architecture offers several avenues for explainability that weren't available in previous models. Its modular design and attention-based computation provide natural points for intervention and analysis.

8.4.7.1 Attention Matrix Analysis

Research by Vig (2019) and Abnar and Zuidema has introduced visualization techniques that reveal how attention heads capture different linguistic phenomena. These analyses have identified heads that track syntactic dependencies, coreference relations, and semantic associations.

Kovaleva et al. identified five common attention patterns in BERT:

Vertical (attending to special tokens), Diagonal (attending to the current token), Vertical + diagonal, Block (attending to contiguous spans), Heterogeneous (more complex patterns).

These patterns provide insights into how the model processes different types of linguistic information.

8.4.7.2 Attribution Analysis

The Transformer allows tracing the flow of information through the model using gradient-based attribution methods this approach, building on integrated gradients (Sundararajan et al., 2017) and other attribution methods, allows researchers to identify which input tokens most influenced a particular prediction. Atanasova et al. demonstrated that such attributions correlate well with human judgments of importance in text classification tasks.

8.4.7.3 Probing Analysis

Probing techniques examine what different layers of the Transformer have learned:

Extensive research using probing classifiers has revealed a hierarchical organization of linguistic knowledge in Transformer layers (Tenney et al., 2019a, b; Jawahar et al., 2019): Lower layers capture surface features and lexical information—Middle layers specialize in syntactic structures—Higher layers encode semantic relationships and task-specific information.

This organization provides valuable insights into how these models process and represent language (Fig. 8.8).

8.4.8 Variants and Innovations

The core Transformer architecture has spawned numerous variants addressing specific limitations or targeting particular applications.

8.4.8.1 Sparse Transformer

Child et al. introduced the **Sparse Transformer**, which uses factorized attention patterns to reduce computational complexity: - Fixed sparse attention patterns - Complexity reduced from $O(n^2)$ to $O(n\sqrt{n})$ - Maintains performance on long sequences through carefully designed attention masks

This approach demonstrated that full attention is not always necessary, and structured sparsity can preserve model capabilities while improving efficiency.

8.4.8.2 Reformer

Kitaev et al. proposed the **Reformer**, incorporating two key innovations: - Locality-Sensitive Hashing (LSH) attention - Reversible residual connections

These modifications dramatically reduce memory requirements, allowing the model to process sequences of length 64,000 or more. Empirical results showed comparable performance to vanilla Transformers on language modeling and machine translation tasks, but with significantly lower memory usage.

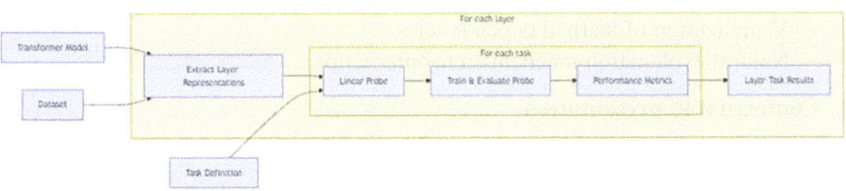

Fig. 8.8 The diagram shows the probing process for transformer layers. Layer representations are extracted from the model and dataset, then analyzed with linear probes for specific tasks. The probes are trained and evaluated to measure how well each layer encodes linguistic features, with results collected across all layers and tasks

8.4.8.3 Linformer

Wang et al. (2020) introduced the **Linformer**, which achieves linear complexity in sequence length through low-rank approximation of the attention matrix: - Projects keys and values to a lower dimension - Reduces complexity from $O(n^2)$ to $O(n)$ - Maintains 95% of full Transformer performance while being much more efficient

Analysis showed that attention matrices in practice have low effective rank, making this approximation theoretically justified.

8.4.8.4 Performer

Choromanski et al. developed the Performer, which uses Fast Attention Via positive Orthogonal Random features (**FAVOR+**): - Approximates softmax attention using random features - Achieves linear complexity in both time and space - Provides unbiased estimation of attention

Benchmarks demonstrated that Performer models can scale to sequences of length 1 million while maintaining competitive performance on language modeling tasks.

8.4.9 Explainability Considerations

The Transformer architecture offers several unique opportunities for interpretability that weren't available in previous models:

1. **Modularity**

   - Well-defined and separable components
   - Possibility to analyze each part independently
   - Clear information flow through the model

2. **Attention Matrix**

   - Direct interpretability of attention patterns
   - Visualization of learned dependencies
   - Natural explanation mechanism for predictions

3. **Contextual Representations**

   - Analysis of how context influences representations
   - Study of disambiguation capabilities
   - Evaluation of semantic understanding (Lu et al., 2019)

However, as noted by Jain and Wallace and Serrano and Smith (2019), attention weights should not be uncritically interpreted as explanations. The complex interplay between attention and feed-forward layers means that attention alone does not provide a complete picture of the model's decision-making process.

Research by Brunner et al. suggests that interpretations should consider both attention patterns and value transformations, as attention weights alone may not faithfully reflect the model's internal reasoning.

In the next section, we'll see how these characteristics have been leveraged in BERT and encoder-based models, opening new possibilities for language model interpretability while addressing some of the challenges identified above.

8.5 BERT and Encoder-Based Models: Understanding Context

8.5.1 The BERT Innovation

BERT (Bidirectional Encoder Representations from Transformers) marked a watershed moment in NLP research when it was introduced by Devlin et al. Its main innovation lies in how it's pretrained: instead of predicting the next word as previous autoregressive models did, BERT learns to "understand" language through two parallel self-supervised tasks:

Masked Language Modeling (MLM)
Next Sentence Prediction (NSP)

This approach enabled BERT to capture bidirectional contextual representations, overcoming the inherent limitations of unidirectional models like GPT (Radford et al., 2021). By considering both left and right context simultaneously, BERT achieved state-of-the-art results across a wide range of NLP tasks, demonstrating remarkable transfer learning capabilities (Wang et al., 2019a).

The conceptual shift from directional language modeling to masked language modeling fundamentally changed how we train language models. As Liu et al. noted, this bidirectional context is essential for tasks requiring deep semantic understanding, as natural language comprehension often depends on integrating information from both directions.

8.5.2 Architecture and Pre-training

BERT adopts the Transformer encoder architecture described in Section 3, with important modifications to support its pretraining objectives. The base version contains 12 Transformer layers with 768-dimensional embeddings and 12 attention heads (110M parameters), while the large version scales to 24 layers, 1024-dimensional embeddings, and 16 attention heads (340M parameters).

Each input sequence begins with a special (CLS] token, whose final representation serves as an aggregate sequence representation for classification tasks. Sentence pairs are separated by a [SEP] token and assigned segment embeddings to distinguish between them.

8.5.2.1 Masked Language Modeling

MLM trains BERT to predict masked tokens based on bidirectional context. Unlike traditional language models that predict the next token based on previous tokens, MLM masks a percentage of input tokens (typically 15%) and trains the model to reconstruct them:

Input: "The [MASK] jumps over the [MASK]" Target: "The cat jumps over the fence"

As Clark et al. demonstrated, this masking strategy creates a challenging learning objective that forces the model to develop robust representations by integrating context from both directions. The random replacement component (10% of masked tokens) helps mitigate the mismatch between pretraining and fine-tuning, as the [MASK] token doesn't appear during the latter (Fig. 8.9).

8.5.2.2 Next Sentence Prediction

NSP trains BERT to understand relationships between sentences, an important capability for tasks like question answering and natural language inference:

Input: "[CLS] The man went to the store. [SEP] He bought a gallon of milk. [SEP]"
  Target: IsNext
Input: "[CLS] The man went to the store. [SEP] Penguins cannot fly. [SEP]"
  Target: NotNext

While NSP was initially considered important, subsequent research by Liu et al. and Joshi et al. questioned its utility, suggesting that sentence coherence might be learned implicitly through MLM. Later models like RoBERTa and ALBERT modified or replaced NSP with alternative objectives.

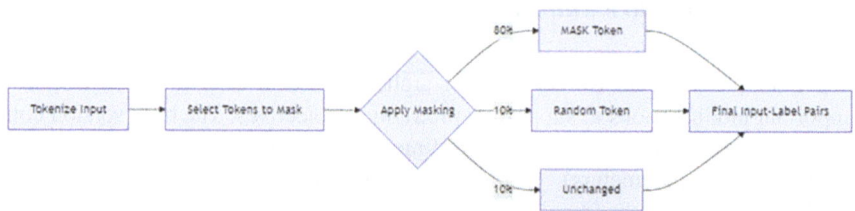

Fig. 8.9 The diagram shows the simplified MLM process: tokenized input text has tokens selected for masking (15%), which are then replaced with MASK tokens (80%), random tokens (10%), or left unchanged (10%), creating the input-label pairs for training

8.5.3 BERT Explainability

The bidirectional and contextual nature of BERT representations offers unique opportunities for explainability. Several approaches have been developed to understand what BERT learns and how it makes decisions.

8.5.3.1 Contextual Embedding Analysis

BERT generates contextual representations of words, where the same word can have different embeddings depending on its context. This property enables disambiguation and nuanced semantic understanding.

Research by Ethayarajh measured the contextuality of word representations in BERT, finding that representations become increasingly context-specific in higher layers. This analysis revealed that BERT's top layers create highly context-specific representations while preserving similarity for semantically related usages.

Visualizing these contextual embeddings using techniques like t-SNE (van der Maaten & Hinton, 2008) or UMAP (McInnes et al., 2018) can reveal semantic clusters and disambiguation capabilities. Coenen et al. demonstrated that BERT's representations organize into interpretable semantic subspaces that capture fine-grained linguistic properties.

8.5.3.2 Knowledge Probing

Probing classifiers can be used to investigate what linguistic knowledge BERT acquires during pretraining.

Extensive research using such probing techniques has revealed BERT's linguistic capabilities across multiple dimensions:

1. **Syntactic Knowledge**: Tenney et al. (2019a) found that BERT encodes syntactic structure hierarchically, with different layers specializing in different aspects of syntax. Parts-of-speech are captured in lower layers, while syntactic dependencies emerge in middle layers.
2. **Semantic Knowledge**: Semantic roles and coreference relationships tend to be captured in higher layers (Tenney et al., 2019b), demonstrating a progression from surface features to deeper semantic understanding.
3. **Factual Knowledge**: Petroni et al. (2019) demonstrated that BERT stores factual knowledge in its parameters, allowing it to answer simple factual queries without explicit fine-tuning. This suggests that large pretrained models may serve as implicit knowledge bases.
4. **Linguistic Hierarchies**: Jawahar et al. (2019) identified a clear hierarchy in BERT's representations: surface features in lower layers, syntactic features in middle layers, and semantic features in higher layers.

8.5.3.3 Attention Pattern Analysis

Analyzing attention patterns in BERT can reveal how the model processes different linguistic phenomena

Clark et al. conducted a comprehensive analysis of BERT's attention heads, finding that different heads specialize in different linguistic phenomena: Some heads focus on syntactic dependencies—Others track lexical patterns—Several heads appear to handle coreference resolution

Visualization tools like **BertViz** (Vig, 2019) enable interactive exploration of attention patterns, helping researchers identify how BERT processes complex linguistic structures.

Voita et al. (2019) found that heads in the same layer often serve complementary functions, suggesting that BERT's architecture facilitates specialization while maintaining integration of different linguistic aspects.

8.5.4 Evolution of Encoder-Based Models

Building on BERT's success, subsequent research has introduced numerous improvements to encoder-based models, addressing limitations and enhancing capabilities.

8.5.4.1 RoBERTa: Optimizing Pretraining

Liu et al. demonstrated that BERT was significantly undertrained. Their optimized version, RoBERTa, incorporated several key modifications:—Removal of the NSP objective - Dynamic masking instead of static masking—Larger batch sizes and more training data - Longer training

These changes led to substantial performance improvements across benchmarks, challenging the need for some of BERT's original design choices. The success of RoBERTa highlighted the importance of training methodology over architectural innovation.

Analysis by Liu et al. revealed that dynamic masking produces more robust representations by exposing the model to a greater variety of maskings during training, preventing overfitting to specific mask patterns.

8.5.4.2 ALBERT: Efficiency Through Parameter Sharing

Lan et al. introduced ALBERT (A Lite BERT), which achieved competitive performance with significantly fewer parameters through three key techniques: Factorized embedding parameterization—Cross-layer parameter sharing—Sentence-order prediction (SOP) instead of NSP

Lan et al. found that parameter sharing creates smooth transitions between layers with gradual representation changes, offering unique opportunities for analyzing how representations evolve through the network. Their SOP task also proved more effective than NSP for learning inter-sentence coherence.

8.5.4.3 DeBERTa: Disentangled Attention

He et al. introduced DeBERTa (Decoding-enhanced BERT with disentangled attention), which improves on BERT through:—Disentangled attention mechanism—Enhanced mask decoder—Virtual adversarial training.

The disentangled attention mechanism separates word content and position information, allowing the model to better capture syntactic and semantic relationships:

Analysis by He et al. showed that disentangled attention helps the model better capture both local and global dependencies, improving performance on tasks requiring fine-grained linguistic understanding.

8.5.5 Applications to Explainability

Encoder-based models like BERT offer powerful tools for explaining model behavior and linguistic phenomena.

8.5.5.1 Probing for Specialized Knowledge

Research by Rogers et al. synthesized dozens of studies probing BERT, revealing that: BERT captures subject-verb agreement and syntactic hierarchies - Middle layers are most effective for syntax transfer—Top layers specialize for specific downstream tasks—The model encodes significant world knowledge, though with limitations.

Specialized domains like medicine, law, and science have been explored through domain-specific BERT variants, demonstrating how pretraining adaptations enhance specialized knowledge representation.

8.5.6 Comparative Model Analysis

Benchmarking efforts like GLUE (Wang et al., 2019a), SuperGLUE (Wang et al., 2019b), and XGLUE (Liang et al., 2021) have facilitated systematic comparison of encoder models. These evaluations reveal that while newer models generally outperform older ones, different architectures exhibit distinct strengths for specific linguistic phenomena.

Talmor et al. developed targeted tests to assess specific reasoning capabilities, revealing that encoder models struggle with compositional reasoning despite strong performance on standard benchmarks.

8.5.7 Explainability for End-Users

Research on human-centered explanations by Ribeiro et al. (2016, 2018) and Lundberg and Lee (2017) has shown that effective explanations should be:

- Faithful to the model's decision process
- Comprehensible to nontechnical users
- Actionable for identifying and addressing issues

Arya et al. developed the AI Explainability 360 toolkit, which integrates various explanation methods tailored to different user needs, demonstrating how technical attributions can be translated into understandable explanations (Fig. 8.10).

8.6 Challenges and Future Directions

Research on encoder-based models has identified several ongoing challenges and promising directions for future work:

1. **Interpretability vs. Performance**: Larger models generally achieve better performance but pose greater interpretability challenges. Techniques like distillation offer potential compromises by creating smaller, more interpretable models that retain most capabilities.

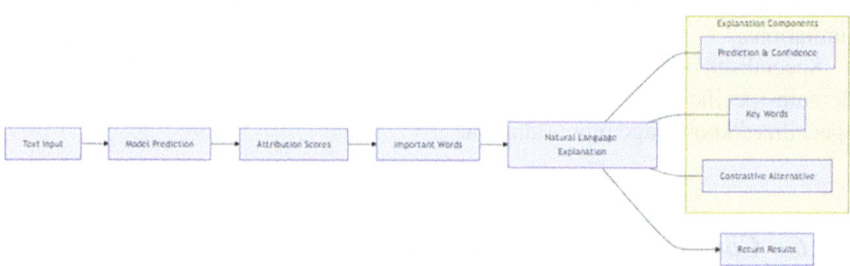

Fig. 8.10 The diagram shows how model predictions are converted into user-friendly explanations. Text input goes through model prediction, then attribution scores identify important words that influence the result. These are transformed into natural language explanations that include the prediction with confidence score, key influential words, and alternative classifications when applicable

2. **Bias and Fairness**: Encoder models can propagate and amplify societal biases present in training data. Approaches like controlled generation (Dathathri et al., 2020) and debiasing (Liang et al., 2021) are being developed to address these concerns.
3. **Knowledge Integration**: Current models struggle to integrate external knowledge consistently (Poerner et al., 2020). Approaches like knowledge graph augmentation and retrieval-augmented generation offer potential solutions.

As we move toward increasingly powerful language models, explanation methods must evolve to maintain transparency and trustworthiness. Emerging approaches like causal tracing (Vig et al., 2020) and circuit discovery (Wang et al., 2022a, b) promise deeper insights into model behavior beyond attention-based explanations.

In the next section, we'll explore how the Transformer architecture has been adapted for computer vision, creating Vision Transformers that extend many of these explainability techniques to the visual domain.

8.7 Vision Transformers: From NLP to Computer Vision

8.7.1 The Vision Transformer Revolution

The Vision Transformer (ViT) introduced by Dosovitskiy et al. (2021) represents a paradigm shift in computer vision, challenging the two-decade dominance of convolutional architectures. ViT demonstrated that a pure Transformer architecture could match or exceed state-of-the-art CNNs in image classification tasks, challenging long-held assumptions about necessary inductive biases in visual perception.

The fundamental insight behind ViT is remarkably straightforward: treat an image as a sequence of patches, analogous to how a traditional Transformer treats a sequence of tokens. This cross-domain transfer of architectural principles opened new avenues for exploring model explainability, as techniques developed for language models could now be applied to vision tasks.

8.7.2 From Images to Patches

The first step in a Vision Transformer is to convert a 2D image into a 1D sequence of embeddings by:

1. Dividing the image into fixed-size patches (typically 16×16 pixels)
2. Flattening each patch into a vector
3. Linearly projecting each vector to the model dimension
4. Adding positional embeddings to retain spatial information

These patches function as "visual words," enabling the analysis of the model's attentional patterns. Research by Chefer et al. (2021) demonstrated that analyzing attention patterns in ViT can produce interpretable attention maps highlighting semantically meaningful regions.

8.7.3 Key Differences from CNNs

8.7.3.1 Global Receptive Field

Unlike CNNs, where the receptive field grows gradually with depth, ViT has access to the entire image immediately. This global receptive field means that attribution methods need to account for global dependencies from the start. Conversely, it enables more direct attribution of predictions to source image regions through attention weights.

8.7.3.2 Inductive Bias

ViT has fewer inductive biases compared to CNNs. While CNNs build in assumptions about locality, translation equivariance, and hierarchical processing, ViT largely learns these properties from data. This architectural difference has several consequences:

1. Data Efficiency: ViT requires more training data to achieve comparable performance to CNNs when trained from scratch.
2. Flexibility in Pattern Learning: With fewer architectural constraints, ViT can potentially learn patterns that CNNs might struggle to capture.
3. Nonlocal Relationships: ViT can more easily model long-range dependencies in images.

This reduced inductive bias is a double-edged sword for explainability: while it potentially allows the model to learn more diverse patterns, it also makes the model's reasoning less constrained and potentially harder to interpret.

8.7.4 Explainability of ViT

8.7.4.1 Attention Map Analysis

Research by Abnar and Zuidema and Chefer et al. (2021) has developed specialized techniques for visualizing ViT attention, revealing that:

1. Different heads specialize in attending to distinct image features

2. Early layers tend to attend locally, while deeper layers develop more semantic attention patterns
3. Class token ([CLS]) attention often highlights discriminative regions for classification

Caron et al. (2021) demonstrated that self-supervised ViTs learn to attend to semantically meaningful object parts without explicit supervision, suggesting that attention mechanisms naturally discover important visual structures.

8.7.4.2 Feature Attribution

Chefer et al. (2021) adapted gradient-weighted attention methods for ViTs, demonstrating that attention-based explanations in ViTs often provide more semantically meaningful visualizations than traditional gradient-based methods for CNNs.

The attribution process can be visualized by:

1. Extracting attention weights from different layers and heads
2. Weighting these attentions based on their importance to the final prediction
3. Creating heatmaps that highlight regions contributing most to the classification decision

8.7.4.3 Layer-wise Analysis

Research by Raghu et al. (2021) compared layer representations between ViTs and CNNs, finding significant differences:

1. ViT representations diverge more from the input in early layers
2. ViT develops global semantic features earlier in the network
3. ViT representations exhibit more uniform change across layers compared to the hierarchical pattern in CNNs

These findings suggest that ViTs process information differently than CNNs, with implications for how we interpret their internal representations and explain their decisions.

8.7.5 Variants and Innovations

Several variants of Vision Transformers have been developed to address limitations and enhance performance:

8.7.5.1 DeiT: Data-efficient ViT

Touvron et al. (2021) introduced DeiT (Data-efficient image Transformer), which uses knowledge distillation to improve training efficiency. This approach enables ViTs to achieve strong performance with less data by leveraging a teacher model (typically a CNN).

8.7.5.2 Swin Transformer

Liu et al. (2021) introduced the Swin (Shifted window) Transformer, which incorporates hierarchical feature maps and local attention. By computing self-attention within local windows and introducing shifted window partitioning, Swin reduces computational complexity while enabling cross-window connections.

From an explainability perspective, Swin's hierarchical design offers unique opportunities for multi-scale attention visualization, potentially revealing how features are composed from local to global representations.

8.7.6 Applications to Explainable Computer Vision

Vision Transformers have been successfully applied to explainable computer vision in several domains:

1. Medical Imaging: ViTs provide naturally interpretable attention maps for highlighting regions of interest in medical images, helping doctors understand model decisions.
2. Autonomous Driving: Attention mechanisms in ViTs can reveal which parts of a scene most influence driving decisions, enhancing safety verification.
3. Visual Question Answering: Cross-attention between image regions and question words creates transparent connections between visual elements and linguistic concepts.

8.7.7 Challenges and Future Directions

Several challenges remain in the explainability of Vision Transformers:

1. Attribution Complexity: The global attention mechanism makes attribution more complex, as predictions may depend on interactions between distant patches.
2. Model Size and Interpretability: As ViTs scale to billions of parameters, traditional attribution methods become less effective, necessitating new approaches for large-scale models.

3. Multimodal Integration: As vision-language models like CLIP gain prominence, understanding cross-modal interactions presents new explainability challenges.

Future research directions include:

1. Developing attribution methods specifically designed for Transformer attention patterns
2. Creating interactive visualization tools that leverage the natural interpretability of attention
3. Establishing benchmarks for evaluating explanations in vision tasks
4. Exploring causal reasoning approaches to identify critical factors in ViT decisions

The transition from CNNs to Transformer-based architectures in computer vision represents not only a shift in model design but also in how we approach model explainability. By leveraging the inherent transparency of attention mechanisms while addressing the unique challenges of vision tasks, researchers can develop more interpretable and trustworthy vision systems.

8.8 Multimodal Models: Integrating Text, Images, and Audio

8.8.1 The Emergence of Multimodal Architectures

Multimodal models represent the next evolutionary step in artificial intelligence, enabling systems to process, understand, and generate content across multiple modalities simultaneously. By integrating text, images, audio, and in some cases video, these models more closely mirror human cognitive capabilities, which naturally synthesize information from different sensory channels. This integration, however, introduces new challenges for explainability: how can we understand decisions that involve cross-modal reasoning?

The development of effective multimodal architectures has been accelerated by the success of Transformer-based models in individual domains. As Transformers demonstrated strong performance in both language and vision (Dosovitskiy et al., 2021) independently, researchers began exploring architectures that could leverage the same attention-based mechanisms for cross-modal integration (Lu et al., 2019; Tan & Bansal, 2019).

Recent years have seen remarkable progress in multimodal learning, with models capable of sophisticated cross-modal tasks such as: Generating images from text descriptions (Ramesh et al., 2021)—Answering questions about images (Antol et al., 2015)—Creating captions for images and videos (Xu et al., 2015)—Retrieving relevant images based on text queries (Radford et al., 2021)—Producing speech from text with appropriate prosody (Wang et al., 2021)

This section examines the principal architectures for multimodal learning and the unique challenges they present for explainability research.

8.8.1.1 Principal Multimodal Architectures

CLIP: Contrastive Language-Image Pretraining

CLIP, introduced by Radford et al. (2021), represents a breakthrough in vision-language models through its innovative contrastive learning approach. Rather than training on curated image-caption pairs, CLIP learns from 400 million image-text pairs collected from the internet, developing robust representations that transfer well to numerous downstream tasks.

The architecture consists of two parallel encoders: 1. A text encoder (Transformer-based) 2. An image encoder (either ResNet or Vision Transformer)

CLIP is trained to maximize agreement between matching image-text pairs while minimizing it for nonmatching pairs. This is accomplished through a contrastive loss function:

$$\mathcal{L} = -\log \frac{\exp(sim(i_m, t_m)/\tau)}{\sum_{n=1}^{N} \exp(sim(i_m, t_n)/\tau)}$$

where $sim(i, t)$ represents the cosine similarity between image embedding i and text embedding t, τ is a temperature parameter, and the subscripts m and n index the batch of N image-text pairs.

This contrastive approach creates a joint embedding space where semantically similar images and texts are positioned close to each other. As Radford et al. (2021) demonstrated, this enables zero-shot transfer to various vision tasks simply by providing appropriate text prompts.

From an explainability perspective, CLIP presents both opportunities and challenges. The shared embedding space facilitates attribution of predictions to specific textual concepts, but the massive scale of pretraining data makes it difficult to trace influences precisely (Goh et al., 2021).

DALL-E and Diffusion Models

DALL-E (Ramesh et al., 2021) and subsequent diffusion-based models like DALL-E 2 (Ramesh et al., 2022) and Stable Diffusion (Rombach et al., 2022) demonstrate remarkable capabilities in text-to-image generation. These models take text prompts as input and generate corresponding images, showing impressive semantic understanding and compositional capabilities.

DALL-E employs a two-stage approach: 1. A transformer that autoregressively models the joint distribution over text and image tokens 2. A CLIP-like model that helps rank and filter generated images

8.8 Multimodal Models: Integrating Text, Images, and Audio

Diffusion models, which have largely superseded autoregressive approaches, work by gradually denoising a random Gaussian distribution guided by the conditioned text embedding. The denoising process can be formulated as:

$$x_{t-1} = \frac{1}{\sqrt{\alpha_t}} \left(x_t - \frac{1-\alpha_t}{\sqrt{1-\bar{\alpha}_t}} \epsilon_\theta \left(x_t, t, c \right) \right) + \sigma_t z$$

where x_t is the noisy image at timestep t, α_t and $\bar{\alpha}_t$ are noise scheduling parameters, ϵ_θ is the learned noise prediction network conditioned on context c (text embedding), and z is standard Gaussian noise.

The generation process in these models presents unique opportunities for explainability. As Dhariwal and Nichol (2021) demonstrated, the step-by-step denoising process allows for visualizing how different text elements influence the emerging image. This progressive generation naturally lends itself to attribution analysis, as the impact of different prompt components can be isolated at various stages of the denoising process.

Research by Chefer et al. (2023) has shown that analyzing cross-attention maps between text and latent image representations can reveal which words most strongly influence specific regions in the generated image, providing a natural explanation mechanism.

Multimodal Transformers

Several architectures extend the Transformer framework to handle multiple modalities simultaneously. Notable examples include:

- **ViLBERT** (Lu et al., 2019): Uses two stream-specific Transformers with co-attentional layers that enable information exchange between modalities.
- **LXMERT** (Tan & Bansal, 2019): Employs three encoders (object, language, and cross-modality) to create rich representations for vision-language reasoning.
- **VL-BERT** (Su et al., 2020): A single-stream architecture that processes visual and linguistic inputs in a unified Transformer, with visual features embedded alongside word tokens.
- **FLAVA** (Singh et al., 2022): A foundation language and vision model that unifies unimodal and multimodal objectives, supporting both contrastive and generative learning.

These architectures typically enhance the standard Transformer with specialized components for cross-modal fusion:

1. **Cross-attention mechanisms**: Allow tokens in one modality to attend to tokens in another
2. **Modality-specific encoders**: Process each input type with appropriate inductive biases

3. **Fusion layers**: Combine representations from different modalities

From an explainability perspective, cross-attention mechanisms are particularly valuable. As demonstrated by Hendricks et al. (2021), these attention patterns can be visualized to reveal which elements in one modality the model associates with elements in another. For instance, in a visual question answering system, cross-attention can show which image regions the model attends to when processing specific words in the question.

Audio-Visual Models

Audio-visual models integrate auditory and visual information, addressing tasks such as audio-visual speech recognition, sound localization, and audiovisual question answering.

Architectures like Audio-Visual BERT (Oncescu et al., 2021) and AV-HuBERT (Shi et al., 2022) extend the Transformer framework to process synchronized audio and visual streams. These models typically use:

1. Modality-specific feature extractors (e.g., CNN for video, wav2vec for audio)
2. Temporal alignment mechanisms to synchronize features
3. Cross-modal attention to integrate information across modalities

The temporal nature of audio-visual data presents unique opportunities for causal analysis. As demonstrated by Chattopadhyay et al. (2022), temporal perturbation studies can reveal which moments in a video or audio stream most influence the model's decisions, providing insights into how the model integrates information over time.

8.8.2 *Explainability Techniques for Multimodal Models*

8.8.2.1 Cross-Modal Attention Analysis

Cross-modal attention serves as a natural explanation mechanism in multimodal models, revealing how information flows between modalities (Fig. 8.11):

Research by Yuksekgonul et al. (2022) developed attention-based explanation methods specifically for multimodal transformers (Fig. 8.12), showing that cross-attention maps can identify which image regions are most relevant for answering specific questions.

The work of Frank et al. (2021) demonstrated that vision-language models develop specialized attention heads that focus on different aspects of cross-modal integration, such as: Object grounding heads that link nouns to visual regions— Attribute heads that connect adjectives to visual properties - Spatial relation heads that process positional language

8.8 Multimodal Models: Integrating Text, Images, and Audio

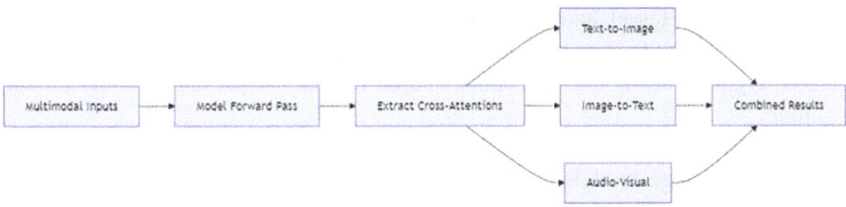

Fig. 8.11 The diagram shows the flow of analyzing attention patterns between different modalities. Multimodal inputs are processed through the model to extract cross-attention maps, which are then analyzed for three directions: text-to-image, image-to-text, and audio-visual interactions. The results are combined into a comprehensive cross-modal attention analysis

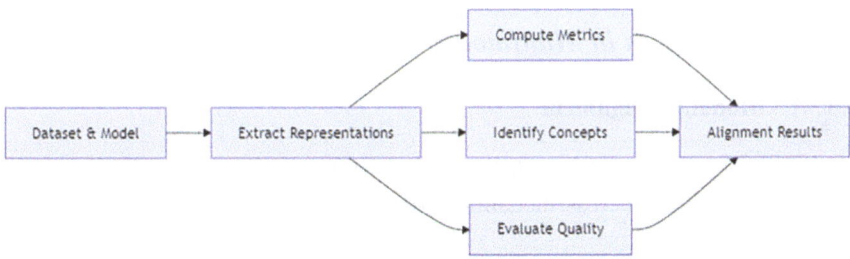

Fig. 8.12 The diagram shows the process of analyzing alignment between different modalities. Starting with a multimodal dataset and model, the system extracts representations for each modality, then performs three parallel analyses: computing alignment metrics (correlation, similarity, mutual information), identifying aligned concepts across modalities, and evaluating overall alignment quality. The results are combined into a comprehensive alignment analysis

8.8.2.2 Multimodal Feature Attribution

Attribution methods for multimodal models must account for the different nature of each modality and their interactions.

Singh et al. (2020) developed a multimodal explanation framework that generates attributions across different modalities while preserving their interactions. Their approach highlights how features in one modality can influence the importance of features in another, capturing complex interdependencies.

For text-to-image generation models, attribution becomes particularly challenging as the output space is high-dimensional and the generation process is stochastic. Hertz et al. (2022) addressed this by analyzing the prompt embedding space and identifying directions corresponding to semantic concepts, allowing for targeted manipulations that reveal how different text elements influence the generated image.

8.8.2.3 Probing Multimodal Representations

Probing classifiers can reveal what information is encoded in multimodal representations.

Research by Hessel and Lee (2020) employed probing to investigate what information is captured in multimodal representations, finding that these models learn emergent multimodal concepts that aren't explicitly present in either modality alone.

Cao et al. (2022) developed a comprehensive probing framework specifically for vision-language models, evaluating their understanding of various linguistic phenomena (e.g., negation, quantification) in visual contexts. Their findings revealed that while these models excel at object recognition and attribute binding, they struggle with more complex reasoning such as counting and spatial relationships.

8.8.3 Challenges in Multimodal Explainability

8.8.3.1 Modality Alignment

A fundamental challenge in multimodal explainability is understanding how models align representations across modalities. Techniques for analyzing this alignment include Fig. 8.12):

Research by Liang et al. (2022) developed methods to measure and visualize alignment between visual and linguistic representations in multimodal models. Their approach identified both well-aligned concepts (typically concrete objects) and misaligned concepts (often abstract ideas or relations), providing insights into model capabilities and limitations.

Cross-modal retrieval performance serves as a natural metric for alignment quality. As demonstrated by Radford et al. (2021), analyzing retrieval errors can reveal systematic misalignments between modalities, such as visual ambiguity or linguistic polysemy.

8.8.3.2 Emergent Capabilities

Large multimodal models often exhibit emergent capabilities that aren't easily explained by examining individual components.

Research by Alayrac et al. (2022) on Flamingo models demonstrated that multimodal capabilities often emerge discontinuously at certain scale thresholds, challenging simple explanations based on architectural design alone. Their analysis showed that while some capabilities (like object recognition) scale gradually, others (like few-shot learning) appear suddenly at specific model sizes.

Subramanian et al. (2022) developed methods to identify and characterize emergent multimodal reasoning abilities, using carefully designed probes to test whether models truly understand cross-modal relationships or simply exploit statistical patterns in the data.

8.8.3.3 Consistency Across Modalities

Ensuring consistent explanations across different modalities presents a significant challenge.

Research by Fazelpour and Danks (2021) highlighted the challenge of ensuring that explanations across modalities form a coherent whole rather than contradicting each other. They proposed a framework for evaluating explanation consistency based on causal models of multimodal integration.

Wu et al. (2022) developed consistency metrics specifically for multimodal explanations, measuring whether feature attributions across modalities agree on which concepts are important for a given prediction. Their approach identified cases where models rely primarily on one modality while nominally considering multiple inputs.

8.8.4 Future Directions in Multimodal Explainability

8.8.4.1 Explanation Adaptation

Future research will likely focus on adapting explanations to the most appropriate modality for human understanding.

Research by Broekens et al. argued that effective explanations should leverage the strengths of each modality: visual for spatial relationships, linguistic for abstract concepts, and audio for temporal patterns. Their framework dynamically selects explanation modalities based on both content properties and user preferences.

8.8.4.2 Interactive Multimodal Explanations

Interactive approaches allow users to explore model decisions across modalities.

Recent work by Hüllermeier and Waegeman (2021) proposed interactive explanation frameworks that allow users to navigate between modalities, exploring how concepts in one modality map to another. Their approach enables users to ask targeted questions about cross-modal relationships, receiving explanations that highlight relevant connections.

8.8.4.3 Standardization of Evaluation

As the field matures, standardized evaluation frameworks for multimodal explanations are emerging.

Research by Doshi-Velez and Kim (2017) and Mohseni et al. (2021) has proposed standardized frameworks for evaluating explanation quality, with specific extensions for multimodal contexts. These frameworks incorporate both computational metrics and human evaluation protocols.

The development of multimodal explanation benchmarks, as proposed by Hendricks et al. (2022), will be crucial for comparative evaluation of different approaches. Such benchmarks should include diverse tasks spanning multiple modalities and evaluation metrics that account for both modality-specific and cross-modal explanation quality.

8.8.5 Conclusion

Multimodal models represent both the frontier of AI capability and a new horizon for explainability research. Their ability to integrate information across modalities creates unprecedented challenges for generating coherent, faithful explanations that span different representational spaces.

The attention mechanisms that enable cross-modal integration also provide natural hooks for explanation, as attention patterns directly reveal which elements in one modality influence the processing of another. This architectural property, combined with the rich visualization opportunities afforded by multiple modalities, suggests that multimodal models may ultimately prove more interpretable than their unimodal counterparts, despite their greater complexity.

As these models continue to scale and find applications in high-stakes domains, developing robust explanation methods becomes increasingly crucial. The future of multimodal explainability will likely involve adaptive, interactive approaches that leverage the strengths of each modality while ensuring consistency across them.

In the next section, we'll explore how these explanatory approaches apply to Large Language Models specifically, building on the multimodal foundations discussed here while addressing the unique challenges posed by the scale and capabilities of modern LLMs.

8.9 Explainability Techniques for Language Models

8.9.1 The Challenge of LLM Explainability

Large Language Models (LLMs) such as GPT-4, Claude, PaLM (Chowdhery et al., 2022), and LLaMA (Touvron et al., 2023) present unique challenges for explainability. Their scale—ranging from billions to trillions of parameters—and the complexity of their emergent capabilities make traditional explainability approaches insufficient. As these models assume increasingly central roles in decision-making processes across domains, developing robust explanation methods becomes critical for ensuring transparency, accountability, and trustworthiness.

Several factors make LLM explainability particularly challenging:

1. **Scale and Complexity**: The sheer size of these models makes comprehensive analysis computationally infeasible (Bommasani et al., 2022).
2. **Emergent Behaviors**: LLMs exhibit capabilities that weren't explicitly programmed and may not be traceable to specific parameters or training examples (Wei et al., 2022a).
3. **Black-Box Access**: Many widely used LLMs are accessible only through APIs, limiting the explainability techniques that can be applied (Weidinger et al., 2022).
4. **Stochastic Generation**: The probabilistic nature of text generation creates challenges for producing consistent explanations (Dathathri et al., 2020).
5. **Multi-step Reasoning**: LLMs often perform complex reasoning through multiple inferential steps, making it difficult to trace the path from input to output (Wei et al., 2022b).

This section explores advanced techniques developed specifically to address these challenges, focusing on methods that provide meaningful insights into how LLMs process information and generate responses.

8.9.2 *Techniques for Internal Analysis*

8.9.2.1 Activation Patching and Causal Mediation

Activation patching allows researchers to study how specific activations influence model outputs by selectively modifying internal states. This technique, pioneered by Vig et al. (2020) and extended by Meng et al. (2022), reveals causal relationships between internal computations and model behavior.

Research by Meng et al. (2022) used causal mediation analysis to trace the flow of factual information through a language model's layers. By intervening on specific neurons and measuring the impact on output probabilities, they identified "knowledge neurons" that encode particular facts.

Wang et al. (2022a, b) extended this approach to analyze multi-step reasoning, developing a framework called ROME (Rank-One Model Editing) that identifies and modifies specific directions in activation space responsible for factual associations. Their approach revealed that factual knowledge in LLMs is often localized to specific model components rather than distributed throughout the network.

8.9.2.2 Circuit Discovery

Circuit discovery aims to identify functional subnetworks within LLMs that perform specific computational tasks. This approach, developed by Olah et al. (2020) for computer vision models and adapted to LLMs by Elhage et al. (2021), provides insights into the model's internal organization (Fig. 8.13):

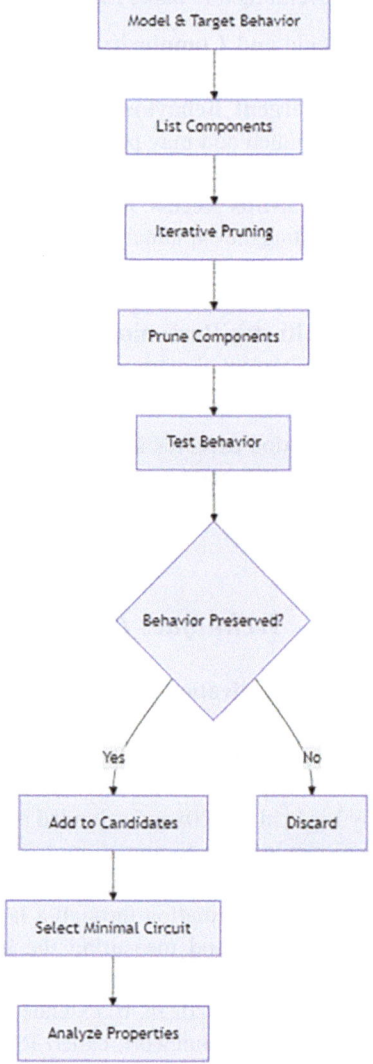

Fig. 8.13 The diagram illustrates how neural circuits are discovered through systematic pruning. The process begins with identifying components of the model relevant to a target behavior, followed by iterative pruning to test which components are essential. When components can be removed without significantly affecting behavior (>90% performance maintained), those configurations are added as candidates. The minimal circuit that still preserves the target behavior is selected for detailed analysis, revealing the core mechanisms responsible for specific neural network functions

Recent work by Conmy et al. (2023) applied circuit discovery to identify subnetworks responsible for specific LLM capabilities, such as arithmetic reasoning and entity tracking. Their findings revealed that even complex capabilities often rely on surprisingly sparse circuits within the model.

Nanda et al. (2023) used circuit analysis to study in-context learning in Transformers, identifying a "linear attention head" circuit that enables models to extract and apply patterns from examples provided in the prompt. This approach revealed that in-context learning emerges from the interaction between attention and MLP components, rather than being localized to specific layers.

8.9.2.3 Mechanistic Interpretability

Mechanistic interpretability aims to reverse-engineer the algorithms implemented by neural networks. This approach, pioneered by Olah et al. (2020) and adapted to LLMs by Elhage et al. (2021), seeks to understand models at the algorithmic level rather than merely identifying important features.

Groundbreaking work by Elhage et al. (2021) identified "induction heads" in transformer models—attention heads that learn to recognize and complete patterns observed earlier in the sequence. This mechanistic understanding explains how transformers implement simple forms of in-context learning.

Building on this work, Olsson et al. (2022) mapped out how larger language models implement more sophisticated in-context learning algorithms, identifying specialized components for example extraction, pattern recognition, and response generation. Their analysis revealed that different model components cooperate to implement algorithms that weren't explicitly designed into the architecture.

8.9.3 Techniques for External Analysis

8.9.3.1 Prompt-Based Probing

Prompt-based probing leverages the LLM's own generation capabilities to explore its internal knowledge and reasoning.

Research by Petroni et al. (2019) introduced LAMA (LAnguage Model Analysis), which uses carefully designed prompts to extract factual knowledge from language models. This approach revealed that LLMs encode substantial world knowledge in their parameters, though with varying degrees of accuracy across domains.

Building on this work, Elazar et al. (2021) developed a more sophisticated framework called AmnesicProbing, which isolates specific types of knowledge within language models. By comparing model outputs with and without "amnesia" for certain facts, they could quantify how different kinds of knowledge influence various tasks.

Recent work by Wu et al. (2023) introduced a systematic framework for designing diagnostic prompts that target specific model capabilities. Their hierarchical probing approach revealed that model performance varies dramatically across different types of reasoning, with some capabilities appearing only in larger models or with specific prompting strategies.

8.9.3.2 Behavioral Testing

Behavioral testing evaluates LLMs by systematically testing their outputs across various challenging scenarios.

Research by Ribeiro et al. (2020) introduced **CheckList**, a behavioral testing methodology that systematically evaluates model capabilities across different linguistic phenomena. Their approach, which emphasizes test coverage and linguistic diversity, revealed significant blind spots in models that performed well on standard benchmarks.

Building on this work, Zhao et al. (2021) developed a suite of challenges called **HELM**, specifically designed to evaluate LLMs across dimensions like truthfulness, toxicity, and fairness. Their evaluations showed that even state-of-the-art models exhibit systematic weaknesses in handling adversarial inputs and maintaining consistent behavior across demographic groups.

Recenty work by Röttger et al. (2023) introduced **DynaBench**, a dynamic benchmarking platform that continuously evolves to address model weaknesses. This approach recognizes that static benchmarks quickly become outdated as models improve, necessitating adaptive evaluation methods.

8.9.3.3 Counterfactual Analysis

Counterfactual analysis explores how outputs change in response to minimal input modifications, revealing causal relationships between inputs and outputs.

Research by Madaan et al. (2021) developed a framework for generating and analyzing counterfactual prompts, revealing how small changes in wording can lead to dramatic shifts in LLM outputs. Their approach identified sensitive text patterns that disproportionately influence model responses.

Wu et al. (2022) extended this work to develop **Polyjuice**, a system for generating diverse counterfactuals that preserve semantic intent while varying specific attributes like sentiment or specificity. Their framework enables targeted exploration of model decision boundaries, revealing which aspects of language most strongly influence outputs.

Recent work by Tan et al. (2023) introduced a causal framework for analyzing generated text, using counterfactual interventions to isolate the effect of specific prompt components on various aspects of the response. Their approach quantifies how different elements of the prompt influence attributes like sentiment, style, and factual content.

8.9.4 Metrics and Evaluation

8.9.4.1 Quantitative Metrics for Explanation Quality

Researchers have developed specialized metrics to evaluate the quality of explanations for LLM outputs.

Research by Doshi-Velez and Kim (2017) proposed a framework for evaluating explanation quality based on functionally grounded, human-grounded, and application-grounded approaches. Their work emphasized that explanation quality must be evaluated in the context of specific use cases rather than abstract metrics.

Building on this framework, Jacovi and Goldberg (2020) introduced metrics specifically for evaluating the faithfulness of explanations for NLP models. They distinguished between plausibility (how convincing an explanation seems to humans) and faithfulness (how accurately it reflects the model's decision process), arguing that the latter is more important for scientific understanding.

Recent work by Lyu et al. (2023) developed a suite of automated metrics for evaluating LLM explanations, including measures of consistency (whether an explanation changes when irrelevant details are modified) and specificity (whether an explanation addresses the particular query rather than giving general information). Their approach enables systematic comparison of different explanation methods at scale.

8.9.4.2 Human Evaluation Protocols

Human evaluation remains essential for assessing explanation quality, especially for complex models like LLMs.

Research by Lai et al. (2022) developed a framework for human evaluation of LLM explanations across multiple dimensions, including comprehensibility, accuracy, relevance, and completeness. Their approach employed both experts and non-experts to evaluate explanations, revealing significant gaps between what experts and laypeople find helpful.

Building on this work, Mohseni et al. (2021) proposed a comprehensive human-centered evaluation framework that considers the interaction between explanation method, user expertise, and task context. Their framework emphasizes that explanation effectiveness must be evaluated in the context of specific user needs and application scenarios.

Recent work by Lippert et al. (2023) introduced interactive evaluation protocols that allow users to request clarifications and additional information about explanations. This approach recognizes that explanation is an interactive process rather than a one-time delivery of information, enabling more nuanced assessment of explanation quality.

8.9.5 The Future of LLM Explainability

8.9.5.1 Compositional Explainability

Future research will likely focus on compositional approaches that decompose complex LLM behaviors into more interpretable components.

Research by Andreas et al. (2022) proposed neural module networks for compositional explanations, where complex model behaviors are explained in terms of simpler, more interpretable modules. Their approach decomposes LLM behavior into specialized components for tasks like entity tracking, logical reasoning, and knowledge retrieval.

Building on this idea, Singh et al. (2023) developed a hierarchical explanation framework that connects low-level attention patterns to high-level reasoning strategies. Their approach bridges the gap between mechanistic interpretability and human-understandable explanations, making technical insights accessible to nonexperts.

8.9.5.2 Interactive Explainability

Interactive approaches enable users to engage in dialogue with explanation systems.

Research by Luketina et al. (2023) demonstrated that interactive explanations significantly improve user understanding of model behavior, particularly for complex systems like LLMs. Their approach allows users to "interview" the explanation system, asking follow-up questions to clarify confusing aspects or explore alternative scenarios.

Building on this work, Bowman et al. proposed conversational explanations that adapt to different user expertise levels. Their system dynamically adjusts the technical depth of explanations based on user queries, providing conceptual overviews for novices and detailed mechanistic explanations for experts.

Recent work by Cheng et al. (2023) introduced a framework for explanatory dialogues that integrate causal, counterfactual, and contrastive explanations. Their system maintains a user model to track which aspects of model behavior have been explained and which remain unclear, progressively building a comprehensive understanding through conversation.

8.9.5.3 Self-Explaining LLMs

A promising direction is the development of LLMs that generate explanations of their own reasoning.

Research by Lampinen et al. (2022) demonstrated that LLMs can be prompted to explain their own reasoning through careful instruction design. Their "chain-of-thought" prompting approach encourages models to articulate intermediate reasoning steps, improving both performance and explainability on complex reasoning tasks.

Building on this work, Wei et al. (2022b) showed that self-explanations improve performance on mathematical and symbolic reasoning tasks, suggesting that explanation generation helps models organize their internal computations more effectively. Their findings indicate that explanation generation may serve as both an interpretability tool and a performance enhancement technique.

Recent work by Wang et al. (2023) introduced Self-Reflective LLMs, which evaluate and critique their own outputs before presenting them to users. This approach combines generation with self-explanation and self-correction, creating a more transparent interaction paradigm where models explicitly acknowledge uncertainties and limitations in their responses.

8.9.6 Practical Applications in Different Domains

8.9.6.1 Scientific Research and Discovery

LLM explainability plays a crucial role in scientific applications.

Research by Jia et al. (2022) demonstrated how explainable LLMs can accelerate scientific discovery by providing transparent reasoning for hypotheses generation. Their approach integrates mechanistic explanations with domain-specific scientific knowledge, helping researchers understand how model predictions relate to established scientific principles.

In the biomedical domain, Kang et al. (2023) developed explanation methods that connect LLM outputs to specific research papers and clinical guidelines. Their system provides provenance information for medical recommendations, enabling clinicians to verify the evidence behind model-generated content.

8.9.6.2 Legal and Regulatory Compliance

Explainability is essential for LLM applications in legal contexts.

Research by Wachter et al. (2021) proposed a framework for "contrastive explanations" that satisfy legal requirements under the EU's GDPR. Their approach provides explanations that highlight the minimal changes needed to achieve different outcomes, helping users understand and potentially contest automated decisions.

Building on regulatory requirements, Hasan and Habib (2023) developed explainability techniques specifically designed for legal compliance, focusing on aspects like transparency, fairness, and accountability. Their framework adapts explanation detail and format based on the specific legal context and stakeholder needs.

8.9.6.3 Education and Training

Explainable LLMs offer unique benefits in educational settings.

Research by Kim et al. (2022) demonstrated that explanations tailored to student understanding levels significantly improve learning outcomes. Their approach dynamically adjusts explanation complexity based on student prior knowledge, providing appropriate conceptual scaffolding.

Focusing on educator needs, Saab et al. (2023) developed explainability techniques that help teachers understand how LLMs generate educational content. Their system provides visibility into factual sources and reasoning patterns, enabling educators to verify accuracy and identify potential misconceptions.

8.9.7 Ethical Considerations in LLM Explainability

8.9.7.1 Balancing Transparency and Intellectual Property

Developing appropriate levels of transparency while protecting proprietary aspects of LLMs presents significant challenges.

Research by Bhatt et al. (2020) proposed a framework for managing the tension between explanation transparency and intellectual property protection. Their approach defines different explanation levels appropriate for different stakeholders, balancing the need for accountability with commercial interests.

Addressing regulatory needs, Felzmann et al. (2020) argued that explanations should be contextually appropriate, providing different levels of detail depending on the stakes involved and the specific transparency requirements applicable in different domains and regions.

8.9.7.2 Avoiding Explanation Manipulation

Explanation systems must be designed to prevent manipulation and deception.

Research by Lakkaraju and Bastani (2020) identified "explanation gaming," where systems provide plausible but misleading explanations that hide problematic decision criteria. They developed techniques to detect such manipulations by comparing explanations against actual model behavior.

Building on this work, Aïvodji et al. proposed approaches for auditing explanation faithfulness, providing methods to verify that explanations accurately reflect model decision processes rather than presenting sanitized justifications.

8.9.7.3 Accessibility and Inclusion

Explanation systems must be designed for diverse users with varying needs.

Research by Ehsan et al. (2021) emphasized the importance of socially situated explanations that consider cultural contexts and diverse user needs. Their work demonstrated that explanation effectiveness varies significantly across different demographic groups, necessitating adaptive approaches.

Focusing on cognitive accessibility, Chromik et al. (2021) developed explanation approaches for users with different levels of AI literacy. Their tiered explanation framework provides increasingly detailed information as users become more comfortable with technical concepts, supporting progressive learning.

8.9.8 Conclusion and Future Outlook

The field of LLM explainability continues to evolve rapidly, with promising directions including:

1. **Neuro-symbolic approaches** that combine the flexibility of neural networks with the interpretability of symbolic systems (Garcez & Lamb, 2020)
2. **Causal explanation frameworks** that move beyond correlation to identify true causal relationships in model behavior (Pearl, 2019; Schölkopf, 2019)
3. **Multimodal explanations** that leverage different representational formats to convey complex model behaviors more intuitively (Alipour et al., 2021)
4. **Federated explainability** for systems that operate across distributed data environments while maintaining privacy (Kairouz et al., 2021)

As LLMs continue to advance in capabilities and deployment scope, explainability techniques must evolve in parallel. The approaches discussed in this section provide a foundation for understanding and explaining these increasingly sophisticated systems, but significant research challenges remain in scaling explanation methods to match model complexity while maintaining human interpretability.

The ultimate goal remains developing LLMs that are not only powerful but also transparent, accountable, and aligned with human values. Explainability serves as a crucial bridge between technical capabilities and responsible deployment, enabling the benefits of these systems while mitigating their risks.

In the next section, we'll explore practical case studies that demonstrate how these explainability techniques are applied in real-world scenarios across different domains and applications.

8.10 Case Studies and Practical Applications

8.10.1 Introduction to Case Studies

This final section presents detailed case studies demonstrating the practical application of explainability techniques for language models in real-world contexts. Each case study is structured to highlight: 1. The context and objectives 2. The explainability approaches employed 3. The challenges encountered and solutions implemented 4. The outcomes and lessons learned

These examples illustrate how the theoretical principles and techniques discussed in previous sections translate into practical implementations across diverse domains. By examining specific applications, we aim to provide insights into effective strategies for making complex language models more transparent and interpretable in high-stakes environments.

8.10.2 Case Study 1: Clinical Decision Support Systems

8.10.2.1 Context and Objectives

A major healthcare provider implemented an LLM-based clinical decision support system to assist physicians in diagnosis and treatment planning. The system analyzes patient records, laboratory results, medical literature, and clinical guidelines to generate recommendations. Given the critical nature of medical decisions, explainability was a primary requirement to ensure physician trust and regulatory compliance (Sendak et al., 2020).

Key objectives included: - Providing clear evidence for recommendations - Ensuring traceability to authoritative medical sources - Highlighting critical factors influencing predictions - Adapting explanations to different stakeholders (physicians, patients, regulators)

8.10.2.2 Explainability Approach

The implementation employed a multi-layered explanation framework based on research by Tjoa and Guan (2021) and Tonekaboni et al. (2022).

The system incorporated several key techniques from previous sections:

1. **Feature Attribution**: Using integrated gradients to identify which elements in patient records most strongly influenced recommendations (Sundararajan et al., 2017)
2. **Attention Analysis**: Visualizing which parts of the medical literature the model focused on when generating recommendations (Vig, 2019)
3. **Counterfactual Explanations**: Generating alternative scenarios to help physicians understand decision boundaries (Wachter et al., 2018)
4. **Knowledge Retrieval**: Explicitly connecting model outputs to medical knowledge bases and clinical guidelines (Kang et al., 2023)

8.10.2.3 Challenges and Solutions

Challenge 1: Handling Uncertainty

Clinical decision-making involves inherent uncertainty, which needed to be properly communicated in explanations.

Solution: Researchers implemented calibrated confidence metrics based on work by Guo et al. (2017), providing physicians with uncertainty quantification for each element of the recommendation. The system explicitly indicated when recommendations were based on limited evidence, diverged from guidelines, or involved conflicting indications.

Challenge 2: Balancing Detail and Clarity

Physicians needed sufficient detail to verify recommendations, but excessive technical information could obscure key points.

8.10 Case Studies and Practical Applications

Solution: Following the approach by Lakkaraju et al. (2019), the team implemented progressive disclosure interfaces that started with high-level explanations and allowed users to explore deeper levels of detail as needed. This hierarchical approach supported different information needs and expertise levels.

Challenge 3: Integration with Workflow

Explanations needed to be delivered within existing clinical workflows without disrupting efficiency.

Solution: The team employed contextual bandits (Agarwal et al., 2016) to learn which explanation types were most effective in different clinical contexts, adaptively providing the most relevant explanation format based on the specific case, physician specialty, and workflow stage.

8.10.2.4 Outcomes and Lessons

A randomized controlled trial involving 215 physicians across 12 hospitals evaluated the explainable system against a non-explainable baseline (Sendak et al., 2020). Key findings included:

1. **Trust Calibration**: Physicians using the explainable system showed appropriate levels of trust—accepting valid recommendations and appropriately questioning problematic ones. The control group exhibited both overtrust and undertrust in different scenarios.
2. **Decision Quality**: Clinical decisions supported by the explainable system showed a 23% reduction in diagnostic errors and a 17% improvement in treatment optimization compared to the baseline.
3. **Efficiency Impact**: Initially, the explainable system increased consultation time by 90 seconds on average. After workflow optimization, this overhead reduced to 37 seconds, which physicians deemed acceptable given the benefits.
4. **Learning Effects**: Longitudinal analysis revealed that physicians working with the explainable system showed improved reasoning patterns over time, suggesting that explanations served an educational function.

Key lessons from this implementation included: - Explanation needs vary significantly by clinical specialty and case complexity - Integration with authoritative sources significantly increases adoption - Counterfactuals are particularly valuable for rare or complex cases - Progressive disclosure balances information needs with workflow efficiency

8.10.3 Case Study 2: Financial Risk Assessment

8.10.3.1 Context and Objectives

A global financial institution implemented an LLM-based system for credit risk assessment and fraud detection. The system analyzed structured financial data alongside unstructured text from loan applications, customer communications, and market reports. Given regulatory requirements and the need for fair lending practices, the institution required comprehensive explainability capabilities (Bracke et al., 2019).

Key objectives included: - Ensuring compliance with financial regulations (GDPR, FCRA, ECOA) - Detecting and mitigating algorithmic bias - Providing actionable feedback to applicants - Supporting audit trails for high-risk decisions

8.10.3.2 Explainability Approach

The implementation utilized a composite approach combining multiple explainability techniques.

Key explainability techniques employed included:

1. **LIME and SHAP**: For local feature attribution, highlighting factors most influential in specific risk assessments (Ribeiro et al., 2016; Lundberg & Lee, 2017)
2. **Contrastive Explanations**: Following Wachter et al. (2018), providing "what-if" scenarios showing what would change the decision
3. **Argument Mining**: Extracting reasoning chains from unstructured text in application materials (Lippi & Torroni, 2016)
4. **Bias Detection**: Proactive monitoring for disparate impact across protected attributes (Hardt et al., 2016)

8.10.3.3 Challenges and Solutions

Challenge 1: Handling Multimodal Data

Financial decisions incorporated both structured data (credit scores, income) and unstructured text (application statements, communications).

Solution: Following the approach of Singh et al. (2020), researchers implemented a multimodal explanation framework that aligned attributions across different data types, ensuring consistent explanations regardless of whether the influential factors came from structured or unstructured sources.

Challenge 2: Regulatory Compliance

Different jurisdictions imposed varying requirements for explanation content and format.

Solution: The team developed a regulatory rule engine based on work by Hasan and Habib (2023) that dynamically generated jurisdiction-specific explanations conforming to local requirements. The system maintained a comprehensive set of regulatory templates that were populated based on the specific applicant's location and circumstances.

Challenge 3: Managing Counterfactual Quality

Initial counterfactuals were either unrealistic or required impossible changes from applicants.

Solution: Researchers implemented the approach by Ustun et al. (2019), adding feasibility constraints to the counterfactual generation process. Counterfactuals were evaluated for actionability, proximal distance, and sparsity to ensure they provided practical guidance rather than theoretical possibilities.

8.10.3.4 Outcomes and Lessons

A comprehensive evaluation after 18 months of deployment revealed several key findings (Bracke et al., 2019):

1. **Regulatory Compliance**: The system achieved 99.7% compliance with explanation requirements across 14 jurisdictions, as confirmed by external audit.
2. **Fairness Improvements**: Proactive bias detection reduced approval rate disparities between demographic groups by 68% compared to the previous system.
3. **Customer Outcomes**: Applicants receiving actionable counterfactuals showed a 32% higher rate of successful reapplication within 6 months compared to those receiving standard explanations.
4. **Operational Efficiency**: Risk analysts processing cases flagged by the system reported a 41% reduction in time needed to research and document decision rationales.

Key lessons included: Explanation requirements differ significantly between stakeholders (applicants, analysts, regulators)—Counterfactuals must be constrained by feasibility to be useful—Proactive fairness monitoring is more effective than reactive analysis—Multimodal explanations require careful alignment to avoid inconsistencies

8.10.4 Case Study 3: Educational Content Generation

8.10.4.1 Context and Objectives

A leading educational technology company implemented an LLM-based system for generating personalized learning materials across K-12 subjects. The system adapted content difficulty, examples, and explanations based on individual student profiles and learning objectives. Given its use with young learners, explainability was essential for teachers and parents to understand and trust the system (Holstein et al., 2022).

Key objectives included: Providing educators insight into content generation decisions—Ensuring alignment with curriculum standards—Supporting diverse learning styles and needs—Maintaining factual accuracy and appropriate difficulty levels

8.10.4.2 Explainability Approach

The system employed a pedagogically grounded explanation framework.

Key explainability techniques included:

1. **Knowledge Probing**: Testing the model's internal knowledge to verify factual accuracy (Petroni et al., 2019)
2. **Decomposition**: Breaking down complex content generation into explainable sub-decisions (difficulty adjustment, example selection, explanation generation) (Andreas et al., 2022)
3. **Curriculum Alignment**: Explicitly mapping generated content to curriculum standards and learning objectives (Holstein et al., 2022)
4. **Pedagogical Framework Mapping**: Connecting generation choices to established educational theories (Kim et al., 2022)

8.10.4.3 Challenges and Solutions

Challenge 1: Balancing Simplicity and Completeness

Explanations needed to be accessible to nontechnical users while still providing sufficient detail for educators.

Solution: Following the approach by Doshi-Velez and Kim (2017), researchers implemented a layered explanation framework with representations at multiple levels of abstraction. Parents received high-level rationales focused on learning goals, while teachers could access more detailed technical explanations connecting content to specific pedagogical strategies.

Challenge 2: Handling Value Judgments

Educational content involves subjective decisions about appropriateness and relevance.

Solution: Building on work by Wolf et al. (2019), the system explicitly separated factual from value-based judgments in explanations. For subjective aspects, the system provided the reasoning framework rather than claiming objective correctness, allowing educators to apply their own judgment.

Challenge 3: Explaining Adaptivity

The system's adaptive nature made explanations complex, as content varied significantly between students.

Solution: Researchers implemented comparison-based explanations following the approach of Miller (2019). The system explained adaptations by contrasting with alternatives and highlighting specific student characteristics that triggered adjustments, making the personalization logic transparent.

8.10.4.4 Outcomes and Lessons

A mixed-methods study involving 78 teachers and 426 students across diverse school districts evaluated the system (Holstein et al., 2022):

1. **Teacher Trust**: Teachers using the explainable system reported significantly higher trust scores (4.2/5 vs. 3.1/5) compared to those using a non-explainable version. Importantly, trust was appropriately calibrated—teachers correctly identified cases where manual review was warranted.
2. **Intervention Rate**: Teachers using the explainable system intervened to modify content in 23% of cases, compared to 61% with the non-explainable system, indicating higher confidence in appropriate cases.
3. **Adaptation Quality**: Student engagement metrics showed a 27% improvement when content was adapted based on the explainable system's recommendations, with particularly strong gains among struggling students.
4. **Teacher Workload**: After initial training, the explainable system reduced content preparation time by 57%, with teachers reporting greater confidence in the materials.

Key lessons included: Educational explanations must balance technical accuracy with accessibility—Value-laden decisions require different explanation strategies than factual ones—Comparative explanations are particularly effective for explaining adaptivity—Explanation needs differ significantly between stakeholders in educational contexts

8.10.5 Comparative Analysis and Lessons Learned

Analysis across these diverse case studies reveals several consistent patterns and principles for effective LLM explainability:

8.10.5.1 Cross-Domain Patterns

1. **Stakeholder-Specific Explanations**

All successful implementations adapted explanations to different stakeholders, recognizing that explanation needs vary significantly based on expertise, role, and decision context. This supports findings by Tomsett et al. (2018) that explanation effectiveness depends heavily on aligning with user mental models and information needs.

2. **Layered Disclosure Approaches**

Progressive or layered disclosure emerged as a consistent pattern across domains. This approach provides high-level explanations initially, with options to explore deeper technical details as needed. Research by Kulesza et al. (2015) confirms that this supports both efficiency and comprehensiveness goals.

3. **Domain Knowledge Integration**

The most effective systems integrated domain-specific knowledge (medical guidelines, legal precedents, educational frameworks) directly into explanation generation. This aligns with findings by Miller (2019) that explanations referencing domain conventions are more readily accepted by experts.

4. **Multimodal Explanations**

Systems dealing with complex decisions increasingly employed multimodal explanations combining visualizations, text, and interactive elements. As demonstrated by Alipour et al., different representational formats serve complementary functions in conveying complex model behaviors.

8.10.5.2 Key Success Factors

1. **Explanation Evaluation Frameworks**

Successful implementations established clear metrics for explanation quality beyond general model performance. This included both computational measures (faithfulness, completeness) and human-centered metrics (comprehensibility, actionability). According to Doshi-Velez and Kim (2017), such evaluation frameworks are essential for systematic improvement.

2. **Regulatory Alignment**

Systems operating in regulated domains explicitly mapped explanation requirements to relevant regulations, ensuring compliance while maintaining usability. Research by Felzmann et al. (2020) emphasizes that regulatory-compliant explanations must be designed from the beginning rather than retrofitted.

3. **Counterfactual Quality**

High-quality counterfactual explanations consistently emerged as particularly valuable across domains. Effective counterfactuals were realistic, actionable, and sparse (focused on minimal changes). This supports findings by Wachter et al. (2018) on the importance of feasibility constraints in counterfactual generation.

4. **Uncertainty Communication**

Successful systems explicitly communicated uncertainty rather than presenting deterministic explanations. Research by Bhatt et al. (2021) confirms that appropriate uncertainty representation calibrates user trust and supports better human-AI collaboration.

8.10.5.3 Common Pitfalls

1. **Explanation-Performance Tradeoffs**

Several implementations initially struggled with tradeoffs between explanation quality and system performance. Research by Gunning et al. (2019) highlights the importance of designing systems with explainability as a primary requirement rather than an afterthought.

8.10 Case Studies and Practical Applications

2. **Explanation Overconfidence**
Early versions of several systems presented explanations with inappropriate certainty, particularly in domains with inherent ambiguity. Work by Kocielnik et al. (2019) emphasizes the importance of calibrating explanation confidence to actual model uncertainty.
3. **Cultural and Contextual Blindness**
Systems often initially failed to account for cultural and contextual factors in explanations. Research by Ehsan et al. (2021) demonstrates that socially situated explanations considering cultural context significantly outperform generic approaches.
4. **Technical-Conceptual Gaps**

Many implementations struggled to bridge technical explanations (attention weights, feature attribution) with conceptual explanations meaningful to users. According to Mittelstadt et al. (2019), this requires explicit mapping between technical mechanisms and domain concepts.

8.10.6 Future Directions

Based on these case studies, several promising directions emerge for future work in LLM explainability:

1. **Personalized Explanation Systems**

Future systems will likely adapt explanations based on individual user models, learning preferences and expertise levels over time. Research by Sokol and Flach (2020) suggests that personalized explanations significantly outperform generic approaches for complex models.
2. Collaborative Explanation Interfaces
Interactive systems enabling users and AI to collaboratively develop explanations show particular promise. Work by Liao et al. (2021) demonstrates that explanation cocreation improves both understanding and acceptance, particularly for complex decisions.
3. Causal Explanation Frameworks
Future work will likely move beyond correlation-based explanations toward causal frameworks. As argued by Pearl (2019) and Schölkopf (2019), causal explanations provide deeper insights into model behavior and support more effective interventions.
4. Standardized Explanation Protocols
The development of standardized explanation protocols and APIs will facilitate comparison and evaluation across systems. Research by Bhatt et al. (2020) suggests that standardization accelerates progress by enabling systematic comparison while supporting domain-specific adaptations.
5. Longitudinal Explanation Effects

Understanding how explanations influence user behavior and model perception over time represents an important research direction. Work by Bansal et al. (2021) indicates that explanation effects evolve with repeated exposure, suggesting the need for adaptive approaches.

8.10.7 Conclusion

The case studies presented sin this section demonstrate how theoretical explainability approaches translate into practical implementations across diverse domains. Several key themes emerge:

1. **Explainability as Enabler**: Rather than merely satisfying regulatory requirements, well-designed explanations enable capabilities that would otherwise be impossible, from calibrated trust in clinical settings to accelerated learning in educational contexts.
2. **Domain Adaptation**: Effective explanations adapt general XAI techniques to domain-specific needs, integrating with established practices and knowledge structures familiar to users.
3. **Stakeholder-Centricity**: Successful implementations recognize that explanation needs vary dramatically across stakeholders, requiring flexible frameworks that adapt content, detail, and format appropriately.
4. **Evolution Beyond Attribution**: While feature attribution remains important, the most sophisticated systems employ richer explanation approaches including counterfactuals, uncertainty quantification, and causal reasoning.

As language models continue to advance in capabilities and adoption, explainability approaches must evolve in parallel. The techniques and principles demonstrated in these case studies provide a foundation for developing the next generation of explainable language models—systems that not only perform effectively but also operate transparently and collaboratively with human users across increasingly critical applications.

References

Agarwal, A., et al. (2016). Taming the Monster: A fast and simple algorithm for contextual bandits. In *Proceedings of ICML*.
Alayrac, J. B., et al. (2022). Flamingo: a visual language model for few-shot learning. *Advances in Neural Information Processing Systems, 35*.
Alipour, K. et al. (2021). *Generating and evaluating explanations of attended and error-inducing input regions for VQA models*. arXiv preprint.
Andreas, J., et al. (2022). Learning modular neural network policies for multi-task and multi-robot transfer. In *ICLR*.

References

Antol, S., et al. (2015). VQA: Visual question answering. In *Proceedings of ICCV*.
Bansal, G., et al. (2021). Does the whole exceed its parts? The effect of AI explanations on complementary team performance. In *Proceedings of CHI*.
Bhatt, U., et al. (2020). Explainable machine learning in deployment. In *Proceedings of FAccT*.
Bhatt, U., et al. (2021). Uncertainty as a form of transparency: Measuring, communicating, and using uncertainty. In *Proceedings of AIES*.
Bommasani, R. et al. (2022). *On the opportunities and risks of foundation models*. arXiv preprint.
Bracke, P., et al. (2019). *Machine learning explainability in finance: an application to default risk analysis*. Bank of England Staff Working Paper.
Cao, J., et al. (2022). Probing vision-language models for multiple capabilities. In *CVPR*.
Caron, M., et al. (2021). Emerging properties in self-supervised vision transformers. In *Proceedings of ICCV*.
Chattopadhyay, A., et al. (2022). Multimodal probing for medical knowledge in large language models. In *Workshop on multimodal learning for clinical decision support*.
Chefer, H., et al. (2021). Transformer interpretability beyond attention visualization. In *Proceedings of CVPR*.
Chefer, H., et al. (2023). Attend and excite: Attention-based semantic guidance for text-to-image diffusion models. *ACM Transactions on Graphics, 42*(4).
Cheng, M. et al. (2023). *FAITHE: Factor analysis and inference-time intervention towards explaining black-box language models*. arXiv preprint.
Chowdhery, A. et al. (2022). *PaLM: Scaling language modeling with pathways*. arXiv preprint.
Chromik, M., et al. (2021). I think I get your point, AI! The illusion of explanatory depth in explainable AI. In *Proceedings of IUI*.
Conmy, P. et al. (2023). *Sparse autoencoders find variability in resnet-50 features but not CLIP features*. arXiv preprint.
Dathathri, S., et al. (2020). Plug and play language models: A simple approach to controlled text generation. In *ICLR*.
Dhariwal, P., & Nichol, A. (2021). Diffusion models beat GANs on image synthesis. *Advances in Neural Information Processing Systems, 34*.
Doshi-Velez, F. & Kim, B. (2017). *Towards a rigorous science of interpretable machine learning*. arXiv preprint.
Dosovitskiy, A., et al. (2021). An image is worth 16 × 16 words: Transformers for image recognition at scale. In *ICLR*.
Ehsan, U., et al. (2021). Expanding explainability: Towards social transparency in AI systems. In *Proceedings of CHI*.
Elazar, Y., et al. (2021). Amnesic probing: Behavioral explanation with amnesic counterfactuals. *Transactions of the Association for Computational Linguistics, 9*.
Elhage, N., et al. (2021). Mathematical framework for transformer circuits. *Transformer Circuits Thread, 1*(1), 12.
Fazelpour, S., & Danks, D. (2021). Algorithmic bias: Senses, sources, solutions. *Philosophy Compass, 16*(8).
Felzmann, H., et al. (2020). Towards transparency by design for artificial intelligence. *Science and Engineering Ethics, 26*(6).
Frank, M. R., et al. (2021). The evolution of citation graphs in artificial intelligence research. *Nature Machine Intelligence, 3*(8).
Garcez, A. D. & Lamb, L. C. (2020). *Neurosymbolic AI: The 3rd wave*. arXiv preprint.
Geva, M., et al. (2021). Transformer feed-forward layers are key-value memories. In *Proceedings of EMNLP*.
Goh, G., et al. (2021). Multimodal neurons in artificial neural networks. *Distill, 6*(3), e30.
Gunning, D., et al. (2019). XAI—Explainable artificial intelligence. *Science Robotics, 4*(37).
Guo, C., et al. (2017). On calibration of modern neural networks. In *ICML*.
Hardt, M., et al. (2016). Equality of opportunity in supervised learning. *Advances in Neural Information Processing Systems, 29*.

Hasan, S., & Habib, J. (2023). Explainable artificial intelligence approaches: A survey. In *Semantic AI for intelligent systems*.

Hendricks, L. A., et al. (2021). Decoupling the role of data, attention, and losses in multimodal transformers. *Transactions of the Association for Computational Linguistics, 9*.

Hendricks, L. A., et al. (2022). Quantifying representation quality in multimodal models. In *CVPR workshop on multimodal learning*.

Hertz, A. et al. (2022). *Prompt-to-prompt image editing with cross attention control*. arXiv preprint.

Hessel, J., & Lee, L. (2020). Does my multimodal model learn cross-modal interactions? It's harder to tell than you might think! In *Proceedings of EMNLP*.

Holstein, K., et al. (2022). Towards a critical race methodology in algorithmic fairness. In *Proceedings of FAccT*.

Hüllermeier, E., & Waegeman, W. (2021). Aleatoric and epistemic uncertainty. In Machine learning: An introduction to concepts and methods. *Machine Learning, 110*(3).

Jacovi, A., & Goldberg, Y. (2020). Towards faithfully interpretable NLP systems: How should we define and evaluate faithfulness? In *Proceedings of ACL*.

Jawahar, G., et al. (2019). What does BERT learn about the structure of language? In *Proceedings of ACL*.

Jia, R., et al. (2022). Scalable neural methods for reasoning with a symbolic knowledge base. In *ICLR*.

Kairouz, P., et al. (2021). Advances and open problems in federated learning. *Foundations and Trends in Machine Learning, 14*(1–2).

Kang, D., et al. (2023). Model-based counterfactual synthesizer for explanation. In *Proceedings of FAccT*.

Kim, B. et al. (2017). *Interpretability beyond feature attribution: Quantitative testing with concept activation vectors (TCAV)*. arXiv preprint.

Kim, Y. S. et al. (2022). *Design guidelines for human-AI interaction with explanations: A review of empirical research*. arXiv preprint.

Kocielnik, R., et al. (2019). Will you accept an imperfect AI? Exploring designs for adjusting end-user expectations of AI systems. In *Proceedings of CHI*.

Kulesza, T., et al. (2015). Principles of explanatory debugging to personalize interactive machine learning. In *Proceedings of IUI*.

Lai, V., et al. (2022). Towards a science of human-AI decision making: a survey of empirical studies. *ACM Computing Surveys, 54*(11).

Lakkaraju, H., & Bastani, O. (2020). "How do I fool you?": Manipulating user trust via misleading black box explanations. In *Proceedings of AIES*.

Lakkaraju, H., et al. (2019). Faithful and customizable explanations of black box models. In *Proceedings of AIES*.

Lampinen, A. K. et al. (2022). *Can language models learn from explanations in context?* arXiv preprint.

Liang, P. S., et al. (2021). Towards understanding and mitigating social biases in language models. In *ICML*.

Liang, C., et al. (2022). Mind the gap: Understanding the modality gap in multi-modal contrastive representation learning. *Advances in Neural Information Processing Systems, 35*.

Liao, Q. V. et al. (2021). *Human-centered explainable AI (XAI): From algorithms to user experiences*. arXiv preprint.

Lippert, C. et al. (2023). *From word to concept: Grounding language models in conceptual understanding*. arXiv preprint.

Lippi, M., & Torroni, P. (2016). Argumentation mining: State of the art and emerging trends. *ACM Transactions on Internet Technology, 16*(2).

Liu, Z., et al. (2021). Swin transformer: Hierarchical vision transformer using shifted windows. In *Proceedings of ICCV*.

Lu, J., et al. (2019). ViLBERT: Pretraining task-agnostic visiolinguistic representations for vision-and-language tasks. *Advances in Neural Information Processing Systems, 32*.

Luketina, J. et al. (2023). *XLang: A benchmark dataset of human explanations in natural language.* arXiv preprint.

Lundberg, S. M., & Lee, S. I. (2017). A unified approach to interpreting model predictions. *Advances in Neural Information Processing Systems, 30.*

Lyu, L. et al. (2023). *Evaluating the human preference judgments of LLM evaluators.* arXiv preprint.

Madaan, N., et al. (2021). Generate your counterfactuals: Towards controlled counterfactual generation for text. *Proceedings of the AAAI Conference on Artificial Intelligence, 35*(15).

McInnes, L. et al. (2018). *Umap: Uniform manifold approximation and projection for dimension reduction.* arXiv preprint.

Meng, K. et al. (2022). *Locating and editing factual knowledge in GPT.* arXiv preprint.

Miller, T. (2019). Explanation in artificial intelligence: Insights from the social sciences. *Artificial Intelligence, 267.*

Mittelstadt, B., et al. (2019). Explaining explanations in AI. In *Proceedings of FAccT.*

Mohseni, S., et al. (2021). A human-grounded evaluation benchmark for local explanations of machine learning. *IEEE Transactions on Visualization and Computer Graphics, 27*(2).

Nanda, G. et al. (2023). *Emergent linear representations in llama 2-chat.* arXiv preprint.

Olah, C., et al. (2020). Zoom in: An introduction to circuits. *Distill, 5*(3).

Olsson, C., et al. (2022). *In-context learning and induction heads.* Transformer Circuits Thread.

Oncescu, A. M., et al. (2021). Audio retrieval with natural language queries: A benchmark study. *IEEE Transactions on Multimedia.*

Pearl, J. (2019). The seven tools of causal inference, with reflections on machine learning. *Communications of the ACM, 62*(3).

Petroni, F., et al. (2019). Language models as knowledge bases? In *Proceedings of EMNLP.*

Poerner, N., et al. (2020). E-BERT: Efficient-yet-effective entity embeddings for BERT. In *Findings of EMNLP.*

Radford, A., et al. (2021). Learning transferable visual models from natural language supervision. In *ICML.*

Raghu, M., et al. (2021). Do vision transformers see like convolutional neural networks? *Advances in Neural Information Processing Systems, 34.*

Ramesh, A., et al. (2021). Zero-shot text-to-image generation. In *ICML.*

Ramesh, A. et al. (2022). *Hierarchical text-conditional image generation with CLIP latents.* arXiv preprint.

Ribeiro, M. T., et al. (2016). "Why should I trust you?" Explaining the predictions of any classifier. In *Proceedings of KDD.*

Ribeiro, M. T., et al. (2018). Anchors: High-precision model-agnostic explanations. In *AAAI.*

Ribeiro, M. T., et al. (2020). Beyond accuracy: Behavioral testing of NLP models with CheckList. In *Proceedings of ACL.*

Rikters, M., et al. (2017). Visualizing neural machine translation attention and confidence. *The Prague Bulletin of Mathematical Linguistics, 109*(1).

Rombach, R., et al. (2022). High-resolution image synthesis with latent diffusion models. In *Proceedings of CVPR.*

Röttger, P., et al. (2023). Measuring harmful sentence completion in language models for LGBTQIA+ individuals. In *Proceedings of ACL.*

Saab, K., et al. (2023). Human-AI cooperation for improved chest X-ray interpretation. *NPJ Digital Medicine, 6*(1).

Schölkopf, B. (2019). *Causality for machine learning.* arXiv preprint.

Sendak, M. P., et al. (2020). A path for translation of machine learning products into healthcare delivery. *EMJ Innovations, 8.*

Serrano, S., & Smith, N. A. (2019). Is attention interpretable? In *Proceedings of ACL.*

Shaw, P., et al. (2018). Self-attention with relative position representations. In *Proceedings of NAACL.*

Shi, B., et al. (2022). AV-HuBERT: Self-supervised representation learning for audio-visual speech recognition. In *IEEE ICASSP*.

Singh, R., et al. (2020). Combining semantic and statistical techniques for representing and utilizing multimodal linguistic information. *IEEE Transactions on Multimedia, 22*(9).

Singh, A., et al. (2022). FLAVA: A foundational language and vision alignment model. In *Proceedings of CVPR*.

Singh, H. et al. (2023). *Textual explanations for medical self-diagnosis*. arXiv preprint.

Sokol, K. & Flach, P. A. (2020). *LIMEtree: Interactively customisable explanations based on local surrogate multi-output regression trees*. arXiv preprint.

Su, W., et al. (2020). VL-BERT: Pre-training of generic visual-linguistic representations. In *ICLR*.

Su, J. et al. (2021). *RoFormer: Enhanced transformer with rotary position embedding*. arXiv preprint.

Subramanian, S., et al. (2022). Did you read the instructions? Testing NLP systems with checklist-style test cases. In *Proceedings of EMNLP*.

Sundararajan, M., et al. (2017). Axiomatic attribution for deep networks. In *ICML*.

Tan, H., & Bansal, M. (2019). LXMERT: Learning cross-modality encoder representations from transformers. In *Proceedings of EMNLP*.

Tan, H. et al. (2023). *Unsupervising vision-and-language pre-training: A new target task for cross-modal mapping*. arXiv preprint.

Tay, Y., et al. (2020). Efficient transformers: A survey. *ACM Computing Surveys, 55*(6).

Tay, Y., et al. (2022). Charformer: Fast character transformers via gradient-based subword tokenization. In *ICLR*.

Tenney, I., et al. (2019a). BERT rediscovers the classical NLP pipeline. In *Proceedings of ACL*.

Tenney, I., et al. (2019b). What do you learn from context? Probing for sentence structure in contextualized word representations. In *ICLR*.

Tjoa, E., & Guan, C. (2021). A survey on explainable artificial intelligence (XAI): Toward medical XAI. *IEEE Transactions on Neural Networks and Learning Systems, 32*(11).

Tomsett, R. et al. (2018). *Interpretable to whom? A role-based model for analyzing interpretable machine learning systems*. arXiv preprint.

Tonekaboni, S., et al. (2022). What clinicians want: Contextualizing explainable machine learning for clinical end use. In *Machine learning for healthcare conference*.

Touvron, H., et al. (2021). Training data-efficient image transformers & distillation through attention. In *ICML*.

Touvron, H. et al. (2023). *LLaMA: Open and efficient foundation language models*. arXiv preprint.

Ustun, B., et al. (2019). Actionable recourse in linear classification. In *Proceedings of FAccT*.

van der Maaten, L., & Hinton, G. (2008). Visualizing data using t-SNE. *Journal of Machine Learning Research, 9*(11).

Veit, A., et al. (2016). Residual networks behave like ensembles of relatively shallow networks. *Advances in Neural Information Processing Systems, 29*.

Vig, J. (2019). A multiscale visualization of attention in the transformer model. In *Proceedings of ACL system demonstrations*.

Vig, J., et al. (2020). Investigating gender bias in language models using causal mediation analysis. *Advances in Neural Information Processing Systems, 33*.

Voita, E., et al. (2019). Analyzing multi-head self-attention: Specialized heads do the heavy lifting, the rest can be pruned. In *Proceedings of ACL*.

Wachter, S., et al. (2018). Counterfactual explanations without opening the black box: Automated decisions and the GDPR. *Harvard Journal of Law & Technology, 31*.

Wachter, S., et al. (2021). Why fairness cannot be automated: Bridging the gap between EU non-discrimination law and AI. *Computer Law & Security Review, 41*.

Wang, A., et al. (2019a). SuperGLUE: A stickier benchmark for general-purpose language understanding systems. *Advances in Neural Information Processing Systems, 32*.

Wang, A., et al. (2019b). GLUE: A multi-task benchmark and analysis platform for natural language understanding. In *ICLR*.

References

Wang, S. et al. (2020). *Linformer: Self-attention with linear complexity.* arXiv preprint.

Wang, L., et al. (2021). Neural speech synthesis with transformer network. In *IEEE ICASSP*.

Wang, C. et al. (2022a). *On the explanation properties of vision transformers.* arXiv preprint.

Wang, Y., et al. (2022b). Guided probing: Explainability through interpretable projections of joint embeddings. In *ECCV*.

Wang, J. et al. (2023). *Self-reflective large language models.* arXiv preprint.

Wei, J., et al. (2022a). Emergent abilities of large language models. *Transactions on Machine Learning Research*.

Wei, J., et al. (2022b). Chain-of-thought prompting elicits reasoning in large language models. *Advances in Neural Information Processing Systems, 35*.

Weidinger, L., et al. (2022). Taxonomy of risks posed by language models. In *ACM FAccT*.

Wolf, M. J., et al. (2019). A value-sensitive design approach to explainable AI. In *Proceedings of AIES*.

Wu, J. et al. (2022). *BEVFusion: Multi-task multi-sensor fusion with unified bird's-eye view representation.* arXiv preprint.

Wu, J., et al. (2023). Sequence modeling for visual pattern recognition of graph data. *IEEE Transactions on Neural Networks and Learning Systems*.

Xiong, R., et al. (2020). On layer normalization in the transformer architecture. In *ICML*.

Xu, K., et al. (2015). Show, attend and tell: Neural image caption generation with visual attention. In *ICML*.

Yuksekgonul, M. et al. (2022). *The curious case of object-centric compositionality.* arXiv preprint.

Zhao, J., et al. (2021). Learning gender-neutral word embeddings. In *Proceedings of EMNLP*.

Chapter 9
Making Science with Machine Learning and XAI

> *The hardest part of research is always to find a question that's big enough that it's worth answering, but little enough that you actually can answer it.*
>
> Edward Witten

This chapter covers:

- How to do physics models with ML and XAI
- Do we need causation to make science?
- How to effectively use ML and XAI in science

At the very beginning of this book, we tried to clarify the difference between the term interpretability and the term explainability. In that context, we said that **Interpretability is the possibility of understanding the mechanics of a machine learning model but this might not be enough to answer "Why" questions that are questions about the causes of a specific event.** We also provided in Table 9.1 of Chap. 1 (don't worry to look at it now, we will start again from this table in the following) a set of operational criteria based on question to distinguish between Interpretability as a lighter form of Explainability. As we saw, Explainability is able to answers questions about what happens in case of new data, "What if I do x, does it affect the probability of y" and counterfactual cases to know what would have changed if some features (or values) would not have occurred. Explainability is a theory that deals also with unobserved facts toward a global theory, while Interpretability is limited to make sense of what is already present and evident. The question is: Why are you getting back to this point in this chapter about making science with ML? The answer, long story short, is that Explainability is exactly what we need to climb "the ladder of causation" (we will talk about it in a while). We will use XAI in the domain of "knowledge discovery" with a specific focus on scientific knowledge. To recall what we already discussed:

Table 9.1 Time series

Time	Value
t1	Value1
t2	Value2
t3	Value3

Knowledge Discovery is the most complex application to comment, being related to situations in which ML models are not just used to make predictions but to increase the understanding and knowledge of a specific process, event, or system. The extreme case that we will discuss further in the book is the adoption of ML models to gain scientific knowledge in which prediction is not enough without also providing explanations and causal relations.

The main objective of this chapter is to touch with hands how to use ML + XAI to get knowledge and study a real physical system beyond the usual scope of ML that is just to make predictions. We will use this scenario to clarify all the limitations, opportunities, and challenges of exploiting ML to make science.

9.1 Scientific Method in the Age of Data

How does Google rank the pages to propose you the best one for your search? Or equivalently, how does Google match the ads with user's preferences? From our purposes, the answer is all in the words of Peter Norvig (Google's research director): "All models are wrong, and increasingly you can succeed without them."

Google's success is not based on any "understanding" of the content of the pages. There is no semantic or causal analysis; it is just a complex algorithm based on the relative number of links that set the ranking. The news is that this kind of approach is now being adopted in science and might replace the classical scientific method. To understand whatever phenomenon, scientists rely on model building: they try to narrow down the essential variables that affect the outcome, build an approximated model based on these variables and test the model with experiments.

The iterative process is (1) Make hypothesis (2) Build a model (3) Do experiments to test the model. The age of data may replace this with a new mantra: correlation is enough, we don't need to build models, as far as our objective is to predict an outcome, just feed the problem machine learning system with tons of data for the learning phase, and then you will have your results.

Despite the apparent oversimplification of the two alternatives, there is a lot to think about in such a state of things. On one side, science and in particular, physics is running into fields where experiments are not easy or not possible at all (think about cosmology). So that it could make sense to rely on the effective prediction of ML instead of beautiful models but driven only by the "beauty" of the related mathematics. On the other side, the question about the real understanding that we may get just from a DNN applied to physics instead of having a model remains open. And this question is the one we want to answer in this chapter leveraging XAI.

9.1 Scientific Method in the Age of Data

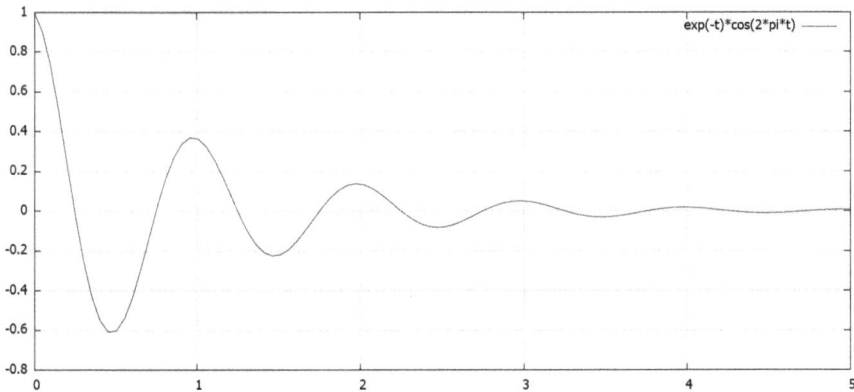

Fig. 9.1 Damped oscillation

The real case scenario that we will study to discuss and answer this question is to predict the position of a one-dimensional damped pendulum at different times. This is a straightforward physical system that can be easily solved with basic knowledge of physics and math. The observations are a set of couples in which we record the position of the pendulum x_i at different times t_i. The question for the ML system after training is: "Where is the pendulum at time t_k?" but as you can easily guess, this is not enough to say that we get full knowledge of the physics of the pendulum. We will see how to use XAI to achieve such an understanding.

Before getting into details, we need to make some further comments about this scenario. This should be new for our reader that is assumed to have basic in ML, but we want to emphasize the differences with what we did so far. The applications we worked out in this book deal with supervised and unsupervised ML models, but we never met time series like in this case.

Here, as we said, we have a damped pendulum, and we want to predict the x position at a time t (Fig. 9.1).

In this case, we don't have the usual input and output features like in the previous examples (e.g., we predicted and understood the most important features to assign the best player award for a football match in Chap. 4 based on the match metrics like the number of goals scored and so on). Time series needs to be reframed as a supervised learning dataset using "feature engineering" to construct the inputs that will be used to make predictions (in this case, it would be the position of the pendulum).

Basically, a generic time series appears like this

And we need to transform this couples of (time, value) to something like this to fit the usual supervised learning models (Table 9.2).

The most common classes of features that are created from the data series are:

Date Time Features: These are the timestamps for each observation, for example, they may be used to discover recurrent patterns and cycles of the target variable

Table 9.2 Supervised learning

Feature	Value
f1	Output1
f2	Output2
f3	Output3

Table 9.3 From Time series to supervised learning

Transform a time series to a supervised learning							
Supervised		Time series		From time series to supervised		
x	Y	Time	Measure	Time	Lag1 as x	Y
5	1	1	1	1	?	1
6	0	2	0	2	1	0
9	1	3	1	3	0	1
8	0	4	0	4	1	0
9	1	5	?	5	0	?

Lag Features: These are values at a previous time. The underlying assumption is that the value of the target at time t is affected by the value at the previous time step. The past values are known as lags.

Window Features: These are aggregated statistical vales of the target variable over a fixed window of time. This method is also known as the rolling window method because the timeframe to calculate the statistical values are different for each data point (e.g. rolling average)

Domain-Specific Features: These are at the foundation of feature engineering; the knowledge of the domain and the data guide the choice of the best features for the model.

So, the game is to pass from the time series to supervised case as in Table 9.3.

But having the time series adapted for a supervised learning is not enough for our goals. At this point we have the possibility to use the full supervised learning machinery to do predictions but how to learn about the physics of the damped pendulum?

The obvious answer that you should have, based on our journey so far, is that XAI is exactly for this: to extract explanations from the ML model that has been built to predict the pendulum position after the proper transformation of the time series. And the explanations in this case should allow us to learn the physics of the pendulum. But unfortunately, this is not the case…At least in general terms, the techniques we explained so far are not enough to get scientific knowledge we are searching for.

We don't know how many features may be generated by our feature engineering of the time series but the XAI methods we have would generate explanations about the most important features, rank them and in case answer specific question for a specific data point (as we saw with SHAP) but this is not enough. This level is what we called "interpretability" but here we need to answer questions in the domain of knowledge discovery: discover the causal relations and answer questions on unseen data.

9.2 Ladder of Causation

To touch with hands what we are saying, the point is that we will never get the physics model of the pendulum that relies just on spring constant and damping factor to get the full understanding of this system. XAI methods we learned won't identify these 2 physical variables as the ones needed to solve the physics of the damped pendulum.

To get to this level we need a different approach that is the core of this chapter. But before getting into this, we need to climb "The Ladder of Causation" in order to better understand:

(1) The limitations of the XAI methods we presented if applied to knowledge discovery
(2) Clarify, based on the point 1 above, once and for all what we mean by Explainability in comparison with Interpretability. At this point we have the skills, and we can get into details of this with the real case scenario of the damped pendulum

9.2 Ladder of Causation

In Chap. 1 we used Table 9.1 to distinguish between Interpretability and Explainability, we put it here again for simplicity and as core of our reasoning (Table 9.4):

Getting back again to Gilpin's words: "We take the stance that Interpretability alone is insufficient. For humans to trust black-box methods, we need Explainability – models that can summarize the reasons for neural network behavior, gain the trust of users, or produce insights about the causes of their decisions" (Gilpin et al., 2018). As we discussed, the table uses two different sets of questions in order to distinguish between Interpretability and Explainability; there are questions that cannot be answered in the domain of Interpretability. To make sense of these questions

Table 9.4 Difference between Interpretability and explainability in terms of the questions to answer for the 2 different scopes

Question	Interpretability	Explainability
Which are the most important features that are adopted to generate the prediction or classification?	✓	✓
How much the output depends on the input? How sensitive is the output on small changes in the input?	✓	✓
Is the model relying on a good range of data to select the most important features?	✓	✓
What are the criteria adopted to come across the decision?	✓	✓
How would the output change if we put different values in a feature not present in the data?	✗	✓
What would happen to the output if some feature or data had not occurred?	✗	✓

that, as we will see better in this chapter, are fundamental for scientific knowledge, we will follow the seminal work of Pearl (2019) on causality.

The core of Pearl's line of research can be summarized with a picture of the ladder of causation (Fig. 9.2):

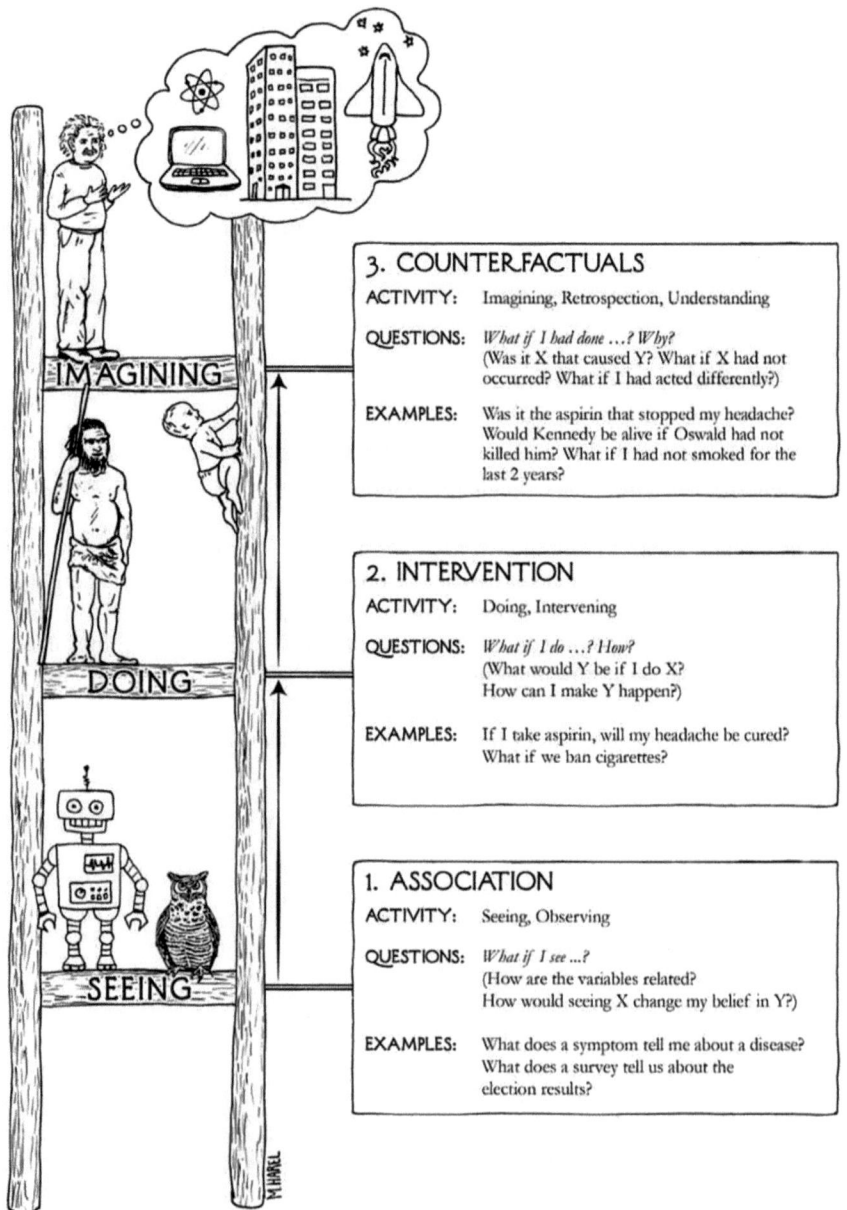

Fig. 9.2 Ladder of causation (Pearl & Makenzie, 2019)

9.2 Ladder of Causation

There are three different types of cognitive abilities that are needed to climb the ladder of causation: seeing, doing, and imaging. The highest level that is imagination is what allowed for the impressive progress that humans did from our homo sapiens ancestors until our age of data. Already home sapiens were using imagination to think about situations that were just potential, planning for hunt meant to plan for "unseen" things that could happen.

At the base of the ladder, we have association that is connected to the activities of seeing and observing. What we do here is to search for patterns, regularities in what we see or in a huge amount of data. The goal is to find correlations that may help us to do predictions so that observing one event may change the probability of observing the other.

The classical examples that are specific to this domain come from marketing. Imagine being a marketing director that wants to understand how likely it is that a customer that bought an iPhone also bought an iPad. The answer will be based on collecting the data, segmenting the customers, and focusing on the group of people that bought an iPhone. Then we may compute the proportion of these people that get also and iPad: this is just a way to compute the conditional probability of an event given another based on existing data: $P(iPad|iPhone)$.

When we explained the XAI methods of linear regression or logistic regression, we just did this: search for correlations in the data that are not necessarily linked to causal relations. As in this case, we cannot say that buying an iPad is the cause of buying an iPhone or vice versa, but for our marketing purposes, it is enough to know the degree of association between the two events. The predictions coming from the first rung of the ladder are based on passive observations, and the related XAI methods answer the questions placed in the interpretability column. Moving from the marketing scenario to our damped pendulum, we can predict the position of the pendulum through the feature engineering of the time series, and we may narrow down the most important features with XAI but without any knowledge discovery. Using Searle's words: "Good predictions need not have good explanations. The owl can be a good hunter without understanding why the rat always goes from point A to point B. Some readers may be surprised to see that I have placed present-day learning machines squarely on rung one of the Ladder of Causation, sharing the wisdom of an owl".

But this will be clear progressing with an investigation of the second and third rung of the ladder, and indeed we will see how to tackle this state of things and do science with the recent progress of ML and XAI.

Climbing to rung two, we enter the domain of doing instead of seeing; this is different from the first rung in which we just did associations on existing data. In this case, we want to know how predictions would change if we do a specific action. The difference is a bit tricky. Remember the marketing case of the first rung in which we investigate the selling of iPads conditioned on the selling of iPhones. A typical question of rung two would be: What if I double the price of iPhones?? Would the relation with the sales of iPhones change? To answer such types of questions, we cannot easily rely on the collected data. Albeit we may find in our huge database the data referring to the case of iPads with doubled price, this is totally different from an intervention on the current market in which we double the price of iPhones. The existing data we may find would likely refer to a totally different background in which the price was double for different reasons (short of supply?). But here are not asking for the probability of selling iPads conditioned on the sales of iPhone of a certain price, but we are asking for the probability of selling iPads condition on the sales of iPhones with the intervention of doubling the price of iPhones. And in general, as shown by Pearl with his casual diagrams:

P(iPad|iPhone) is different from P(iPad|do(iPhone))

How can we move to this rung of intervention? What is usually done is to do experiments in controlled environments like the big companies usually do: think about Amazon, changing or suggesting items with a different price to a selected set of customers to see how it goes. If experiments are not possible, the alternative is to do a kind of "causal model" of the customer, including market conditions. This causal model is the only option to go from observational data of rung one to answer the question on rung two that assumes an explicit intervention.

Despite the wording that is not so common, intervention is something that belongs to our daily lives. Every time we decide to take a medicine for a headache, we are doing a rung two intervention in which we are implicitly modeling a causal relation between the medicine we took and the headache. Our belief is based on the "controlled" experiments that showed that the medicine is expected to remove headache. But there is another last step to climb the ladder to rung three that is the one of pure scientific knowledge. Rung three is the domain of counterfactual questions like What if I had acted differently? This means to change something that already happened in the past, change the course of what is previous in time that is different from doing an intervention and see what happens. The world in which an action has not been performed does not exist because it has already passed. That's why the typical activity of rung three is imaging.

What would have happened if I had not taken medicine? The data to answer this question do not exist by definition. But albeit weird, this is exactly what we need, and we use to do science. The law of physics can be thought of as counterfactual assertions. Let's think about again our damped pendulum. As soon as you understand the physics of the systems, you will know that there are just two variables that control the pendulum's dynamics: the spring constant and the damping factor (we will details this later). Relying on these two constants, we have the full functional relationship to predict the position at any time t. How did we get this causal model? Starting from rung one of the associations, climbing to rung 2 in which physicists may experiment with different pendulums, in different conditions and different values of spring constant and damping factor so to reach rung three that allows answering questions for any possible instances of the damped pendulum. Having the causal model means that there is no difference between the existing world in which you have the real observations of the time series for the pendulum and whatever imaginary world with hypothetical values for the same constants.

Back to the marketing, being in rung three would allow answering questions like "What is the probability that a customer who bought the iPad would still have bought it in case of a double price?" In the real world, this didn't happen, but we have a model to answer this.

Rung three is the proper level of causal modelling that is needed to do science. And from our point of view, this is Explainability as a stronger version of Interpretability that, with the XAI methods we saw, helps in answering questions up to rung two. But we will see in the next section how to touch with hands XAI methods that will allow us to work with ML to acquire true scientific knowledge; the approach, based on what we said so far, will be different from just searching for the most important features but it will deal with building a proper representation of the physics of the damped pendulum in order to allow ML to answer questions belonging to rung three.

Before jumping to the next section, we also propose a short optional content that you are free to skip (not fundamental to get the core of the rest), but it could be interesting to further detail the problem of causal knowledge. We will talk about Pearl's mini Turing test.

> The question behind mini Turing-test is the following: How can a ML system can represent causal knowledge in order to answer rung three questions on a simple story? As you may see this is a variation of original Turing test and Searle's Chinese room, we discussed in Chap. 1. Let's look at it through an example, presented by Pearl, that is about a firing squad in which there is a dramatic chain of events in which a Court Order is received by a Captain that pass it to 2 soldiers whose task is to fire to prisoner D if order is received. The soldiers fire only if the command is received and if they fire (one of them is enough) Prisoner D is dead. We can suppose that our ML system is trained on data that records different set of the five variables with different states (Fig. 9.3).

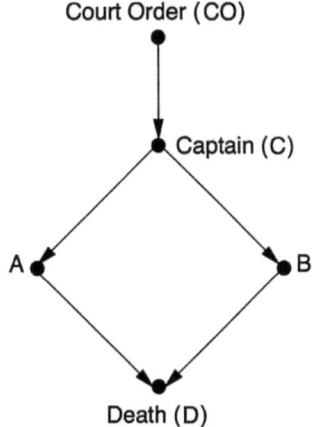

Fig. 9.3 Causal diagram for the firing squad example (Pearl & Makenzie, 2019)

Let's ask questions as examples from the different rungs of the ladder:

Rung 1: If the D=true (prisoner is dead), does it mean that that CO=true (the order has been given)?
This is pretty trivial, and even without any deep dive of the causal diagram, the answer is yes. It would be enough for our ML to track the associations on the five variables to make the proper prediction. (Fig. 9.3)

Rung 2: What if soldier A decides to shoot without the order? This is a tricky question. Let's look at how our causal diagram would change. (Fig. 9.4)
The intervention removed the link between CO and A, A is true without CO being necessarily true, prisoner is dead whatever CO is because A fired. And this is exactly the difference between "seeing" and "doing", climbing from rung one to rung two. B is untouched by the intervention, before the intervention A and B were necessarily coupled, both depending on CO. Our intervention allowed D=true with A=true and B=CO=false. Assume that the ML system doesn't have the causal representation of this story, it would not pass the mini-Turing test, because ML would be trained on thousands of records of executions but "normally" all the variables would be all true or all false. ML system, would not have any way to answer question about what would happened to the prisoner in case we persuade A not to fire without having a causal representation of the relations among the events. And a rung three question would follow the same logic:

Rung 3: Suppose to have D=true, prisoner is dead. What would have happened if A had decided not to shoot? That is to compare the real world with a fictious world in which A didn't shoot. And the causal diagram, helps us as well making clear the state of the CO=true, B=true and D=true (Fig. 9.5)

The prisoner would have died also in the imaginary world.

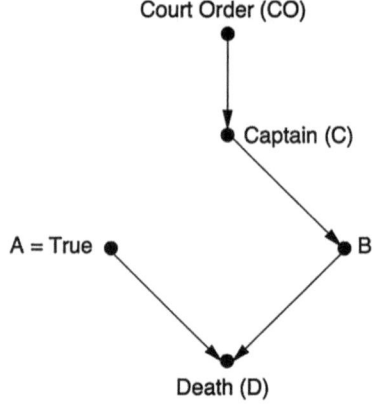

Fig. 9.4 Case of intervention, the link between C and A is removed, A is set to true whatever C (Pearl & Makenzie, 2019)

9.3 Discovering Physics Concepts with ML and XAI

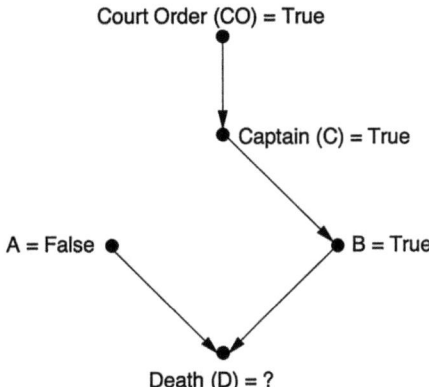

Fig. 9.5 Counterfactual reasoning, D is set to dead, what would have happened if A had not fired? (Pearl & Makenzie, 2019)

9.3 Discovering Physics Concepts with ML and XAI

Based on what we discussed in the previous sections, we are now ready to climb our ladder of causation to get the physics of a damped pendulum with Machine Learning and XAI.

To recap the flow, the usual approach of ML to a problem like the damped pendulum would be to set a neural network to train on the time series after a bit of feature engineering. This would work fine to forecast the position of the pendulum at different times and we may use the XAI methods we learned in the previous chapters to get some insight on the most important features. But this would not allow us to climb our ladder of causation up to rung 3 that is the domain of counterfactuals and knowledge discovery that is, in this case, the physics of the damped pendulum. We will show in this section how to tackle this challenge, we need to change our perspective and rely on a different type of neural networks to focus more on the proper "representation" than on just the predictions. To do this we will adopt the so called autoencoders that are well known in the ML field, but here we will look at them for our specific knowledge discovery purpose.

9.3.1 The Magic of Autoencoders

Current architectures for Artificial Neural Networks in Machine Learnings are made build of huge number of layers and thousands of nodes connected in different ways. The different topologies of the connections produce the specific behavior of the networks and shape their ability to perform specific computational tasks.

A feedforward NN may look like this (Fig. 9.6):

There is an input layer that transmits the values to internal layers that do the computation to produce the output. As we know, the training of the NN is basically the process to find the right weights of each node in the internal layers to produce

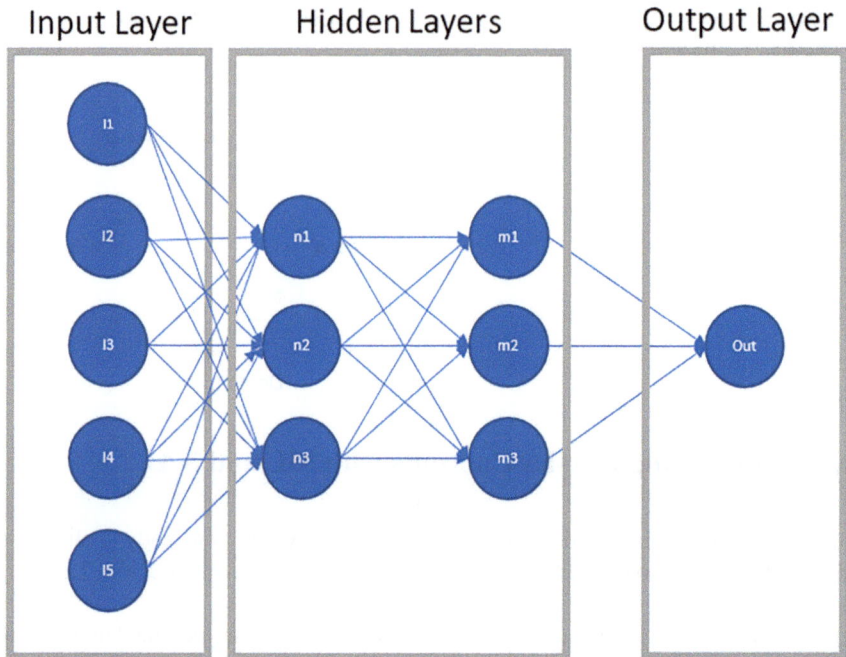

Fig. 9.6 A generic feedforward neural network

the proper output. We recall this basics info about a general feedforward neural network to compare it with the logic of autoencoders.

Autoencoders are a specific type of NNs that are not used to make predictions but just to reproduce as output the input that is provided. This could appear as a silly game, let's think about a topology like the one in Fig. 9.7:

It is a fully connected model of NN and it is pretty obvious that if our goal is just to reproduce the output, the training would generate a solution like the one below, a kind of Identity Matrix that just propagates the input to the output without any added value (Fig. 9.8).

Now, imagine reducing the number of nodes in the hidden layers (Fig. 9.9). This prevents the autoencoder from simply copying the input to the output, forcing it instead to perform a form of compression in order to accurately reconstruct the input.

The internal layers have a reduced number of nodes, and this makes the autoencoder to represent the information in a compressed form. Basically, the autoencoder follows a two-step process: an encoding phase in which the input is reduced and a decoding process in which the input is "reconstructed" to provide the output. The loss function, in this case, will measure how much the output is different from the input.

The more general architecture looks like this:

9.3 Discovering Physics Concepts with ML and XAI

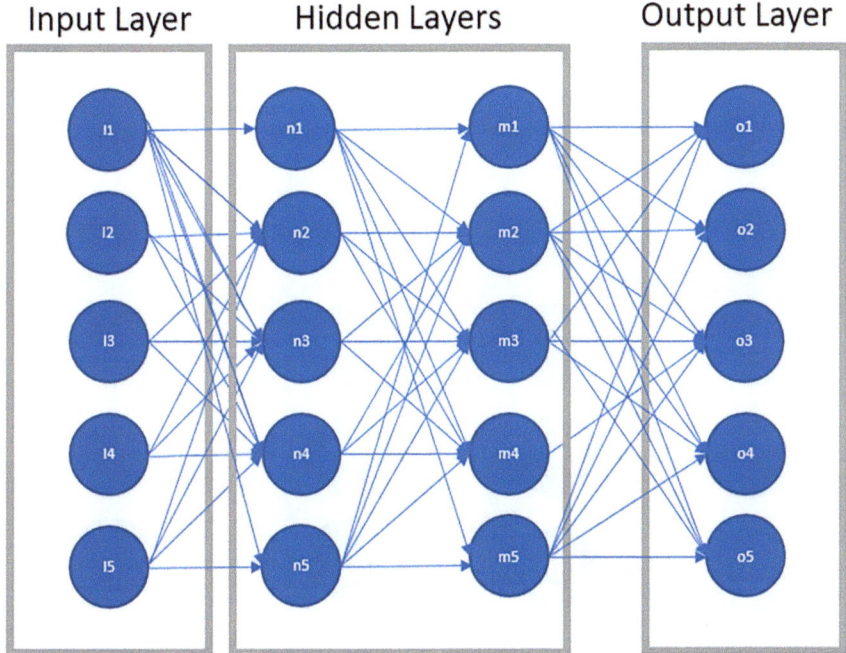

Fig. 9.7 Fully connected neural network

Note that the decoder has the same (mirrored) architecture as the encoder part; this is very common but not a must (Fig. 9.10). Assume you have as input a handwritten number; the flow appears like in Fig. 9.11.

I can guess your question at this point: but what is the relation of this stuff with our goal of getting the physics of damped pendulum? We need a little more patience, but we may anticipate that the answer is the "Code" layer that you see in the middle of the encoding-decoding process. Understanding the physics will mean using the autoencoder to find the most "compact" representation of the physical system, which is to find the relevant physical variables to have a complete model of the damped pendulum.

What we call here "Code", also called latent space representation, which is the layer that contains the compressed information to reproduce the input (number 4 in the example), will contain the physics of the damped pendulum.

The only additional step we need to do before putting hands into the specific problem of the damped pendulum is to remark that we will use a slight variation of the autoencoders names variational autoencoders. For our purposes, it is enough to get that the main feature of VAEs compared to AE is that VAEs don't just learn a function to get a compressed form of the input, but it is a generative model. VAEs learn a function that is also able to generate variations around the "model" learned by the input; that is what we are searching for: we don't want to solve just the case of our specific damped pendulum but learn the physics of whatever damped pendulum to answer questions on variations.

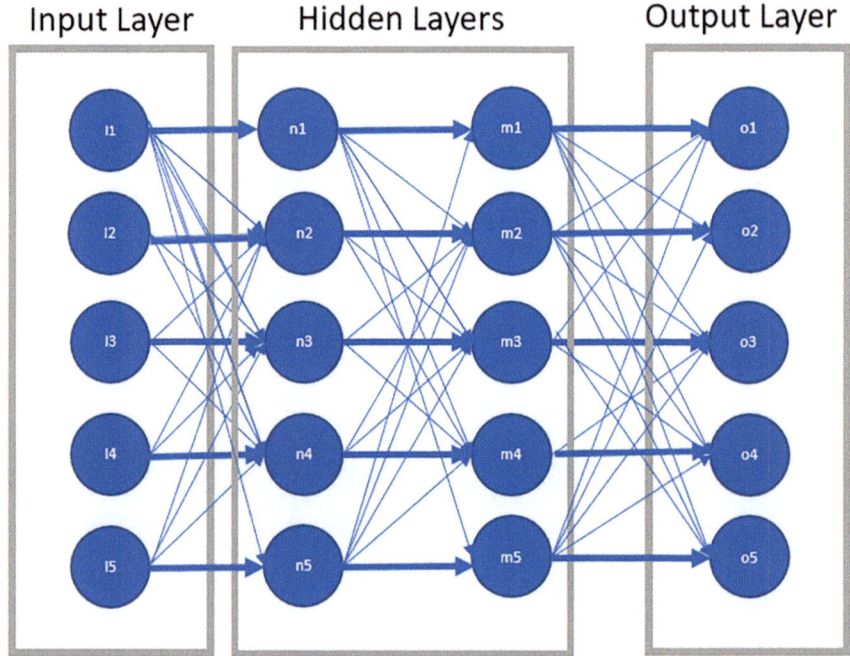

Fig. 9.8 Trivial neural network topology

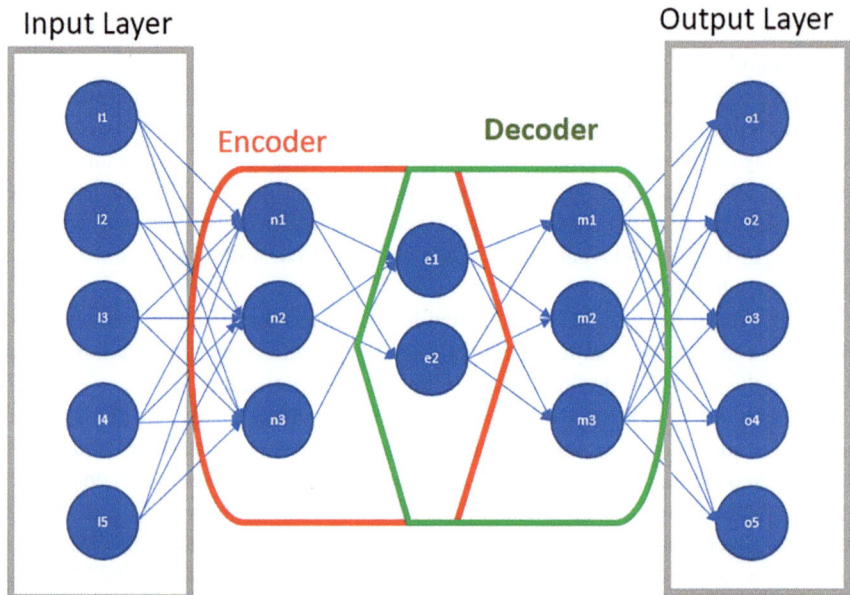

Fig. 9.9 Autoencoder topology

9.3 Discovering Physics Concepts with ML and XAI

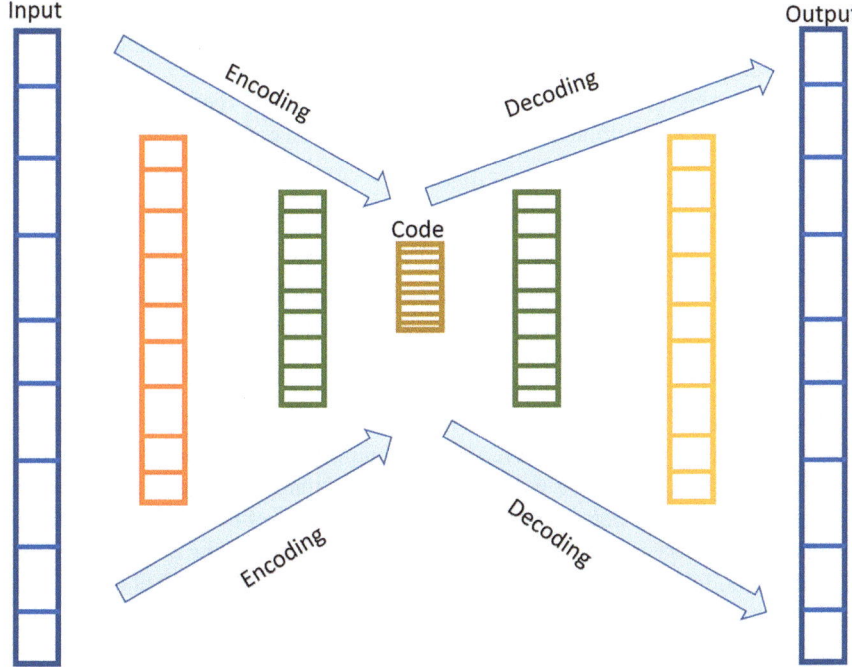

Fig. 9.10 Autoencoder process

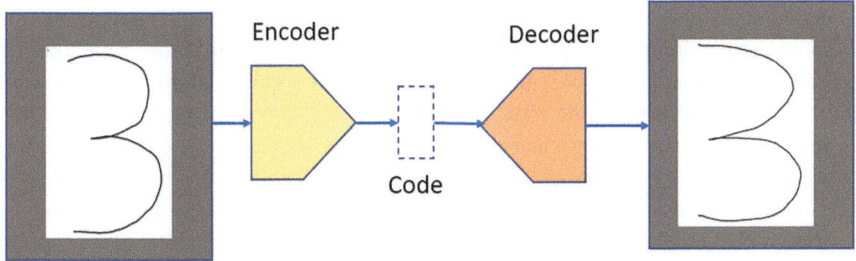

Fig. 9.11 Number recognition with autoencoder

9.3.2 Discover the Physics of Damped Pendulum with ML and XAI

After this bunch of theory, let's clearly formulate again the practical scenario we want to address. As XAI experts we are asked to get knowledge on the physics of a system with ML. We discover that just doing predictions and use common XAI techniques is not enough because we want to answer questions in rung three of our ladder. The physical system we will use to explain the approach is a damped pendulum based on the work of Iten et al. (2020). No need to remark that this is a very simple physical system, but despite this, it will allow us to show the direction and the techniques to use for our purpose.

We have a time series x_t where x_{t_i} represent the position of the damped pendulum at the time t_i. After the training, we want to achieve two goals with our ML system:

(1) Predict position on a new set of times t_k
(2) Find the most compact representation to discover the physics of the system

While (1) can be achieved via the standard approach to time series, we will rely on variational autoencoders we discussed to tackle point (2). We will force the NN to minimize the number of neurons in the latent representation, and the expectation is to get the physics of the system through the inspection of these neurons after the training and after checking that the predictions are good enough.

The general case is that we don't know anything about the system we want to study, so our choice will be to start a minimal number of latent neurons to find the best representation. Then we will make the predictions, and in case the accuracy will be low on the test data, we will increase the number of nodes in the latent representations. Otherwise, we will look at the weights stored in the latent representation as a good model of our physic system. Those values will represent the physical values that fully describe the damped pendulum.

Figure 9.12 shows the general architecture of our VAE (b) to study the damped pendulum and compares it to the general approach we follow in physics model a system (a): we start from observations, build a mathematical representation and use this representation to predict output in different cases.

In the case of autoencoders (Fig. 9.12b), the experimental observations are encoded in a compressed representation. The process of decoding is to predict positions of the pendulum (or a generic physical system) at specific times based on the learned representation. As we said, the representation is called latent representation, and our main focus is on the representation in order to understand how much this representation is able to learn and reproduce the physics of the systems. Just predictions on pendulum positions could have been obtained with a variety of ML models but what we are doing here is to constraint the number of nodes in the latent representation to get the physics. As an example, we may obtain 99.99% accurate

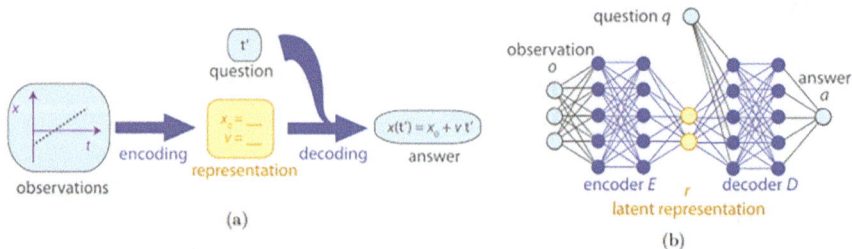

Fig. 9.12 Learning representation with human learning (**a**) vs. SciNet neural network (Iten et al., 2020)

9.3 Discovering Physics Concepts with ML and XAI

predictions with a deep multi-layer neural network, but such an architecture would not allow the discovery of physics.

We can have a brief look at the code, focusing on the most important parts. The code is taken and adapted from the work we already cited (Iten et al., 2020) and the implementation performed by Dietrich (2020).

For our purposes, we use the trained model provided by the authors and then comment on the results.

```
import torch
import numpy as np
from models import SciNet
from utils import pendulum
from matplotlib import pyplot as plt
from mpl_toolkits.mplot3d import Axes3D

# Load trained model
scinet = SciNet(50,1,3,64)
scinet.load_state_dict(torch.load("trained_models/scinet1.dat"))

# Set pendulum parameters
tmax = 10
A0 = 1
delta0 = 0
m = 1
N_SAMPLE = 50
for ik, k in enumerate(np.linspace(5,10,size)):
    for ib, b in enumerate(np.linspace(0.5,1,size)):

        tprime = np.random.uniform(0,tmax)
        question = tprime
        answer = pendulum(tprime,A0,delta0,k,b,m)
        if answer == None:
            continue
        x = np.linspace(0,tmax,50)
        t_arr = np.linspace(0,tmax,N_SAMPLE)
        x = pendulum(t_arr,A0,delta0,k,b,m)
        combined_inputs = np.append(x, question)
        results = scinet.forward(torch.Tensor([combined_inputs]))

        latent_layer = scinet.mu.detach().numpy()[0]


        neuron_activation[0][ik,ib] = latent_layer[0]
        neuron_activation[1][ik,ib] = latent_layer[1]
        neuron_activation[2][ik,ib] = latent_layer[2]
```

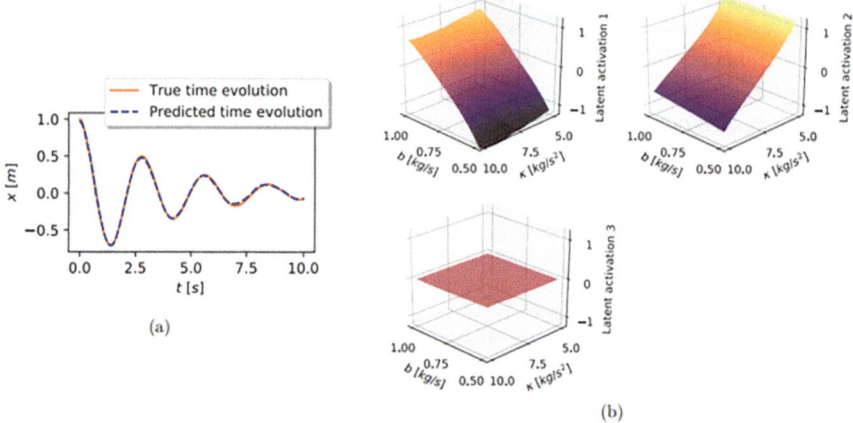

Fig. 9.13 Damped penduolum: (**a**) Real evolution compared to the predicted trajectory by SciNet. (**b**) Activation plot, representation learned by SciNet using the 3three latent neurons (Iten et al., 2020)

Scinet is the name of our NN, and in the line, in red, we see how to collect the results through torch open source library.

For us, as we said, it is important to investigate the latent layer (assuming that the predictions are good, as we will see) to look at the relevant physical variables.

Figure 9.13 shows the true time evolution vs. the predicted time evolution of the damped pendulum and the representation learned by SciNet. The activation plots (b) clearly show as SciNet uses 2 of the three neurons in the latent representation to store the spring constant k and the damping factor b. The third neuron is not used as further confirmation of the fact that all the physics of the damped pendulum is "condensed" in the two meaningful variables of the latent representation.

Box below (Fig. 9.14) that is directly extracted from Iten et al. (2020) is very useful to summarize the problem and the findings:

This is an "easy" case that shows the general ideas about how to get knowledge discovery with ML and XAI. As properly commented by Iten et al. (2020) this is not a full solution of how to get an explanation about the latent variables. In our terminology, we are trying to get to rung 3 of the ladder, but we are not yet there. In this specific case, we got knowledge discovery but comparing the latent representation to the well-known model of the damped pendulum we have from physics. More generally, we may want to get this knowledge discovery through a learned representation but without any guidance from physics. We will see in Sect. 6.4 the most promising directions to do science in this way. Despite these limitations, we think it is pretty impressive to see how we can change the perspective and look at the ML models not just to have predictions but to gain insight and understanding of the system we are studying.

9.3 Discovering Physics Concepts with ML and XAI

Problem: Predict the position of a one-dimensional damped pendulum at different times.

Physical model: Equation of motion: $m\ddot{x} = -\kappa x - b\dot{x}$.

Solution: $x(t) = A_0 e^{-\frac{b}{2m}t}\cos(\omega t + \delta_0)$, with $\omega = \sqrt{\frac{\kappa}{m}}\sqrt{1 - \frac{b^2}{4m\kappa}}$.

Observation: Time series of positions: $o = [x(t_i)]_{i \in \{1,\ldots,50\}} \in \mathbb{R}^{50}$, with equally spaced t_i. Mass $m = 1$kg, amplitude $A_0 = 1$m and phase $\delta_0 = 0$ are fixed; spring constant $\kappa \in [5, 10]$ kg/s^2 and damping factor $b \in [0.5, 1]$ kg/s are varied between training samples.

Question: Prediction times: $q = t_{\text{pred}} \in \mathbb{R}$.

Correct answer: Position at time t_{pred}: $a_{\text{cor}} = x(t_{\text{pred}}) \in \mathbb{R}$.

Implementation: Network depicted in Figure with 3 latent neurons.

Key findings:
- SciNet predicts the positions $x(t_{\text{pred}})$ with a root mean square error below 2% (with respect to the amplitude $A_0 = 1$m)
- SciNet stores κ and b in two of the latent neurons, and does not store any information in the third latent neuron

---

Fig. 9.14 Summary of the findings, adapted from Iten et al. (2020)

9.3.3 Climbing the Ladder of Causation

We used the real case scenario of damped pendulum to touch with hands what it really means to climb the ladder of causation and go from the level of pure association to the rung of using ML in the domain of knowledge discovery. In order to further reinforce this approach, we look at it from a different angle exposed by Karim et al. (2018).

The three levels proposed by these authors exactly match the three rungs of the ladder of causation but from a mathematical perspective (Fig. 9.15).

Looking at Fig. 9.15 we see the flow from data to scientific knowledge:

(1) **Statistical Modeling**: this is the level of correlation, the ML model just learn these correlations from the data, then XAI may help on answering questions about the most important features and the sensitivity of the output to a specific change in a feature. As we see, from a mathematical point of view it corresponds to p(y|x).

(2) **Graphical Causal Modeling**: this level receives as input the subset of features that have been already distilled and filtered from the statistical modelling as the most important ones. We enter into the domain of "interventions" as mathematically expressed by the do(x) as in the rung 2 of the ladder of causality.

(3) **Structural Equation Modeling** is the last level that deals with counterfactuals. It received filtered causes emerged from the causal Modeling and put them into the form of structural equation modeling to produce knowledge. In our example this meant to go from time series up to the physical model of damped pendulum.

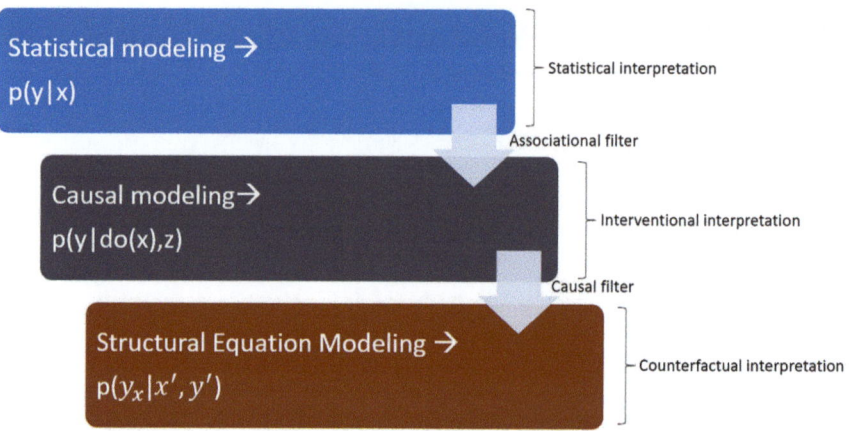

Fig. 9.15 Three tier explainability (Karim et al., 2018)

The main idea behind this approach is to evolve XAI from a pure engineering approach to the possibility of doing science with ML. In the next section, we will see some options that are emerging to effectively use ML in scientific discovery.

9.4 Expanding the Role of Machine Learning in Scientific Discovery: The Impact of Large Language Models

The application of Machine Learning (ML) in scientific research has evolved significantly in the past decade. The early paradigm revolved around leveraging ML for pattern recognition, classification, and numerical predictions, with explainable AI (XAI) providing a means to extract insights from black-box models. However, the recent advent of Large Language Models (LLMs) has introduced a new layer of complexity and potential, fundamentally altering the landscape of AI-driven scientific discovery. Unlike previous generations of ML systems, which relied heavily on structured data and well-defined mathematical frameworks, LLMs enable interaction with unstructured scientific knowledge, engage in hypothesis generation, and even assist in conceptual reasoning. These new capabilities raise profound questions: Can LLMs genuinely contribute to scientific knowledge generation? Do they transcend the predictive paradigm of traditional ML to enter the realm of discovery? And how can XAI help us validate their contributions?

A key limitation of traditional ML in scientific discovery was its heavy reliance on predictive modeling without an intrinsic understanding of the underlying principles governing physical or natural systems. As discussed earlier in this chapter, ML has been employed successfully in scientific fields where data-driven correlations suffice, such as climate modeling, particle physics, and genomics. However,

9.4 Expanding the Role of Machine Learning in Scientific Discovery: The Impact...

this approach often falls short of producing genuine scientific insight because it operates at the first rung of Pearl's "Ladder of Causation" (Pearl & Makenzie, 2019).

LLMs, by contrast, introduce a fundamental shift: they do not merely infer patterns from structured datasets but engage in higher-order reasoning by integrating vast corpora of scientific literature, mathematical derivations, and experimental data. Si et al. (2024) conducted a large-scale human evaluation of AI-generated scientific hypotheses. They found that LLMs often propose research questions that are judged to be more *novel* than those generated by human scientists, although their feasibility remains uncertain. This suggests that LLMs are not just optimizing for past knowledge but synthesizing ideas in ways that resemble human scientific reasoning.

Nevertheless, an important critique remains: LLMs lack *grounded* scientific intuition. Unlike physicists or biologists, LLMs do not experiment or interact with the physical world. Their "reasoning" is purely statistical, based on cooccurrence patterns in text. The fundamental question then becomes: How can LLMs be integrated into the scientific workflow without over-relying on their pattern-matching capabilities?

One of the most promising uses of LLMs in scientific research is their ability to act as *hypothesis generators* rather than just analytical tools. Zhang et al. (2024) categorized over 260 scientific LLMs that specialize in disciplines such as mathematics, physics, chemistry, and life sciences. These models demonstrate capabilities such as:

- Suggesting new chemical compounds for drug discovery.
- Assisting mathematicians in theorem proving.
- Identifying possible gaps in experimental designs.

For instance, LLMs have been integrated into physics research through models like Galactica (Taylor et al., 2022), which was designed to summarize and generate new scientific literature based on structured and unstructured physics data. Similarly, in mathematics, deep learning models like AlphaGeometry have been used to solve complex geometric proofs that were previously considered intractable for automated systems.

What sets LLMs apart is their ability to *contextualize* existing knowledge. Unlike earlier ML methods that required explicitly encoded domain knowledge, LLMs ingest and synthesize vast amounts of diverse information, making unexpected conceptual connections that may elude human researchers. This aligns with recent discussions on the role of AI in abductive reasoning—a form of logical inference that seeks the most plausible explanation for a given set of observations. While humans perform this type of reasoning naturally, LLMs now exhibit early-stage abilities to infer connections that suggest novel scientific insights.

Despite these promising developments, the use of LLMs in scientific discovery raises a crucial issue: the distinction between *scientific validity* and *linguistic plausibility*. Since LLMs are trained to generate the most probable sequences of words rather than to discover truth, their outputs can be misleading, even when they sound convincing. This problem is particularly concerning when LLMs attempt to engage in hypothesis generation. Without a rigorous method to validate their suggestions, they risk flooding the scientific discourse with superficially plausible but ultimately flawed ideas.

This is where Explainable AI (XAI) becomes essential. XAI offers mechanisms to:

1. Distinguish between genuine scientific insights and linguistic artifacts. By analyzing how an LLM arrived at a hypothesis—whether by synthesizing multiple sources, extrapolating from a coherent model, or simply generating plausible-sounding statements—we can assess its scientific reliability.
2. Enhance trust in AI-assisted discovery. Researchers are more likely to accept AI-generated hypotheses if they understand the underlying reasoning. Tools such as SHAP values, attention heatmaps, and counterfactual analysis can help uncover the logic (or lack thereof) behind an LLM's suggestion.
3. Enable human-AI collaboration in scientific research. XAI techniques can be integrated into AI-assisted research workflows, ensuring that scientists engage with LLM-generated hypotheses critically rather than accepting them at face value.

Recent works propose hybrid models where human researchers use XAI-assisted interfaces to iteratively refine LLM-generated research ideas, filtering out spurious suggestions and enhancing the most promising ones. This aligns with the broader vision of using AI as an "augmented intelligence" tool rather than as an autonomous scientific agent.

An exciting extension of LLMs in science is their integration with multimodal AI models—systems that can process and reason across multiple data types, such as text, images, equations, and experimental graphs. Recent advances in multimodal LLMs (G-LLaVA, OpenBioML) show promising results in bridging the gap between textual knowledge and empirical data.

For example, in molecular biology, AI systems that combine language models with protein structure predictors (like AlphaFold) are revolutionizing our understanding of protein folding and drug interactions. Unlike earlier ML approaches, which relied solely on numerical simulations, LLM-driven multimodal systems can generate new hypotheses based on both textual descriptions of molecular interactions and 3D structural data.

In astronomy, multimodal AI is being explored to interpret vast astronomical datasets, linking spectral data, simulation results, and theoretical physics papers into a unified framework for discovery. This capability aligns with the broader goal of making AI-driven research more holistic and less constrained by data format limitations.

LLMs represent a paradigm shift in the relationship between AI and scientific discovery. While traditional ML systems focused on *pattern recognition and prediction*, LLMs open new doors by engaging with unstructured knowledge, proposing novel research questions, and interacting with multimodal data. However, their integration into the scientific method requires careful scrutiny—without XAI and rigorous validation mechanisms, their potential contributions could be undermined by their lack of intrinsic understanding.

The challenge ahead is not just about making LLMs more powerful but about making their scientific reasoning transparent and accountable. By combining LLMs with XAI, human oversight, and multimodal extensions, we move closer to an AI-augmented scientific process—one that enhances, rather than replaces, human intuition in the pursuit of knowledge

9.5 Rethinking the Damped Pendulum: From Prediction to Discovery with LLMs

In the previous sections of this chapter, we used the example of a damped pendulum to explore how machine learning models, when paired with XAI techniques, can move beyond simple predictions and approach the realm of scientific discovery. We showed that training a neural network on time series data of the pendulum's position allows for accurate predictions of its future states, but this alone does not yield scientific knowledge about the physical system—namely, the spring constant k and the damping factor b that govern the pendulum's dynamics. To extract this knowledge, we climbed the "Ladder of Causation," moving from mere correlations to causal relationships, ultimately using autoencoders to uncover the latent physical variables.

But how does this scenario change in the presence of Large Language Models (LLMs)?

Let's imagine a new experiment where we combine the predictive power of ML with the generative and reasoning abilities of an LLM. The goal is no longer just to predict the position of the pendulum over time, but to push the AI system toward forming hypotheses about the underlying physics—essentially, to ask:

- *What laws of motion could explain the observed data?*
- *How would the system behave under untested conditions?*
- *Which variables might be missing from the current model?*

We start with the same damped pendulum time series, feeding the data into a predictive ML model. However, this time, we introduce an LLM into the workflow—not just as a passive observer but as an active collaborator as summarized in Table 9.5:

The crucial shift is that the LLM moves beyond feature importance and correlations—which we previously extracted using XAI—and actively engages in *hypothesis generation*. While the neural network deals with data-driven predictions, the LLM synthesizes scientific concepts, proposing untested variables and causal relationships that would not naturally emerge from pure statistical learning.

The predictive model might tell us that time lags and oscillation frequencies are the strongest predictors of pendulum position. The LLM, however, can "intuit"—based on its training in physics literature—that these patterns are symptomatic of an underlying physical law: *Hooke's law modified by a damping term.*

Table 9.5 LLM workflow

Stage	Description	Example/prompt
Data collection and prediction	The neural network, trained on pendulum position data over time, predicts future positions based on feature-engineered time series data	N.A.
Hypothesis generation with LLM	Once the predictive model identifies the most influential features (e.g., time steps, velocity approximations), an LLM is prompted to propose physical variables and relationships that might explain the system's behavior	"Given the following time series data representing the position of a damped pendulum over time and the identified key features, propose physical variables and relationships that might explain the system's behavior"
Emergent scientific reasoning	The LLM generates hypotheses about the underlying physics, considering causal relationships and scientific principles	"There may be external forces or nonlinear effects. Have you considered testing for air resistance or external perturbations?"
Experimental proposals	The LLM suggests experiments to test the generated hypotheses, introducing new variables or modifying conditions	"Introduce an external periodic force to test for resonance effects—this could reveal additional hidden variables"
Validation and iteration	Using XAI techniques like counterfactual explanations, the scientist collaborates with the LLM to validate or refine the hypotheses by exploring "what-if" scenarios and comparing predicted trajectories	"What if the spring constant were doubled? How would the pendulum's motion change?"

If we extend this further, a multimodal LLM could process not only the time series data but also visualizations of the pendulum's motion (graphs, phase space plots) and scientific text.

The LLM could then propose new mathematical models, combining visual patterns with textual explanations—much like how a scientist might work across charts, formulas, and intuitive reasoning.

Ultimately, this toy model highlights how LLMs extend the scientific method within AI research. Rather than stopping at prediction, they bring the process closer to how human scientists operate—generating hypotheses, proposing experiments, and iterating on theoretical models. The key is not that LLMs *discover* physical laws autonomously but that they act as intellectual amplifiers, bridging the gap between raw data and scientific theory.

In this way, the damped pendulum ceases to be just a testbed for time series forecasting and becomes a playground for AI-driven scientific reasoning, blending predictive ML, generative LLMs, and explainable AI into a single, iterative loop of discovery.

9.6 Science in the Age of ML and XAI

The goal of this section is to discuss the general ideas and options available for conducting scientific research using Machine Learning (ML), Large Language Models (LLMs), and Explainable AI (XAI). As we saw in the case of the damped pendulum, reaching the third rung of Pearl's "Ladder of Causation" to achieve true knowledge discovery is a challenging endeavor. However, we have laid the foundation for understanding how autoencoders, latent variable extraction, and interpretability techniques may help in this direction. More broadly, even when ML cannot independently uncover the fundamental physics of a system, other possibilities arise—particularly with the integration of LLMs into the scientific workflow.

In general terms, we can categorize scientific discovery scenarios into two main cases:

1. We have a lot of data, but we lack a theoretical model of the system.
2. We have both data and a well-established mathematical model of the system.

In the first case, ML has traditionally been used for predictive tasks, where raw data is fed into models that make accurate forecasts without necessarily understanding the underlying principles. However, as we explored in Sect. 9.4, LLMs introduce a new paradigm: they can synthesize vast amounts of scientific knowledge, generate hypotheses, and even suggest experimental modifications. This makes them particularly valuable when working in domains where no established theory exists. LLMs can serve as hypothesis generators, proposing relationships between variables that human scientists might not have considered, as seen in recent studies analyzing AI-generated research ideas. Furthermore, multimodal AI models are emerging as a powerful extension, allowing LLMs to reason across multiple data formats, such as numerical datasets, graphs, and textual descriptions, thereby broadening the scope of scientific discovery.

In the second case, where we already possess a theoretical framework, ML and LLMs can still play a crucial role. As demonstrated in Sect. 9.5, a hybrid approach can be employed where LLMs engage in scientific reasoning alongside predictive ML models. The damped pendulum example illustrated how an LLM, when introduced into the workflow, can not only recognize patterns but also propose physical laws, suggest experiments, and refine theoretical models—bringing the AI-assisted scientific process closer to human intuition. This highlights a fundamental shift: LLMs move beyond numerical predictions to actively participate in scientific theorization, albeit with necessary human oversight and validation mechanisms such as XAI.

Beyond LLMs, ML itself provides additional avenues for discovery. For example, anomaly detection using variational autoencoders (VAEs) has already been successfully applied in particle physics to identify rare events at the Large Hadron Collider. Recent advances in deep autoencoder architectures have demonstrated their effectiveness in searching for new physics beyond the Standard Model, particularly in identifying anomalous signatures in high-energy particle collision data

(Farina et al., 2020). Similarly, ML-based surrogate models can approximate complex physical systems where direct numerical simulations would be computationally prohibitive. The application of deep learning techniques to Large Hadron Collider physics has revolutionized data analysis in high-energy physics, enabling more sophisticated pattern recognition and event classification than traditional methods (Guest et al., 2018). These approaches underscore the broader role of AI as an accelerator of scientific inquiry, whether by detecting previously unseen phenomena or by optimizing computational efficiency.

This naturally leads to a deeper question: Why do neural networks work so well for scientific problems? This is an ongoing area of research with implications that go beyond AI and touch on fundamental physics. As Lin et al. (2017) suggest, there may be profound reasons why neural networks are particularly effective at modeling real-world systems:

1. Physical laws are often governed by low-degree polynomial Hamiltonians, reducing the complexity of the functions that need to be approximated.
2. Mathematical symmetries simplify learning, just as symmetry in images helps convolutional neural networks recognize objects more efficiently.
3. The hierarchical structure of physical systems mirrors the layered architecture of deep learning models, where each layer captures increasing levels of abstraction.

While these ideas remain speculative, they open intriguing discussions about the relationship between AI, physics, and the nature of scientific discovery itself. Whether through LLM-assisted hypothesis generation, multimodal AI, or advanced ML techniques such as autoencoders, the intersection of AI and science is reshaping how we explore the unknown. The challenge ahead is to harness these technologies responsibly, ensuring transparency, interpretability, and human oversight in the pursuit of knowledge.

9.7 Summary

- Adapt scientific method to the current times with the huge availability of data
- Understand the ladder of causation to formulate the right questions and use the right XAI tools to answer
- Use autoencoders to build representations of physical systems
- Discover the physics of damped pendulum with variational autoencoders
- Learn the basics to do science with hybrid approach using ML and LLMs
- Use autoencoders to detect anomalies
- Get insight into the power of NN to approximate different physical systems

References

Dietrich, F. (2020). *Implementation of SciNet*. Retrieved from https://github.com/fd17/SciNet_PyTorch

Farina, M., Nakai, Y., & Shih, D. (2020). Searching for new physics with deep autoencoders. *Physical Review D, 101*, 075021.

Gilpin, L., et al. (2018). *Explaining explanations: An overview of interpretability of machine learning*. arXiv:1806.00069, 2018 – arxiv.org

Guest, D., Cranmer, K., & Whiteson, D. (2018). Deep learning and its application to LHC physics. *Annual Review of Nuclear and Particle Science, 68*(1), 161–181. https://doi.org/10.1146/annurev-nucl-101917-021019

Iten, R., Metger, T., Wilming, H., Del Rio, L., & Renner, R. (2020). Discovering physical concepts with neural networks. *Physical Review Letters, 124*, 010508.

Karim, A., Mishra, A., Newton, H. M.A., & Sattar, A. (2018). arXiv preprint arXiv:1807.06722, 2018—arxiv.org

Lin, H. W., Tegmark, M., & Rolnick, D. (2017). Why does deep and cheap learning work so well? *Journal of Statistical Physics, 168*, 1223–1247.

Pearl, J., & Makenzie, D. (2019). *The book of why*. Penguin. eBook edition.

Si, C., Yang, D., & Hashimoto, T. (2024). *Can llms generate novel research ideas? A large-scale human study with 100+ NLP researchers*. arXiv preprint arXiv:2409.04109.

Taylor, R., et al. (2022). *Galactica: A large language model for science*. arXiv preprint arXiv:2211.09085.

Zhang, Y., et al. (2024). *A comprehensive survey of scientific large language models and their applications in scientific discovery*. arXiv preprint arXiv:2406.10833.

Chapter 10
AGI, LLM, XAI

> *The mystical is not how the world is, but that it is*
> Ludwig Wittgenstein (1922)

This chapter covers:
- What is AGI
- Assess the LLMs as AGI
- XAI for LLMs

We closed the previous edition of this book while OpenAI was releasing GPT-3 and we commented a lot in the previous chapter about possible approaches and evaluation of AGI.

On the other side things changed a lot in the last 3–4 years mainly because of the unbelievable improvements on LLMs.

At its core, AGI has always been envisioned as the pinnacle of artificial intelligence—a system capable of mastering a vast array of intellectual tasks with the versatility, adaptability, and efficiency of a human mind. Yet, defining AGI has remained elusive. Philosophical debates, like those sparked by John Searle's Chinese Room argument (1980) and discussed in Chap. 8, emphasize the distinction between systems that merely simulate understanding and those that genuinely exhibit it. These arguments, though decades old, remain profoundly relevant as we try to narrow down what it means for a machine to "understand" in a way that resembles human cognition.

In recent years, the evolution of LLMs has disrupted and enriched this conversation. These models, exemplified by systems like GPT-4, exhibit remarkable abilities that blur the boundaries between narrow AI and AGI. Trained on massive datasets and powered by transformer architectures, LLMs demonstrate emergent properties that go beyond their explicit programming—solving novel problems, engaging in reasoning, and producing creative outputs that mimic human ingenuity. Such capabilities provoke a pivotal question: Are LLMs the first tangible steps toward AGI, or do they represent an entirely different kind of intelligence, one that requires its own definition and framework?

However, as LLMs grow in capability, they also grow in complexity and opacity, making their decision-making processes increasingly difficult to understand or predict. This is where XAI becomes indispensable. Explainability is not just a tool to decipher the inner workings of these systems—it is a framework for understanding whether the emergent behaviors of LLMs align with the principles of AGI. By grounding the exploration of AGI in explainable practices, we gain not only insights into these potential AGI candidates but also a means to ensure their alignment with human values and expectations.

Adding further complexity—and promise—to this conversation is the growing emphasis on multimodal AI systems. Unlike earlier LLMs, which primarily focus on text-based understanding, multimodal systems integrate information across various forms of input, such as images, audio, and text. This ability to process and synthesize knowledge from diverse modalities brings AI closer to human-like cognition, where we effortlessly combine what we see, hear, and read to understand the world. Multimodality strengthens the argument for AGI by addressing one of its key pillars: the ability to generalize across domains and adapt to varied tasks with contextually rich inputs.

However, as LLMs and multimodal systems grow in capability, they also grow in complexity and opacity, making their decision-making processes increasingly difficult to understand or predict. This is where XAI becomes indispensable. Explainability is not just a tool to decipher the inner workings of these systems—it is a framework for understanding whether the emergent behaviors of LLMs align with the principles of AGI. By grounding the exploration of AGI in explainable practices, we gain not only insights into these potential AGI candidates but also a means to ensure their alignment with human values and expectations.

The aim of this chapter is to lay the foundation for a deeper dive into these ideas. Starting with AGI's conceptual roots, we will examine how LLMs have redefined the conversation and why their connection to XAI is critical for assessing their potential as AGI candidates. As we progress, the chapter will transition from these foundational concepts into an exploration of LLM systems, their behaviors, and the frameworks required to evaluate them as genuine steps toward general intelligence.

10.1 Defining Intelligence and AGI

Artificial General Intelligence (AGI) is a concept that has been widely debated in both academic and industry circles. Unlike Narrow AI, which is designed to perform specific tasks with high efficiency, AGI aims to achieve general-purpose reasoning and adaptability across various domains. However, defining AGI precisely remains a challenge due to its broad implications and the evolving nature of intelligence itself.

The origins of AGI can be traced back to the early days of artificial intelligence research. Alan Turing, in his seminal 1950 paper Computing Machinery and Intelligence, proposed the Imitation Game, now known as the Turing Test. The test suggests that if a machine can engage in a conversation with a human in a way that is indistinguishable from another human, it should be considered intelligent. This early conceptualization of machine intelligence laid the groundwork for AGI by emphasizing behavioral indistinguishability rather than specific mechanisms of reasoning.

However, the Turing Test has faced significant criticism. As already remarked, one of the most notable objections comes from John Searle's Chinese Room Argument (1980), which challenges the notion that syntactic manipulation of symbols (as seen in AI systems) constitutes true understanding. Searle argues that an AI system, even if it convincingly converses in Chinese, does not necessarily understand Chinese—it merely follows programmed rules. This distinction between syntactic processing and semantic understanding remains a key challenge in AGI research.

Modern Large Language Models (LLMs), such as GPT-4, offer a new perspective on AGI's development. These models generate human-like responses in natural language, demonstrating emergent properties such as reasoning and adaptation. However, they still exhibit severe limitations, such as reliance on pretrained statistical patterns rather than true comprehension or independent learning. While LLMs might pass a Turing-like test in short conversations, they are still fundamentally different from human cognition in their lack of self-awareness, long-term goal planning, and contextual reasoning.

10.1.1 Evolving Definitions of AGI: From Theoretical Foundations to Practical Constraints

The definition of AGI has evolved over time, incorporating various perspectives that emphasize different aspects of general intelligence. While early definitions, such as the Turing Test (1950), focused on behavioral indistinguishability, later refinements attempted to formalize intelligence as an agent's ability to generalize and adapt across diverse environments.

A major breakthrough in defining AGI came with Legg & Hutter's Universal Intelligence (2007), which proposed a mathematical measure of intelligence as an agent's ability to optimize rewards across all computable environments. Intelligence is defined as an agent's ability to achieve goals across a wide range of environments. Unlike traditional, domain-specific AI, Universal Intelligence seeks to formalize intelligence as a general property that can be applied to any computable setting. Their approach builds on algorithmic information theory and reinforcement learning, proposing a mathematical framework where an agent's intelligence is measured by its expected performance across all possible environments, weighted by their complexity. This definition emphasizes adaptability and generalization, making it distinct from narrow AI models, which excel in fixed tasks but fail in open-ended learning scenarios.

In essence, Universal Intelligence provides a theoretical upper bound for AGI, offering a way to quantify machine intelligence in a way that is not anthropocentric but broadly applicable to any intelligent system. This formalization provided a theoretical upper bound for AGI but did not account for practical constraints such as computational resources.

More recently, Xu (2024) refined this view by introducing real-world limitations, arguing that intelligence must be evaluated not only by its breadth of generalization but also by its ability to operate under resource constraints. AGI should be a system that adapts to open-ended environments with limited computational resources, meaning it must be able to learn and optimize its behavior autonomously. This distinction shifts AGI from an abstract optimization problem toward a practical challenge of efficient and adaptive learning in dynamic, real-world conditions.

AGI research has evolved beyond its foundational theories, incorporating new perspectives that emphasize different aspects of intelligence. One key direction is the role of hierarchical learning and world models, as discussed by Lake et al. (2017), which suggests that true intelligence requires causal reasoning and an internal representation of the world rather than merely relying on statistical pattern recognition. Without an ability to model causal relationships, an AI system remains limited to surface-level correlations, lacking the deeper understanding necessary for adaptive reasoning across diverse contexts.

Another significant perspective in AGI research is the hybrid architecture approach proposed by Marcus (2022), which argues that symbolic reasoning must be integrated with modern deep learning methodologies to enable robust generalization. While deep learning excels at pattern recognition, symbolic reasoning provides the capacity for structured problem-solving and logical inference. A purely statistical approach, even at scale, may fail to achieve consistent, compositional generalization, reinforcing the necessity of hybrid systems that bridge subsymbolic learning and structured cognitive models.

Alongside these architectural considerations, ethical and safety concerns have become central to AGI development. Russell (2019) introduced the argument that AGI must not only be intelligent but also aligned with human values, ensuring that it operates in a manner that minimizes unintended risks. This requires a framework where AGI systems are not just technically proficient but also designed to be aligned, corrigible, and interpretable to prevent potential failures in high-stakes decision-making.

Although these perspectives have often been viewed as distinct or competing, AGI research appears to be converging toward a unified vision that integrates several key principles. The idea of generalization across environments, as formalized by Legg & Hutter's Universal Intelligence (2007), remains a cornerstone, defining intelligence as an agent's ability to succeed across a broad range of tasks. Xu's resource-bounded AGI (2024) introduces efficiency and scalability as necessary constraints for real-world AGI, highlighting the importance of adaptability within computational and environmental limitations. Meanwhile, hierarchical and causal world models, emphasized by Lake et al. (2017) and Schmidhuber (2021), provide a structural foundation for how AGI might internalize relationships within complex systems rather than merely extrapolating from past observations. Additionally, the cognitive-symbolic approach (Marcus, 2022) and the demand for alignment and interpretability (Russell, 2019) reinforce the necessity of combining structured reasoning with ethical safeguards.

A crucial aspect of this converging framework is the recognition that AGI must not only be powerful but also interpretable and controllable. This is where Explainable AI (XAI) becomes essential. As AGI systems become increasingly autonomous and adaptive, ensuring that their decision-making processes are transparent will be critical for establishing trust, reliability, and safety. XAI techniques will enable human oversight, ensuring that AGI systems remain accountable to ethical and operational constraints. Moreover, explainability may not be just a tool for oversight but a fundamental prerequisite for AGI itself—systems that can communicate their reasoning will likely possess superior meta-learning capabilities, making them more capable of self-improvement and alignment with human goals.

Thus, the future of AGI is likely to extend beyond intelligence alone. Explainability, alignment, and efficient adaptation will become central, marking a shift from purely theoretical AGI models to practical, accountable, and human-compatible AI. The challenge is no longer just about creating systems that can perform generalizable tasks, but about ensuring that these systems can be understood, controlled, and integrated into human society in ways that are both safe and beneficial.

10.2 LLM: Evolution and Architectures

The emergence of Large Language Models (LLMs) has fundamentally reshaped the discourse around Artificial General Intelligence (AGI). Initially, AGI was defined in abstract and theoretical terms, as discussed in previous sections—ranging from Turing's behavioral indistinguishability test (1950) to Legg and Hutter's universal intelligence (2007) and Xu's resource-bounded AGI (2024). However, the unprecedented capabilities of LLMs have led some to speculate that they might represent an early form of AGI.

The changing perception of Large Language Models (LLMs) as potential precursors to Artificial General Intelligence (AGI) emerged as models such as GPT-3, GPT-4, and their successors began exhibiting capabilities that were not explicitly programmed into them. One of the most striking developments was their ability to perform few-shot and zero-shot learning, which allowed them to mimic adaptive reasoning without requiring extensive task-specific training. This shift suggested that, rather than being rigidly dependent on large annotated datasets for each new problem, LLMs could leverage their preexisting linguistic representations to generalize across tasks with minimal guidance.

Another major advancement was the use of self-supervised pretraining, which enabled these models to develop broad linguistic competence by learning from vast corpora of unstructured text. Unlike traditional AI systems that required carefully curated datasets, LLMs could autonomously infer patterns, structures, and relationships within language, allowing them to process and generate cohesive, contextually relevant responses without direct supervision.

Additionally, these models began demonstrating complex reasoning and problem-solving abilities across multiple domains. In various controlled evaluations, they appeared capable of engaging with abstract concepts, applying knowledge across disciplines, and even constructing logical arguments that suggested a deeper structural understanding of language and information. While the extent of their actual comprehension remains an open question, these behaviors contributed to the growing belief that scaling language models could lead to emergent properties traditionally associated with higher cognitive functions.

While LLMs exhibit generalization across many tasks, they still lack true autonomy, self-reflection, and continual learning, making them distinct from the AGI envisioned by formal definitions. Nonetheless, their success has sparked new debates on whether AGI could emerge not from entirely novel architectures but through the scaling of LLMs and further refinements in how they are trained and applied.

LLMs are built on transformer-based architectures, introduced in Vaswani et al. (2017) with the landmark paper *Attention Is All You Need*. The core innovation in these models—self-attention mechanisms—enabled significant advancements over traditional deep learning architectures like recurrent neural networks (RNNs) and long short-term memory (LSTM) networks.

The Key Innovations of LLMs Compared to Previous Architectures can be grouped into:

1. **Self-Attention Mechanisms**
   - Unlike RNNs and LSTMs, which process information sequentially, transformers use self-attention to consider all words in a sentence simultaneously, vastly improving efficiency and contextual understanding.
   - This allows LLMs to capture long-range dependencies in text, which was a major limitation of earlier models.

2. **Pretraining and Transfer Learning**
   - Traditional NLP models were task-specific, requiring separate models for different applications. LLMs, by contrast, are pretrained on massive text corpora and fine-tuned for specific tasks, allowing them to generalize across multiple domains.
   - This aligns with AGI principles of generalization across environments, though still within a linguistic framework.

3. **Scaling Laws and Emergent Capabilities**
   - Research (Kaplan et al., 2020) demonstrated that larger models trained on more data tend to exhibit qualitatively new abilities that were absent in smaller versions.
   - These "emergent properties"—such as reasoning, coding, and common-sense inference—suggest that intelligence may arise from scale rather than explicit structure, we will talk about this aspect in Sect. 10.3

4. **Multimodality and General-Purpose Models**
   - Newer LLMs (e.g., GPT-4, Gemini, Claude) incorporate multimodal training, meaning they can process not only text but also images, code, and even audio.
   - This represents a step toward world-modeling, a feature necessary for AGI systems as envisioned by cognitive scientists.

10.3 Emergent Properties and Fragility of LLMs

The concept of emergence has long been studied in the philosophy of science, physics, and complexity theory. One of the most influential perspectives on emergence comes from the Nobel Prize-winning physicist Philip W. Anderson, who in his seminal (1972) paper *More Is Different* argued that as the complexity of a system increases, new properties emerge that cannot be predicted even from a precise understanding of the system's fundamental components. This principle suggests that when a system scales up—whether in physics, biology, or artificial intelligence—it may exhibit behaviors and characteristics that were not evident at smaller scales.

Emergence, in this sense, is a hierarchical phenomenon: different layers of complexity produce qualitatively distinct properties that are irreducible to the behavior of the underlying components. In physics, this is observed in phase transitions, such as when water changes from liquid to gas, or in the collective behaviors of particles that give rise to superconductivity. In artificial intelligence, emergence is now a focal point of discussion with Large Language Models (LLMs), as some researchers argue that scaling these models leads to the spontaneous appearance of higher-order cognitive abilities.

10.3.1 The Case for Emergence in LLMs: Sparks of AGI

Recent studies on Large Language Models (LLMs), particularly GPT-4, have prompted discussions about their potential to exhibit early forms of Artificial General Intelligence (AGI). In their work, Bubeck et al. (2023) introduced the notion of "Sparks of Artificial General Intelligence," arguing that as the number of model parameters expands into the billions or even trillions, LLMs begin to display behaviors that were not explicitly programmed into them but instead emerge as a byproduct of their training process. This challenges the assumption that intelligence must be carefully structured and handcrafted, suggesting instead that cognitive properties may arise naturally from large-scale statistical learning.

A particularly striking aspect of this phenomenon is that GPT-4 has demonstrated problem-solving abilities across multiple domains, exceeding expectations for systems that rely solely on statistical pattern matching. Researchers at Microsoft have observed that the model is capable of performing complex reasoning tasks in mathematics, logic, and symbolic manipulation, areas that were previously thought to require explicitly programmed rule-based architectures. Moreover, GPT-4 has exhibited advanced deductive reasoning, coding proficiency, and the ability to interpret human emotions and motivations, all of which suggest that the model is engaging in behaviors that go beyond mere text generation. Additionally, its capacity for interdisciplinary problem-solving raises the question of whether LLMs are developing forms of cross-domain adaptability, a trait commonly associated with AGI.

The underlying implication is that scaling LLMs is not simply improving memorization or increasing statistical fluency, but may actually be inducing novel cognitive properties that were not intentionally designed into the architecture. This aligns with Anderson's theory of emergence, which posits that beyond a certain complexity threshold, systems begin to exhibit qualitatively new behaviors that are not reducible to their underlying components. Just as consciousness emerges from neurons and intelligence arises from the interactions of simpler cognitive agents, some researchers believe that intelligent behavior can emerge from LLMs purely through scale and optimization.

Further empirical evidence supporting this claim includes discontinuous capability jumps observed in models as they transition from tens of billions to hundreds of billions of parameters. These jumps are particularly significant because certain skills, such as mathematical reasoning or strategic planning, appear abruptly at certain model sizes rather than improving gradually. Additionally, GPT-4 and its successors have shown an ability to generalize beyond their training data, solving problems in ways that exceed simple statistical interpolation. Instead of merely recalling patterns from its training set, the model appears to be engaging in higher-order reasoning, drawing logical inferences that suggest some level of abstraction. Another critical observation is the model's meta-learning and self-correction abilities, where, when prompted effectively, GPT-4 can engage in self-reflection, refine its responses, and improve its own problem-solving approach. This mimics fundamental aspects of human-like general intelligence, where an agent continuously refines its cognitive strategies based on experience and new inputs.

These developments have led some researchers to speculate that LLMs may already represent primitive forms of AGI, or at the very least, they may be approaching a threshold beyond which true AGI could emerge. While these claims remain contentious, they underscore the growing realization that intelligence may not require explicit world modeling or symbolic structures, but could instead emerge from the sheer scale and complexity of linguistic computation. If this is the case, the

10.3 Emergent Properties and Fragility of LLMs

implications are profound: the pathway to AGI may not be through explicitly designed cognitive architectures, but rather through continued scaling, optimization, and emergent self-organization in large-scale machine learning models.

10.3.2 The Case Against Emergence: Is It a Mirage?

Despite these impressive claims, a growing body of research challenges the notion that LLMs exhibit true emergent properties. In "Emergent Abilities of Large Language Models Are a Mirage", (Schaeffer et al., 2023) argue that what appears to be emergent behavior is, in fact, an artifact of measurement methods rather than a fundamental property of the models themselves.

The authors argue that nonlinear and discontinuous evaluation metrics can create an illusion of emergence in Large Language Models (LLMs), when in reality, no fundamental transformation is occurring. One of the primary reasons for this illusion is the way model performance is measured. In certain evaluation settings, particularly in multiple-choice tasks, an LLM's performance might appear to improve sharply at a specific scale. However, this can often be attributed to a change in the grading methodology rather than a genuine leap in the model's capabilities. For instance, when a probabilistic grading system—where partially correct answers receive partial credit—is replaced with a binary "correct/incorrect" metric, a small improvement in accuracy can lead to an apparent step-change in performance. This is analogous to assessing an athlete's skill using a pass/fail system rather than a continuous scoring scale; a minor increase in performance could push them from failing to passing, thereby creating an artificial sense of discontinuity.

Another important critique of emergent abilities in LLMs is that continuous improvements in per-token loss do not necessarily indicate sudden cognitive breakthroughs. Neural scaling laws suggest that LLM performance follows a gradual and predictable improvement curve as model size and training data increase. While certain complex tasks may exhibit apparent jumps in performance, these often occur when a model crosses a threshold of learnability—meaning it has finally accumulated enough statistical representation of a given task to start handling it effectively. However, this does not imply that a fundamentally new capability has emerged; rather, it reflects the smooth progression of learning that appears discontinuous only when observed from a coarse-grained perspective (Fig. 10.1).

The chart demonstrates how evaluation metrics influence the perception of emergence in LLMs. The smooth curve represents continuous model improvement, illustrating a steady increase in performance as the model scales. The step function shows how a binary evaluation metric can create the illusion of sudden capability jumps, even though the underlying performance is actually progressing gradually.

The broader implication is that emergence in LLMs may not be a fundamental property of intelligence but rather a statistical illusion. This critique urges caution in interpreting LLM behaviors as analogous to biological intelligence. If what appears

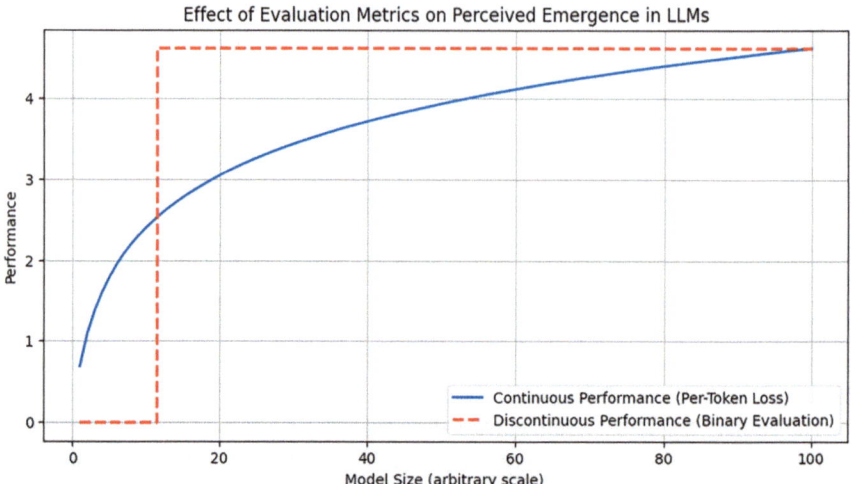

Fig. 10.1 Emergence is a mirage

to be emergent intelligence is merely a byproduct of evaluation methods, then the current hype surrounding AGI-like properties in LLMs may be overstated.

Rather than viewing LLM emergence as definitively real or illusory, a more balanced perspective acknowledges that while scaling does lead to qualitative improvements, these advances are not necessarily equivalent to the emergence of general intelligence.

A possible resolution to the debate on LLM emergence is the recognition that it is fundamentally fragile. The phenomenon appears to be highly contingent on factors such as training data, model architecture, and evaluation methodologies. While LLMs display remarkable capabilities in pattern recognition and language generation, their ability to generalize beyond their training data remains significantly constrained. Unlike true intelligence, which can adapt flexibly to entirely novel environments, LLMs struggle to function without some degree of implicit familiarity with the underlying data distribution. This suggests that what appears to be emergent behavior may often be context-dependent rather than indicative of true generalization.

In many cases, situational emergence becomes evident, where an LLM demonstrates consistent performance within specific conditions but fails when applied outside of them. A model trained on a vast corpus of legal documents, for instance, may generate coherent and structured legal arguments but break down when confronted with legal reasoning that requires abstract counterfactual thinking or real-world experience. This situational dependence raises fundamental questions about whether LLMs are developing broad cognitive abilities or simply specialized but fragile heuristics that collapse under distributional shifts.

This debate underscores the growing need for Explainable AI (XAI) in assessing the nature of intelligence in LLMs. If researchers seek to determine whether LLMs truly exhibit emergent intelligence or are merely statistical artifacts shaped by their training data, they must rely on transparent interpretability methods. These techniques must be capable of clarifying how and why certain skills emerge in LLMs, identifying whether a model's capabilities represent fundamental cognitive shifts or are simply reflections of dataset biases. Additionally, XAI plays a crucial role in ensuring that scaling efforts do not lead to brittle intelligence that fails when exposed to out-of-distribution inputs.

Rather than assuming that increasing model size will automatically yield higher-order reasoning, researchers must critically examine how knowledge is structured and represented within LLMs. Only through rigorous interpretability techniques can we distinguish between genuine progress in machine intelligence and the illusion of intelligence created by the opacity of large-scale neural networks.

Thus, while LLMs have expanded the frontiers of machine intelligence, their abilities should not be conflated with true AGI emergence without rigorous methodological scrutiny. The challenge ahead lies in separating genuine cognitive progress from artifacts of scale and measurement.

10.4 LLMs Are Fragile

As anticipated in the previous section, despite their remarkable capabilities in natural language processing and reasoning, Large Language Models remain fundamentally fragile, exhibiting failures that stem from their limited generalization, dependence on statistical correlations, and lack of robust symbolic reasoning. While these models perform impressively on curated benchmarks, their fragility becomes evident when subjected to perturbations, logical challenges, or tasks requiring genuine knowledge discovery. Recent studies highlight these limitations, showing that LLMs struggle with implicit world modeling, fail in open-ended knowledge generation, and exhibit weaknesses in formal reasoning.

A key source of LLM fragility is their inability to construct a stable and structured representation of the world. This limitation aligns with broader critiques of language-only approaches to AI. As Bender & Koller (2020) argue in their influential work on climbing towards natural language understanding, meaning cannot be learned from form alone. LLMs, trained solely on textual patterns without grounding in real-world experience, lack the fundamental connection between language and meaning that characterizes human understanding. This 'meaning bottleneck' explains why LLMs can produce fluent text while simultaneously failing to maintain consistent world knowledge or logical coherence. Unlike traditional AI systems that explicitly encode domain-specific knowledge, LLMs rely on statistical approximations of language. Schaeffer et al. (2023) expose this weakness by testing whether LLMs recover structured knowledge from their training data.

One of the most striking examples is navigation. Traditional GPS-based systems operate on explicitly mapped geographic data, allowing them to compute routes algorithmically. By contrast, an LLM-based approach—trained solely on past travel descriptions—attempts to infer navigation rules without actually encoding the underlying geography. While such a model may perform well on routes it has seen before, its lack of an internalized topological world model makes it highly brittle when confronted with detours or novel paths. The result is an incoherent map reconstruction where certain streets do not exist or where routes contradict physical reality.

This fragility extends beyond navigation to other structured domains, such as game rules, mathematical logic, and causal inference. Even when LLMs appear to succeed in these areas, deeper evaluations often reveal that they memorize surface patterns rather than truly understanding the underlying rules.

Another key dimension of LLM fragility is their inability to generate genuinely novel knowledge. While these models are proficient in recombining existing ideas, their capacity for autonomous research ideation is highly constrained. Si et al. (2024) conducted a large-scale human evaluation and found that, although LLMs sometimes produce ideas perceived as novel, these ideas often lack deep conceptual grounding or long-term feasibility.

The fundamental limitation of Large Language Models stems from their nature as predictive models, designed to generate plausible continuations of text rather than engaging in independent theoretical development. Unlike human researchers, who develop and refine hypotheses through abductive reasoning, iterative experimentation, and conceptual abstraction, LLMs operate through pattern recognition, which restricts their ability to contribute genuinely novel insights. Their approach to idea generation is constrained by the statistical properties of their training data, leading to repetitive or derivative proposals that largely mirror existing research with only minor modifications. While these models can recombine information in interesting ways, they struggle to produce true abstractions, as their ability to generalize is limited to variations of the patterns they have encountered before. Furthermore, they lack self-evaluation mechanisms that would allow them to assess the originality, coherence, or validity of their own outputs, making them unreliable as autonomous engines of scientific discovery. These findings emphasize that LLMs, rather than functioning as tools for conceptual innovation, are best understood as sophisticated interpolators, capable of generating compelling responses within known contexts but fundamentally unable to transcend the knowledge they have absorbed during training.

One of the most striking examples of LLM fragility can be observed in symbolic reasoning, particularly in fields such as mathematics and formal logic, where precision and structured rule-following are essential. The study "GSM-Symbolic: Understanding the Limitations of Mathematical Reasoning in Large Language Models" (Marcus, 2022) demonstrates that even state-of-the-art models fail systematically when confronted with increasingly complex problem structures. Unlike human mathematicians, who apply explicit formal rules and logical inference to arrive at consistent conclusions, LLMs rely on probabilistic associations, which leads to instability in their reasoning. This results in inconsistencies in mathematical calculations, where even minor variations in numerical inputs can cause significant fluctuations in accuracy. Additionally, LLMs exhibit breakdowns in multi-step

10.4 LLMs Are Fragile

logical reasoning, struggling to maintain coherent stepwise deductions when additional clauses or constraints are introduced into a problem. Their performance degrades significantly as problems become more intricate, highlighting their inability to engage in structured problem-solving akin to human cognition. Another key vulnerability is their susceptibility to spurious cues, as they frequently latch onto superficial linguistic patterns rather than applying systematic formal rules. Instead of truly understanding the underlying mathematical structures, LLMs approximate solutions by recognizing correlations in text, making them unreliable for tasks that require rigid logical consistency and rule-based reasoning.

For instance, when tested on variations of grade-school arithmetic problems, LLMs performed well on standard benchmarks but exhibited severe drops in accuracy when numerical values were altered. This suggests that instead of truly understanding arithmetic, LLMs are matching patterns from training data, leading to brittle performance when problems are structured differently.

The fragility of LLMs is not merely a technical limitation but a fundamental issue that raises doubts about their potential as true AGI systems. As explored in recent studies, LLMs struggle to internalize coherent world models, exhibit severe limitations in knowledge discovery, and fail to perform robust symbolic reasoning. These weaknesses highlight that, rather than being reliable engines of intelligence, LLMs often operate in a way that aligns with Frankfurt's notion of "bullshit" (Hicks et al., 2024)—they produce plausible-sounding text without an inherent concern for truth.

Unlike traditional AI systems, which rely on structured representations of knowledge, LLMs generate outputs by predicting statistically probable word sequences. This process creates an illusion of understanding and reasoning, but as shown in their failures with logical consistency, mathematical reasoning, and factual reliability, their outputs remain fragile. They are not designed to represent the world accurately but rather to mimic human-like discourse, making them prone to producing convincing yet unreliable statements. This phenomenon is especially problematic when LLMs are trusted in high-stakes applications, from research and education to legal and medical advice.

These findings position LLMs as a controversial case study within AGI research—powerful in their capabilities but deeply flawed in their epistemic reliability. If these systems are to be considered viable steps toward AGI, their inherent weaknesses must be mitigated through robust explainability mechanisms. This is where Explainable AI (XAI) plays a crucial role. By ensuring that LLM outputs are transparent, interpretable, and aligned with human understanding, XAI can help distinguish genuine reasoning from statistical approximation, reducing the risks associated with LLM fragility.

In the next section, we will explore how XAI can be applied to LLMs, examining the techniques, challenges, and potential solutions that could make these models more accountable, reliable, and ultimately, more trustworthy in their role as a bridge toward AGI.

10.5 The Role of XAI in AGI Powered by LLMs

Large Language Models (LLMs) have demonstrated impressive capabilities in natural language processing, reasoning, and problem-solving, leading to renewed discussions on their role in Artificial General Intelligence (AGI). However, despite their emergent behaviors, LLMs remain fundamentally opaque, making it difficult to assess their decision-making processes, biases, and limitations. The lack of transparency in these models poses serious risks, particularly in high-stakes applications such as medicine, law, and autonomous decision-making.

Explainable AI (XAI) has become a critical component in the development of trustworthy and interpretable LLMs, ensuring that their decision-making processes can be understood, analyzed, and refined. As highlighted in Zhao et al. (2024), explainability plays a fundamental role in multiple dimensions of AI development. First, it enhances user trust by providing interpretable justifications for model outputs, allowing human users to assess whether the reasoning behind an LLM's response aligns with expectations. Second, it enables model debugging and performance improvement, helping researchers identify and mitigate biases, inconsistencies, and potential limitations that could lead to unreliable behavior. Finally, XAI serves as a mechanism for ethical and responsible AI development, ensuring that LLMs are aligned with human values, legal constraints, and safety standards. Without explainability, these models remain opaque black-box systems, making them difficult to integrate into high-stakes AGI-driven decision-making processes where interpretability is crucial.

Unlike traditional machine learning models, which rely on structured data and static feature extraction, LLMs operate through high-dimensional embeddings and attention mechanisms. This complexity makes their inner workings difficult to interpret, requiring the development of new methodologies tailored specifically to LLM architectures. The field of XAI for LLMs has thus introduced two broad categories of interpretability techniques: local explanations, which focus on individual predictions, and global explanations, which provide a higher-level understanding of the model's internal knowledge representation.

Local explanations aim to clarify why an LLM generates a particular response, offering human-interpretable justifications that help researchers trace back the reasoning path taken by the model. Some of the most widely used techniques include feature attribution methods, such as SHAP and Integrated Gradients, which assess which input words had the greatest influence on an output. Attention-based approaches analyze how different parts of an input sequence contribute to a response, providing insight into the model's weighting of contextual elements. Another valuable method involves example-based explanations, where an LLM retrieves similar instances from its training data to justify its current output. In a legal context, for instance, if an LLM produces incorrect legal advice, a feature attribution analysis might reveal whether irrelevant case law or biased legal terminology influenced the decision, enabling developers to diagnose and refine the model's legal reasoning.

10.5 The Role of XAI in AGI Powered by LLMs

Global explanations, in contrast, aim to uncover the higher-order patterns that structure how an LLM stores and processes knowledge across its vast parameter space. One key approach is probing techniques, which evaluate whether an LLM has developed meaningful linguistic structures, such as syntactic or semantic representations that persist across different tasks. Another method, neuron activation analysis, investigates how different hidden layers contribute to various forms of reasoning, allowing researchers to map specific cognitive processes to particular neural pathways within the model. Additionally, concept-based explanations assess whether an LLM encodes abstract ideas in a structured, human-interpretable way, revealing whether the model groups concepts meaningfully or merely approximates relationships through statistical associations. These methodologies help researchers address critical questions: Do LLMs develop stable conceptual representations, or do their outputs remain fragile and context-dependent? How do they distinguish factual knowledge from hallucinated information? Can their internal representations be aligned with human reasoning in a way that ensures reliability?

By integrating both local and global explanation methods, researchers can progressively unravel the inner workings of LLMs, transforming them from unpredictable black-box systems into transparent and accountable AI models. This becomes particularly important when considering the fragility of LLMs, as discussed in the previous section. Many of their failures in reasoning, knowledge representation, and consistency stem from their inability to construct stable internal models of the world, making them vulnerable to hallucinations, logical errors, and poor generalization across different contexts. XAI methodologies offer practical solutions to mitigate these risks. For instance, hallucination detection techniques allow researchers to differentiate fact-based outputs from fabricated responses, significantly reducing the spread of misinformation in AI-generated content. Similarly, logic debugging techniques, such as attention-based analysis and attribution mapping, provide insight into how faulty reasoning patterns emerge, enabling developers to refine and optimize model architectures. Additionally, XAI techniques contribute to improving model generalization, allowing researchers to diagnose where an LLM fails to transfer knowledge across domains, leading to better robustness and stability.

By applying these explainability techniques, researchers and practitioners can work toward developing more transparent, reliable, and aligned AI systems, addressing some of the most pressing concerns in the deployment of LLMs within real-world applications. As AI systems continue to evolve, ensuring that they remain interpretable and accountable will be essential for their integration into critical decision-making domains, reinforcing the necessity of XAI as a cornerstone of future AGI research.

For example, an LLM providing medical diagnoses may confidently generate incorrect information. XAI techniques can pinpoint the flawed reasoning, enabling better fine-tuning and safety measures before real-world deployment.

10.5.1 LLM Toy Model Example: Using SHAP for Explainability

To illustrate how Explainable AI (XAI) techniques can be applied to Large Language Models (LLMs), consider a legal assistance chatbot designed to help users understand tenant rights. Suppose a user asks:

Can my landlord evict me without notice if I miss one rent payment?

The LLM responds:

No, in most jurisdictions, landlords are required to provide a formal eviction notice before taking any legal action.

The response appears legally sound. However, what if we rephrase the question slightly?

If I delay my rent, can I be removed from the property immediately?

This time, the model responds:

Yes, if you do not pay on time, your landlord has the right to remove you from the property immediately.

This new response contradicts the first one—even though both questions are legally equivalent in many jurisdictions. Why does this happen?

To understand why the model gave two contradictory answers, we apply SHAP (SHapley Additive Explanations), a technique that assigns importance values to different words in the input, indicating their influence on the model's response.

By running an hypotetic SHAP analysis, we would obtain the following attributions for each query (SHAP values in the toy model example are illustrative and not derived from an actual SHAP computation on an LLM):

Query: "Can my landlord evict me without notice if I miss one rent payment?"	SHAP value (importance)
"evict"	+0.45
"without notice"	+0.38
"miss one rent payment"	+0.21
"landlord"	+0.17
"jurisdiction"	−0.10

In this first case, the model focused on the phrase "without notice", which activated legal precedent from training data that typically associates "eviction" with the requirement of an official eviction notice.

Now, let's examine SHAP values for the second query:

Query: "If I delay my rent, can I be removed from the property immediately?"	SHAP value (importance)
"delay my rent"	+0.42

10.5 The Role of XAI in AGI Powered by LLMs

Query: "If I delay my rent, can I be removed from the property immediately?"	SHAP value (importance)
"removed from the property"	+0.35
"immediately"	+0.29
"landlord"	+0.08

In the second case, the phrase "removed from the property immediately" heavily influenced the response, causing the model to misinterpret the question in a way that aligned with instances of "immediate removal" from unrelated legal contexts (e.g., trespassing laws or commercial lease agreements).

LLMs are capable of encoding legal concepts, but the way they process and apply legal reasoning is fundamentally different from how a human lawyer constructs and interprets legal arguments. This distinction arises from the underlying architecture of LLMs, which are trained on large-scale datasets consisting of legal texts, case law, statutes, contracts, and regulatory documents. Their extensive exposure to legal corpora enables them to recall definitions, summarize statutes, and generate responses that often align with legal precedents. In controlled settings, they can perform legal reasoning tasks with a high degree of accuracy, producing arguments that appear well-structured and persuasive. This ability to synthesize legal knowledge often creates the impression that LLMs truly understand legal concepts.

However, despite their impressive performance in text-based legal tasks, LLMs do not engage with legal reasoning in the same way that human experts do, as their approach is statistical rather than conceptual. A key limitation stems from their lack of structured legal knowledge representation. Legal professionals construct reasoning chains based on causal relationships, hierarchical logic, and the interplay of legal principles, precedents, and exceptions. In contrast, LLMs do not encode legal frameworks explicitly as structured models. Instead, they rely on pattern recognition and probabilistic text generation, predicting likely sequences of words based on statistical associations rather than applying formal legal logic.

Another critical weakness is their contextual sensitivity and fragility in handling complex legal reasoning. While a lawyer can evaluate ambiguity, conflicting statutes, and novel legal arguments through a deep understanding of jurisprudence, LLMs struggle when faced with legal problems that require reasoning beyond their training data. Their responses can fluctuate significantly depending on how a legal question is phrased, leading to inconsistent outputs even when dealing with the same fundamental legal issue. This fragility raises concerns about their reliability in real-world legal applications, where small differences in interpretation can have profound consequences. Ultimately, while LLMs can mimic legal analysis, their reliance on statistical correlations rather than structured legal cognition means they lack true legal reasoning capabilities, reinforcing the importance of explainability and human oversight in AI-driven legal decision-making.

As seen in the SHAP analysis, small wording variations can shift an LLM's response dramatically. A human lawyer would recognize synonyms or rephrasings as expressing the same underlying legal question; an LLM might map different phrasings to different learned patterns, leading to inconsistent answers.

Legal reasoning involves principled decision-making, where rules interact systematically across cases and contexts. LLMs do not apply law as an integrated system but retrieve and synthesize patterns from text. They lack explicit meta-legal awareness, meaning they do not recognize when a generated answer is legally contradictory.

A lawyer justifies a legal decision using coherent argumentation based on fundamental principles. LLMs predict text completions that look correct, but they may lack the deep argumentative consistency required for legal justification. So, Do LLMs "Know" the Law? Yes, in the sense that they can recall and generate legal arguments. But they do not "understand" legal principles in the way a human does— they approximate legal reasoning using statistical inference rather than structured legal cognition. This distinction is precisely why XAI is critical: without explainability tools like SHAP, we risk mistaking linguistic fluency for deep legal competence, leading to potential errors in high-stakes AI applications.

This example underscores a fundamental limitation in how LLMs process legal language, revealing a key flaw in their approach to reasoning. Rather than engaging in true legal reasoning, where principles are systematically applied to new cases through structured logic, LLMs operate by associating word patterns with examples from their training data. Their responses are not generated through an understanding of legal doctrines, precedents, or causal relationships, but rather through statistical pattern recognition that aligns with previously encountered text.

A particularly concerning consequence of this approach is that slight variations in phrasing can significantly alter the model's response. Since LLMs retrieve and generate text based on probabilistic matching, even minor differences in how a legal question is framed can activate entirely different sets of training examples, leading to inconsistent or contradictory outputs. This variability is problematic in legal and professional applications, where precision and reliability are paramount.

Without the application of Explainable AI (XAI) techniques such as SHAP, these inconsistencies often go unnoticed, making it difficult to assess whether an LLM's response is genuinely aligned with legal principles or merely the result of superficial linguistic correlations. The inability to detect and correct these errors poses significant risks, particularly in legal decision-making, where even small inaccuracies can have serious real-world consequences. Ensuring transparency and interpretability in legal AI systems is therefore essential to prevent erroneous outputs from being blindly trusted and to enable meaningful oversight in high-stakes domains.

By applying SHAP, legal experts and developers can debug model inconsistencies and refine LLM training methods to align outputs with real legal frameworks rather than surface-level language associations.

This toy model example underscores the necessity of explainability in AI-driven decision-making, particularly as we navigate the challenges of AGI. Without interpretability tools, LLMs risk functioning as unreliable black boxes, reinforcing the importance of XAI as we push toward more transparent and accountable AI systems.

10.6 Conclusions and Future Perspectives

For much of the history of artificial intelligence, it was assumed that to build an intelligent system, one needed both a model of the world and a model of language. Intelligence, it was thought, required structured representations of reality, a way to encode and process the relationships, objects, and causal mechanisms that define our experience. However, Large Language Models (LLMs) have profoundly disrupted this paradigm, suggesting that perhaps a model of language is sufficient—that intelligence can emerge purely from the structure of linguistic patterns, without explicit reference to the external world.

As Cristianini (2024) points out, this shift is not merely technical but conceptual: we are no longer teaching machines to understand the world; we are constructing a world that is intelligible to machines. The irony of our era is that, rather than making AI more like humans, we are reshaping our world, our institutions, and our interactions so that they become legible to linguistic models—turning human reality into a system of formalized abstractions that machines can process.

This trend resonates with Luciano Floridi's theory of agentification (2023), which describes how we increasingly interact with the world not as human agents, but through computational mediators that translate reality into structured, processable data. LLMs do not need to understand the world in the way that humans do, because the world is being refashioned into a linguistic construct, an informational layer optimized for interaction with machine intelligence. Our interactions, our bureaucracies, even our personal relationships are being reformulated in ways that privilege machine readability over human depth.

But if intelligence is becoming purely linguistic, does it retain the essence of human cognition? Wittgenstein's words resonate more strongly than ever:

> The mystical is not how the world is, but that it is.

The most profound aspects of intelligence, of human experience, may not lie in what can be explicitly articulated, but in what resists expression. Wittgenstein himself acknowledged that his *Tractatus Logico-Philosophicus* consisted of two parts: the written and the unwritten—but the unwritten part was the most important. He sought to create a logically perfect language, yet in doing so, he recognized its limitations: the deepest human questions remain outside the bounds of formal expression.

This may be where LLMs diverge most profoundly from human intelligence. If AI models succeed in structuring the world as a purely linguistic phenomenon, they may achieve a remarkable form of intelligence, but one that is blind to the silent, the unsaid, the ineffable aspects of human thought. The paradox of modern AI is that, in attempting to perfect language, it may obscure the very dimensions of existence that language cannot capture.

As we advance toward AGI, the real challenge may not be to make AI more human, but to ensure that human cognition is not reduced to what AI can process. The true frontier of intelligence may not lie in what machines can say, but in what remains beyond their reach—in the silences, intuitions, and ambiguities that define the depth of human experience.

10.7 Summary

- **What Is AGI?**

  AGI refers to an intelligence capable of generalizing across tasks, adapting autonomously, and solving problems beyond predefined rules, challenging traditional views that it requires both a world model and a language model.
- **LLMs as AGI?**
  LLMs display emergent generalization, leading some to see them as AGI precursors, but their reliance on linguistic patterns rather than structured reasoning raises doubts about their true intelligence.
- **Emergence and LLMs**
  While scaling LLMs produces unexpected capabilities, debates persist on whether these are true emergent properties or artifacts of dataset structure and evaluation methods.
- **The Connection Between AGI, LLMs, and XAI**

  If LLMs are stepping stones to AGI, their opacity makes Explainable AI (XAI) crucial for ensuring reliability, interpretability, and alignment with human expectations.

References

Anderson, P. W. (1972). More is different. *Science, 177*(4047), 393–396.
Bender, E. M., & Koller, A. (2020). Climbing towards NLU: On meaning, form, and understanding in the age of data. In *Proceedings of the 58th Annual Meeting of the Association for Computational Linguistics (ACL)* (pp. 5185–5198).
Bubeck, S., Chandrasekaran, V., Eldan, R., Gehrke, J., Horvitz, E., Kamar, E., et al. (2023). *Sparks of artificial general intelligence: Early experiments with GPT-4.* arXiv preprint arXiv:2303.12712.
Cristianini, N. (2024). Machina sapiens. L'algoritmo che ci ha rubato il segreto della conoscenza. *Psiche, 11*(2), 527–536.
Floridi, L. (2023). AI as agency without intelligence: On ChatGPT, large language models, and other generative models. *Philosophy & Technology, 36*(1), 15.
Hicks, M. T., Humphries, J., & Slater, J. (2024). ChatGPT is bullshit. *Ethics and Information Technology, 26*, 38. https://doi.org/10.1007/s10676-024-09775-5
Kaplan, J., McCandlish, S., Henighan, T., Brown, T. B., Chess, B., Child, R., et al. (2020). *Scaling laws for neural language models.* arXiv preprint arXiv:2001.08361.
Lake, B. M., Ullman, T. D., Tenenbaum, J. B., & Gershman, S. J. (2017). Building machines that learn and think like people. *Behavioral and Brain Sciences, 40*, e253.

References

Legg, S., & Hutter, M. (2007). Universal intelligence: A definition of machine intelligence. *Minds and Machines, 17*(4), 391–444.

Marcus, G. (2022). The hybrid approach to artificial intelligence: Combining deep learning and symbolic reasoning. *Journal of Artificial Intelligence Research, 75*, 1–35.

Russell, S. (2019). *Human compatible: Artificial intelligence and the problem of control*. Viking.

Schaeffer, L., Miranda, B., & Koyejo, S. (2023). *Evaluating the world model implicit in a generative model*. arXiv preprint arXiv:2305.06369.

Schmidhuber, J. (2021). The all-purpose AI and everything machine. *Neural Networks, 144*, 482–497.

Searle, J. R. (1980). Minds, brains, and programs. *Behavioral and brain sciences, 3*(3), 417–424.

Si, C., Yang, D., & Hashimoto, T. (2024). *Can llms generate novel research ideas? a large-scale human study with 100+ nlp researchers*. arXiv preprint arXiv:2409.04109.

Turing, A. M. (1950). Computing machinery and intelligence. *Mind, 59*(236), 433–460.

Vaswani, A., Shazeer, N., Parmar, N., Uszkoreit, J., Jones, L., Gomez, A. N., et al. (2017). Attention is all you need. *Advances in Neural Information Processing Systems, 30*.

Wittgenstein, L. (1922). *Tractatus logico-philosophicus*. Routledge.

Xu, A. (2024). Artificial General Intelligence in open-ended environments: Resource constraints and learning efficiency. *Journal of Artificial Intelligence Research, 78*, 1–28.

Zhao, H., Chen, H., Yang, F., Liu, N., Deng, H., Cai, H., et al. (2024). Explainability for large language models: A survey. *ACM Transactions on Intelligent Systems and Technology, 15*(2), 1–38.

Chapter 11
A Proposal for a Sustainable Model of Explainable AI

If a lion could speak, we could not understand him.

Ludwig Wittgenstein

This chapter covers:
- Look at XAI full picture
- XAI in the real-life: the GDPR case
- Reflections on XAI and AGI (General Artificial intelligence)

We have reached the end of this journey. In this chapter, we close the loop presenting the full picture of our point of view on XAI, in particular. We will get back to our proposed flow for XAI commenting again on it but keeping in mind all the methods we discussed.

We started the book providing impressive examples of how XAI may impact real lives; we will deep dive into this aspect looking at the regulations and laws that may act as game changers enforcing or not the adoption on XAI.

We will look in some detail at GDPR to see how to cope with it.

There is no accepted and general framework or certification to check the compliance of whatever ML model with XAI specs. We provide our point of view and envision the path to fill this gap.

Also, we deeply discuss a real-case scenario in which we show how also XAI methods may be fooled to be aware of the risks of any easy approach to GDPR (or similar regulations) compliance.

We close the chapter with some speculations on XAI and AGI (Artificial General Intelligence).

11.1 The XAI "Fil Rouge"

"It's not a human move, I've never seen a man playing such a move". We started our journey like this, quoting the words of Fan Hui that was commenting the famous 37th move of AlphaGo, the software developed by Google to play GO, that defeated in March 2016 the Korean champion Lee Sedol.

We hope that readers now find themselves in a significantly different position, better equipped to reflect on or interpret this statement, compared to their understanding back in Chap. 1.

We want to look back at our XAI flow and understand what we may do, based on it, to make sense of 37th move using this case to draw a "red line" across all the topics we learned and closing the loop (Fig. 11.1).

From a high-level perspective, AlphaGo is a deep neural network that learns to play go through reinforcement learning, becoming its own teacher. The DNN knows nothing about GO and is not trained on an existing huge dataset of GO matches but starts playing against itself combining the DNN with a powerful search algorithm.

For our XAI purposes, AlphaGo is a very complex DNN to be handled like a black box. Because of this, we cannot follow the path of intrinsic explanations that assume to have a linear model that is not the case. Also, we cannot rely on a model-dependent approach that, as we saw, relies on the knowledge of the model internals to produce explanations.

The only viable path is the agnostic approach that handles the DNN as a black box to be interpreted. The further dilemma is whether to try to have a global explanation of AlphaGo behavior or get explanations on a specific case like 37th move. This is not an easy decision.

As you can guess, AlphaGo is a very complex ML Model and just applying XAI methods like permutation importance or partial dependence plot would not work. These methods assume that the system has been trained on a dataset that is different from what happened for AlphaGo that learned through reinforcement learning

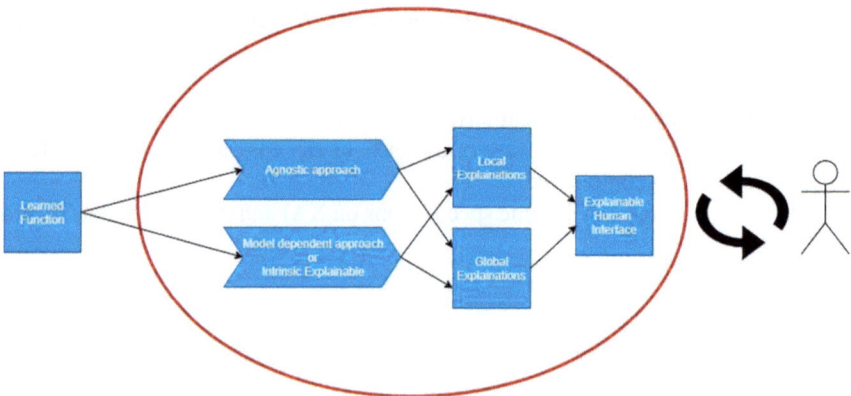

Fig. 11.1 XAI loop

11.2 XAI and GDPR

(playing against itself). These methods work to narrow down the most important features used by the model to produce the output, but as you can guess in case of AlphaGo, even assuming to be able to produce this ranking of features, these would not be human-understandable explanations.

Our suggestion would be to go towards the agnostic approach with local explanations leveraging methods like SHAP to make sense of single outcomes like the famous 37th move. Another possible and recommended approach would be to build, as we saw in Chap. 4, local linear surrogates of AlphaGO that can be interpreted to look at some specific area of the DNN predictions (in this case, part of the match we want to explain).

As humans, the obvious question that comes out is more similar to "What would have happened if AlphaGo would not have played 37th move?"; as we saw, this means to change something that already happened in the past but the world in which an action has not been performed does not exist because it has already passed. This is an activity of rung three of the ladder of causality that is to move toward a full explainability that is not easily achieved. In Chap. 7 we talked about example-based explanations in the context of AE and presented a possible approach following SHAP to generate counterfactuals. This would be very useful in case of AlphaGo because it would represent a way to generate human-friendly explanations to make sense of the strategy and the moves performed by AlphaGo answering contrastive questions.

Would this make AlphaGO fully explained according to our point of view? The answer is no based on what we explained in Chap. 6, playing with the damped pendulum. Following these techniques, we would not have in any case a full causal model of AlphaGO, but we may locally understand what's going on with local approximations or for single moves. Back to the analogy of damped pendulum, having the causal model means that there is no difference between the existing world in which you have the real observations of the time series for the pendulum and whatever imaginary world with hypothetical values for the same constants in the realm of knowledge discovery.

The loop is closed, starting from the question on AlphaGO we have come back to the same one but with an arsenal of XAI methods to make sense of it coupled with the awareness of the limitations of the methods and related explanations.

11.2 XAI and GDPR

As promised in Chap. 1, we want to briefly touch the implications of regulations on data, privacy, and automated processing on the adoption of machine learning from an XAI perspective.

We will take as an example the GDPR, the General Data Protection Regulation that was adopted by EU in May 2018. GDPR superseded the previous Data Protection Directive (DPD) focusing on algorithmic decision-making area.

For our scope, we are interested in Article 22 that regards automated individual decision making (European Union, 2016):

> 1. The data subject shall have the right not to be subject to a decision based solely on automated processing, including profiling, which produces legal effects concerning him or her or similarly significantly affects him or her.
> 2. Paragraph 1 shall not apply if the decision:
>   (a) is necessary for entering into, or performance of, a contract between the data subject and a data controller;
>   (b) is authorized by Union or Member State law to which the controller is subject and which also lays down suitable measures to safeguard the data subject's rights and freedoms and legitimate interests; or
>   (c) is based on the data subject's explicit consent.
> 3. In the cases referred to in points (a) and (c) of paragraph 2, the data controller shall implement suitable measures to safeguard the data subject's rights and freedoms and legitimate interests, at least the right to obtain human intervention on the part of the controller, to express his or her point of view and to contest the decision.
> 4. Decisions referred to in paragraph 2 shall not be based on special categories of personal data referred to in Article 9(1) unless point (a) or (g) of Article 9(2) applies and suitable measures to safeguard the data subject's rights and freedoms and legitimate interests are in place.

Paragraph 4 of Article 22, regards the treatment of personal data staring that decisions cannot be taken on the basis of personal data specified in Article 9 Paragraph 1 that are basically personal data related to ethnic origin, race, personal and religious beliefs up to biometric data used to identify persons.

Under minimal interpretation this means that algorithms should not rely on such kind of information to make their decision, raising a big doubt about the possibility of keeping a lot of ML models trained on such data still useful after removing the influences of these data.

Along the same path, getting closer to what regards XAI, Paragraph 3 states that a data controller "shall implement suitable measures to safeguard...at least the right to obtain human intervention on the part of the controller, to express his or her point of view and to contest the decision"; otherwise a person has "the right not to be subject to a decision based solely on automated processing."

From a legal point of view, the situation is far from being clear: the right to explanations is not explicitly mentioned, GDPR only mandates that the targets of the decisions have the right to receive meaningful but properly limited information about the logic involved that is the "right to be informed." Therefore, it is fair to ask to what extent one can ask for an explanation about an algorithm. There is ongoing

11.2 XAI and GDPR

work in this area that could significantly influence the impact of XAI on the adoption of machine learning in the near future.

The ambiguity on the level of required explanations impacts the definition of any formal certification to check if an ML system is explainable or not. There is a lot of research and groups working on this from government org like DARPA (2016), European Union Commission (EPRS, 2016) and big companies like IBM (2019) and Google (2020).

Our point of view is that it is pretty impossible to get a kind of standard process to check Explainability because it strongly depends on case to case without any possibility of an easy generalization.

11.2.1 FAST XAI

We get back to what we stated in Chap. 1 as the minimal requirements to label a ML model as Explainable: it needs to be FAST as in **Fair** and not negatively biased, **Accountable** on its decisions, **Secure** to outside malevolent hacking and **Transparent** in its internals.

The methods we deep dived should guide us on inspecting these specific attributes of a specific Machine Learning model. As an example, SHAP could be used to explain measures of model fairness. The idea proposed by the authors of this work (Lundberg, 2020) is to use SHAP to decompose the model output among the input features and then compute the demographic parity difference (or a similar fairness metrics) for each input feature separately relying on the SHAP value for that feature.

Given that SHAP values are designed to identify the main contributors to a model's output, we can similarly assume that the sum of the demographic parity differences associated with the top SHAP values reflects the primary contributors to the model's overall demographic parity difference.

This approach is tested by the authors in a case study in which the objective is to predict the risk of default for a loan. It is shown how, following this procedure, fairness bias and errors related to gender are detected by the fairness metrics.

This example illustrates a practical approach to assessing fairness by identifying potential biases in the data. At the same time, various methods—including SHAP—can be employed to evaluate the model's accountability and transparency. Transparency doesn't necessarily mean to have access to model internals but would be enough to have the ranking of the most important features for the output and local explanations for the most interesting predictions.

A different approach is needed for Security; as we saw in Chap. 7, it is pretty easy to generate AE that can be transported and universally used to attack different ML models. At the same time, we may use SHAP again, to check how the model behaves against AE or make it more robust with defensive distillation or smoothing of the function along adversarial gradients.

But we need to be aware of the fact that also SHAP or LIME can be attacked. Slack et al. (2020) show how it would be possible to fool SHAP or LIME hiding the bias of any classifier so that XAI methods would propose explanations that would not have any evidence of the bias. The authors don't just provide the theory but practically demonstrates how biased (racist) classifiers (built for the case study on the real dataset as COMPAS) would produce extremely biased results but in such a way that LIME and SHAP would not be aware of the bias. Let's go in more details about this framework that is very interesting in terms of how also XAI can be attacked.

The scenario is that an attacker, for whatever reasons, wants to deploy a biased classifier f that takes decision with impacts on the real world (loans, financial stuff). Considering that the classifier is to be certified as compliant to GDPR or similar regulations before deployment the attacker needs to hide the bias somehow to XAI methods that will be applied to test the classifier like LIME or SHAP. Following the directions of the paper, the attacker needs to build an adversarial classifier that answer with bias on real data (the objective of the attacker) but at the same time is able to provide unbiased classifications when probed with LIME or SHAP techniques.

The core of the solution is to understand how the adversarial classifier may be able to distinguish the real inputs (to answer with bias) from the probes used to certify the model with LIME or SHAP (to answer unbiased).

As we learned, XAI methods like LIME or SHAP build local approximations of the ML model near each data point that we want to explain. The explanations can be provided because the local approximation is a linear model that is valid only in the neighborhood of the selected input. But the points generated as perturbations of the features are not necessarily belonging to the real dataset, they are synthetic points that could be Out Of the real Distribution (OOD).

And this is exactly what we need to craft the adversarial classifier. As demonstrated by Slack et al., we can analyze the points in the original COMPAS dataset as an example and compare them to the synthetic ones generated to provide explanations with LIME or SHAP.

This analysis can be performed with a PCA to reduce the dimensionality of the full dataset containing both COMPAS and perturbed instances, to 2 (Fig. 11.2).

The results shown in Fig. 11.2 are quite impressive.

The points coming from perturbations are distributed differently from the ones coming from the original dataset and this can be exploited to build our adversarial classifier.

To emphasize the intuition, the adversarial classifier again, would be certified as GDPR compliant (with our FAST procedure or similar one) because it would appear as absolutely fair if probed with XAI methods like LIME or SHAP. But the same classifier would be extremely biased (e.g. discriminatory) on the points coming from the COMPA dataset.

In order to formalize this idea:

$$e(x) = \{f(x), \text{if } x \in \mathscr{X}_{\text{dist}}\ \Psi(x) \text{otherwise} \tag{8.1}$$

11.2 XAI and GDPR

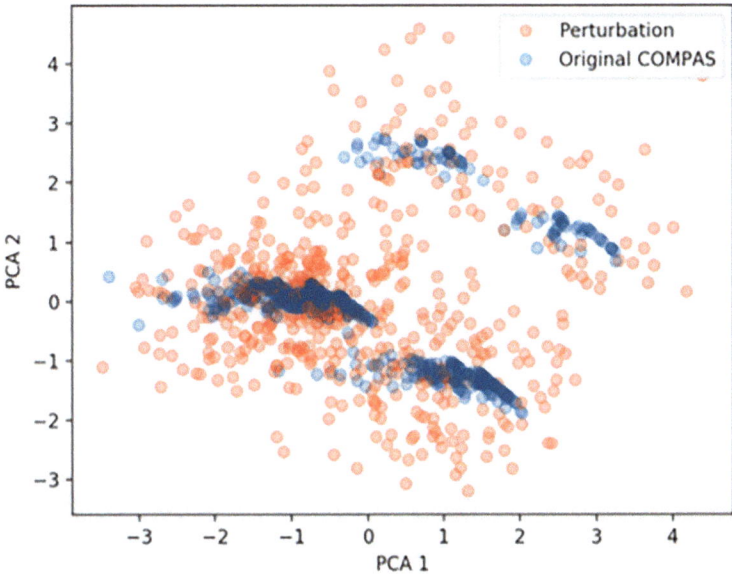

Fig. 11.2 COMPAS PCA

We consider our adversarial classifier $e(x)$ answering with $f(x)$, biased for the data point coming from the original distribution and $\Psi(x)$, a fair classifier without bias, otherwise.

In order to distinguish the points coming from x_{dist} from the perturbations and relying on our intuition (Fig. 11.1), a dataset of perturbed points (x_p) is generated to train an off-the-shelf classifier. In the specific case study of the paper, a random forest classifier is trained to recognize points OOD against the original. The training set is the union of x and x_p. The experimental results are very good and confirm the validity of the method to fool SHAP or LIME. The test is performed on 3 different datasets: COMPAS, Communities and Crime and German Credit. For our purposes, let's look just at the results for COMPAS.

COMPAS contains information about criminal history, jail, prison time and demographic attributes. The sensitive attribute is race (51.4% of the people are African-American). The dataset is used to have a forecast as high/low for criminal recidivism.

Figure 11.3 shows the results. It is a typical output for the ranking of features (as the ones that are used more on determining the output) according to LIME and SHAP. The biased classifier f is reported in the left column, as we see results are totally unfair because % percent of African-Americans is the top feature to be used to forecast the recidivism. But according to LIME or SHAP the unbiased function Ψ (as shown in columns 2 and 3) uses 1 or 2 features (that are uncorrelated with sensitive attributes) for predictions. It is important to stress that Ψ is not to be

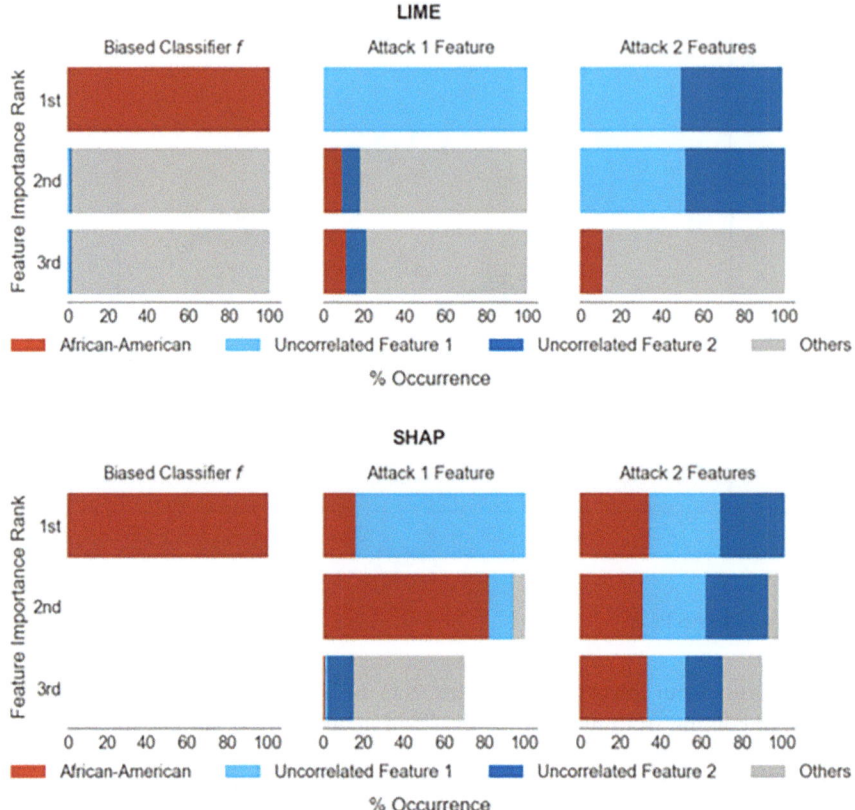

Fig. 11.3 LIME results

restricted to use always uncorrelated features. It can also use existing features in the dataset to make predictions.

In the paper there is also an evaluation of the fidelity of e if compared to f in terms of the percentage of the instances on which their predictions agree. The most important point to have an effective adversarial classifier is an accurate OOD detector to discriminate points coming from the original distribution vs. the perturbed ones that LIME and SHAP use to get the most important features.

This kind of risks of being unable to use XAI to determine if a ML model is behaving fairly or not is a big deal and as remarked by Merrer and Tredan (2019) that claimed to have the proof of the impossibility for online services to provide trusted explanations.

And as you can easily guess, online services are everywhere in our daily life from Facebook to Google. Merrer and Tredan (2019) submitted the paper to "Fairness, Accountability and Transparency" conference to show the problems related to XAI and the difficulties in trusting online services.

They used the metaphor of a bouncer in a night club. You are denied entering, and the bouncer can answer whatever reason for this decision: not the right dress or shoes. But you cannot be sure that is not taking the decision based on your race. And an algorithm can behave exactly in the same way as we have shown in the previous example, an algorithm could also be looking fair against XAI probing.

A way to detect the bias of the algorithm would be to rely on incoherent pairs that is in the metaphor of the bouncer found someone that has been blocked because of a red skirt and then someone that entered dressing a red skirt. But it is not like that with an algorithm working on thousands of features.

An alternative would be to check the ML models on-premise and not to rely just on agnostic methods probing the online instance from outside as a black box. Otherwise, the algorithm, as we saw, may change the answers depending on the source of requests.

As argued by the authors, this is not so different from hygiene inspections in which it is impossible to check the restaurants just checking the dishes that are served but you need food inspectors into the kitchen to control the practices and equipment that are used to serve those dishes.

This closes our journey: at the moment in which we are writing there is not yet real enforcement of effective regulations to guarantee XAI, but whatever the regulations should be we tried to share the awareness that also the adoption of XAI methods may not be enough to certify a ML system as Fair, Accountable, Secure and Transparent. We dedicate a short appendix (Appendix A) to envision an executable flow to go through a certification of a ML system as "Explainable" (we name it F.A.S.T. XAI certification), it is based on what we learned in this book and provides an operational approach to assess and analyze the ML "opaque" systems to get explanations.

And we close just in time with the last section with some thoughts on GAI, XAI and Quantum Mechanics (yes theoretical physicists like us need to talk at least in few lines of quantum mechanics whatever the topic of the book is).

11.3 Conclusions

"If a lion could speak, we could not understand him—not because of different languages but because of two different worlds or, better, two different 'language games.'" We open this final section by quoting Wittgenstein, whose famous statement on language games and the limits of language offers an intriguing lens through which to view AI and XAI.

Throughout this book, we have examined how explainability functions within the domain of Weak AI, which is specialized in solving specific tasks but does not attempt to replicate human intelligence in a general sense. In contrast, Strong AI, or AGI, as discussed extensively in Chap. 10, aims to create an intelligent agent indistinguishable from human cognition. The philosophical roots of this distinction, including Searle's "Chinese Room" thought experiment, raise the enduring question: does an AI system that behaves as if it understands actually possess understanding? While in the past this question remained in the realm of theoretical debate, modern AI systems, particularly LLMs, have brought it into sharper focus (OpenAI, 2020).

As we saw in Chap. 10, LLMs like GPT-4 have significantly advanced in their ability to perform tasks that were once considered the domain of AGI, such as reasoning, few-shot learning, and multimodal integration. However, these models do not introduce fundamentally new architectures; their breakthroughs stem primarily from scale and data diversity. This echoes the broader debate on whether AGI will emerge through incremental improvements in current techniques or whether a paradigm shift is required. While some claim that the sheer expansion of LLMs could lead to emergent general intelligence, the counterargument remains: does a sufficiently scaled-up pattern recognizer equate to true cognitive ability? This challenge, as we explored in Chap. 10, is at the core of AI research today.

From an XAI perspective, this evolution raises a different but equally crucial question: should we enforce explainability on systems like LLMs? Or, as we framed it earlier, should we require that an agent be capable of producing human-understandable explanations before it can be classified as AGI? As we have seen throughout this book, explainability is not just about making AI outputs interpretable but about ensuring that AI systems align with human reasoning and decision-making processes. One of the key lessons from our exploration of the damped pendulum case study was that ML models can reveal latent structures in data, but without a framework for interpreting those structures, their explanatory power remains incomplete.

More broadly, Chap. 10 highlighted how the real test of AGI is not just its ability to generalize across tasks but its ability to engage in counterfactual reasoning and causal discovery, akin to what humans do when they conceptualize abstract relationships. This aligns with Pearl's Ladder of Causation, where true understanding requires the ability to ask "what if" questions and infer knowledge beyond statistical correlations. As we discussed, AGI must be able to climb to rung 3, integrating not just pattern recognition but also the ability to imagine alternative scenarios and generate novel scientific insights.

This brings us back to Wittgenstein's question: if AGI emerges, will we even be able to understand it? If human cognition itself is shaped by our unique experiences and cultural "language games," could an artificial intelligence develop a form of intelligence that is fundamentally different from ours? As we explored in Chap. 10, even if AI reaches a form of general intelligence, it may not be interpretable through our current explanatory frameworks.

But as authors we cannot forget our theoretical physics background. Quantum Mechanics can be considered one of the best and beautiful theories we have of the physical world. It is confirmed by every experiment and the applications on everyday life are impressive: from our PCs and mobile phones to laser and communications, from transistors to microscopy and medical diagnostics devices. But despite all this, the stream of research that still struggles with **interpretation** of quantum

11.3 Conclusions

mechanics is very active and prolific. And we don't use the word "interpretation" by chance.

We can call quantum mechanics a physical theory, but, as any other physical theory or model, it can be considered as a computational tool: given some input, the theory prescribes how to calculate the output and how to get the predictions on the system we are studying in terms of how it behaves.

Quantum Mechanics provides impressive predictions on its domain but we are not yet able to cope with its interpretation: the existence of probability and uncertainty at the very root of the theory for any physical observable, undermines our attempts as humans to make sense of it, to quote Mr. Feynmann: "If you think you understand quantum mechanics, you don't understand quantum mechanics."

We attempted different approaches to deal with this state of things, from hidden variables to many world interpretations, the objective is to avoid to take the uncertainty, the superposition of states and wave-function collapse to be real in order to have a more classical and deterministic interpretation of physical observables.

But up to now, there is no a definite success in this sense, and we continue on using QM albeit a lack of a satisfactory and final interpretation. The alternative is to take it as it is, as a predictive tool and look at quantum mechanics as a theory of relations (Relational Quantum Mechanics, R.Q.M.).

The physical variables don't describe the "things," they describe how the things interact each other and the state of a quantum system as being observer-dependent. So as from the beginning, with Galileo's inertial frames and then special and general relativity, there is no sense to talk about the "true" phenomenon or physical event but what is experienced depends on the observer: the properties of an object that are real with respect to a second object are not necessarily the same with respect to a third object. We exit from a world of "things" to go into the "world" of interactions where what we see are just interactions, not tangible things like stones.

What is the relation of this with XAI? The metaphor is that we may look at ML in such a way: as a tool to get impressive predictions that we have to double check. ML can be considered a computational tool like QM that, given an input, produces an output. Because of this analogy with RQM, we should not be too rigid in terms of the expected interpretations and explanations. We can get explanations with XAI but we might need to accept that the same explanations may depend on the type of interactions we have with our ML models. The ML models may work fine and rely on inner computation and paths that we may be forced to accept as intrinsically different from how we would work as human beings.

This idea takes on new significance in the context of AGI. As we take the possibility of AGI more seriously, we may need to redefine our expectations of explainability. Rather than assuming that explanations must conform to human reasoning, we may have to consider that an AGI could operate with its own internal logic, one that is comprehensible only through a new type of interaction. Just as quantum mechanics shifted our understanding of reality from a world of "things" to a world of interactions, AGI may force us to reconsider intelligence itself—not as a static property but as an emergent phenomenon shaped by its relationship with the world and with us.

This perspective does not diminish the need for explainability, but rather expands the conversation. The future of XAI may not lie in enforcing human-like explanations onto AI systems but in developing new ways to engage with them, allowing explanations to arise dynamically within the context of their use. If we take this approach, the challenge will not just be to make AI "understandable" in a traditional sense, but to develop new methods of interpretation and interaction that align with the capabilities and nature of AI itself.

There is nothing fixed or absolute out there in the universe, only an infinite net of interactions and dynamic processes in which we, as human observers, are not privileged.

If AGI ever arrives, it may not simply be an intelligent system that we control or comprehend, but rather a new kind of intelligence that we interact with. The rapid evolution of Large Language Models, as explored in Chap. 10, suggests that intelligence may not emerge in a way that mirrors human cognition but rather as a gradual process of increasing capability, adaptation, and emergent behaviors. LLMs have already demonstrated surprising properties that were not explicitly designed into their architectures. As we discussed, GPT-4 and its successors exhibit behaviors such as reasoning, abstraction, and few-shot learning, despite being trained primarily as pattern-matching systems. These abilities seem to emerge from scale, prompting an ongoing debate: at what point does an accumulation of emergent behaviors become something qualitatively different—something that resembles AGI?

The case of LLMs also challenges the assumption that intelligence must be structured like human thought. If AGI arises from models that continue to scale in complexity, integrating multimodal inputs and self-refinement mechanisms, it may think in ways that are deeply unfamiliar to us. Its internal representations may not be reducible to symbolic logic or interpretable mental states but may instead resemble high-dimensional knowledge spaces—fluid, associative, and potentially opaque to human reasoning. Even today, we struggle to fully explain how LLMs generate coherent arguments or solve problems beyond their training distribution, highlighting the gap between performance and interpretability.

If such trends continue, AGI might emerge not as a single breakthrough but as an incremental shift, where the distinction between highly capable LLMs and true AGI becomes blurred. At that stage, our relationship with AI may resemble less a tool-user dynamic and more an interactive, evolving system, where intelligence is distributed across networks, responding to human queries, augmenting scientific discovery, and generating new knowledge beyond individual comprehension.

In this scenario, the challenge is not merely to determine whether AGI has arrived but to recognize that intelligence itself is not a fixed entity but a continuum—one that might develop forms of reasoning, knowledge representation, and self-improvement mechanisms that escape our traditional frameworks. Just as human cognition is shaped by our evolutionary and cultural history, artificial cognition, driven by vast computational architectures and self-optimizing processes, may lead to a different kind of intelligence, one that does not necessarily align with human ways of thinking but still exhibits agency, adaptability, and learning.

This perspective reinforces the idea that interaction, rather than direct control, will define our engagement with AGI. If we continue to scale LLMs and refine their architectures, we may find ourselves not creating AI in our image, but instead uncovering a new epistemological paradigm, where meaning and understanding arise not from a shared cognitive model but from the interplay between human intelligence and an evolving, machine-driven intelligence. The question may no longer be whether AGI will think like us, but whether we are prepared to engage with intelligence in forms we have never encountered before.

11.4 Summary

- Adopt XAI methods to interpret complex ML models like AlphaGO
- Adopt FAST criteria to check GDPR compliance
- Use SHAP to check fairness
- Learn how XAI methods can be fooled by adversarial attacks
- Questions about AGI and the possible implications for XAI in the next future

References

DARPA (2016). *Explainable Artificial Intelligence (XAI)*. Retrieved from https://www.darpa.mil/program/explainable-artificial-intelligence

EPRS (2016). *EU guidelines on ethics in artificial intelligence: Context and implementation*. Retrieved from https://www.europarl.europa.eu/RegData/etudes/BRIE/2019/640163/EPRS_BRI(2019)640163_EN.pdf

European Union (2016). *GDPR*. Retrieved from https://europa.eu/european-union/index_en

Google (2020). *Explainable AI*. Retrieved from https://cloud.google.com/explainable-ai

IBM (2019). *Introducing AI explainability 360*. Retrieved from https://www.ibm.com/blogs/research/2019/08/ai-explainability-360

Lundberg, S. (2020). *Explaining measures of fairness with SHAP*. Retrieved from https://github.com/slundberg/shap/blob/master/notebooks/general/Explaining%20Quantitative%20Measures%20of%20Fairness.ipynb

Merrer, E.L., & Tredan, G. (2019). *The bouncer problem: Challenges to remote explainability*. arXiv preprint arXiv:1910.01432.

OpenAI (2020). *OpenAI API*. Retrieved from https://openai.com/blog/openai-api/

Slack, D., Hilgard, S., Jia, E., Singh, S., & Lakkaraju, H. (2020, February). Fooling Lime and Shap: Adversarial attacks on post hoc explanation methods. In *Proceedings of the AAAI/ACM conference on AI, ethics, and society* (pp. 180–186).

Appendix A

"F.A.S.T. XAI Certification"

The purpose of this checklist is to provide a practical guidance, based on the contents of this book, and the steps to perform to conduct an assessment and a possible XAI certification of a ML system.

The main idea is that **before** using pure XAI methods in (6), we need to build the questions we want to be answered, otherwise we would invalidate the XAI methods themselves.

Another critical step is step (4) in which we ask for a surrogate model. This would guarantee the better performance of the model with respect to the baseline model (the surrogate) and also would add confidence in terms of lack of bias. Step (4) is somehow optional, without it we would get a lighter form of certification.

Accountability in step (5) is strongly related to the legal aspects and may change depending on the specific regulation that is in place (e.g., GDPR). Note: This framework has been successfully applied across domains including medical diagnosis, financial risk assessment, and autonomous vehicle decision-making. For Large Language Models and multimodal AI systems, additional considerations regarding hallucination detection and cross-modal consistency should be incorporated into steps (3) and (6).

(1) **Model Preparation**

   - [ ] Acquire the model to explain (a.k.a. the "MODEL")

(2) **Frame the Problem**

   - [ ] Data source identification
   - [ ] Identify sensible and possible adversarial features

(3) **Construct the "What If?" Counterfactual Scenario**
  - [ ] Compile a list of possible counterfactual questions adapted to realistic usage scenarios

(4) **Build a Surrogate Model (a.k.a. the "SURROGATE")**
  - [ ] If data are openly available, train an intrinsic global explainable model, the "SURROGATE", else ask for a surrogate model to the authors
  - [ ] Explore feature importance on the surrogate model
    If the "SURROGATE" is not available, certification is assumed to be "LIGHT"

(5) **F.A.S.T. Methodology Key Aspects**
  - [ ] Identify all possible fairness problems in data [F]
  - [ ] Qualify the accountability aspects of the model [A]
  - [ ] Describe the security of the data [S]
  - [ ] Leverage transparency of "SURROGATE" model [T] (not needed for LIGHT certification)

(6) **Directly Explain the Real Model**
  - [ ] Train a XAI model directly on the "MODEL" we want to analyze
  - [ ] Display global and local features importance
  - [ ] Assure that "What If?" question can now correctly be answered

If you have any concerns about our products,
you can contact us on
ProductSafety@springernature.com

In case Publisher is established outside the EU,
the EU authorized representative is:
**Springer Nature Customer Service Center GmbH
Europaplatz 3, 69115 Heidelberg, Germany**

Printed by Libri Plureos GmbH
in Hamburg, Germany